To Wanda, Marek, and Darek

ATM Network Resource Management

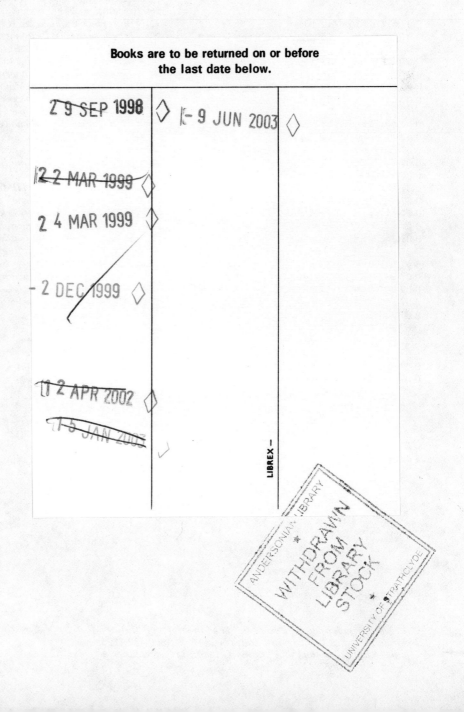

The McGraw-Hill Series on Computer Communications (Selected Titles)

ISBN	AUTHOR	TITLE
0-07-005147-X	Bates	*Voice and Data Communications Handbook*
0-07-005560-2	Black	*TCP/IP and Related Protocols*, 2/e
0-07-005590-4	Black	*Frame Relay Networks: Specifications and Implementation*, 2/e
0-07-011769-1	Charles	*LAN Blueprints: Engineering It Right*
0-07-011486-2	Chiong	*SNA Interconnections: Bridging and Routing SNA in Hierarchical, Peer, and High-Speed Networks*
0-07-016769-9	Dhawan	*Mobile Computing: A Systems Integrator's Handbook*
0-07-018546-8	Dziong	*ATM Network Resource Management*
0-07-020359-8	Feit	*SNMP: A Guide to Network Management*
0-07-021389-5	Feit	*TCP/IP: Architecture, Protocols and Implementation with IPv6 and IP Security*, 2/e
0-07-024563-0	Goralski	*SONET: A Guide to Synchronous Optical Networks*
0-07-024043-4	Goralski	*Introduction to ATM Networking*
0-07-031382-2	Huntington-Lee/Terplan/Gibson	*HP's OpenView: A Manager's Guide*
0-07-034249-0	Kessler	*ISDN: Concepts, Facilities, and Services*, 3/e
0-07-035968-7	Kumar	*Broadband Communications*
0-07-041051-8	Matusow	*SNA, APPN, HPR and TCP/IP Integration*
0-07-060362-6	McDysan/Spohn	*ATM: Theory and Applications*
0-07-044435-8	Mukherjee	*Optical Communication Networks*
0-07-044362-9	Muller	*Network Planning Procurement and Management*
0-07-046380-8	Nemzow	*The Ethernet Management Guide*, 3/e
0-07-051506-9	Ranade/Sackett	*Introduction to SNA Networking*, 2/e
0-07-054991-5	Russell	*Signaling System #7*
0-07-057724-2	Sackett/Metz	*ATM and Multiprotocol Networking*
0-07-057199-6	Saunders	*The McGraw-Hill High Speed LANs Handbook*
0-07-057639-4	Simonds	*Network Security: Data and Voice Communications*
0-07-912640-5	Simoneau	*Hands-On TCP/IP*
0-07-060363-4	Spohn	*Data Network Design*, 2/e
0-07-069416-8	Summers	*ISDN Implementor's Guide*
0-07-063263-4	Taylor	*The McGraw-Hill Internetworking Handbook*
0-07-063301-0	Taylor	*McGraw-Hill Internetworking Command Reference*
0-07-063639-7	Terplan	*Effective Management of Local Area Networks*, 2/e

To order or receive additional information on these or any other McGraw-Hill titles, in the United States please call 1-800-722-4726. In other countries, contact your local McGraw-Hill representative.

ATM Network Resource Management

Zbigniew Dziong

McGraw-Hill

New York   San Francisco   Washington, D.C.   Auckland   Bogotá
Caracas   Lisbon   London   Madrid   Mexico City   Milan
Montreal   New Delhi   San Juan   Singapore
Sydney   Tokyo   Toronto

Library of Congress Cataloging-in-Publication Data

Dziong, Zbigniew
    ATM network resource management / Zbigniew Dziong.
       p. cm.—(McGraw-Hill series on computer communications)
    Includes index.
    ISBN 0-07-018546-8
    1. Asynchronous transfer mode. 2. Telecommunication—Traffic—
Management. I. Title. II. Series.
TK5105.35.D885   1997
004.6—dc21   97-15265
                                                                                  CIP

McGraw-Hill

A Division of The McGraw·Hill Companies

Copyright © 1997 by The McGraw-Hill Companies, Inc. All rights reserved. Printed in the United States of America. Except as permitted under the United States Copyright Act of 1976, no part of this publication may be reproduced or distributed in any form or by any means, or stored in a data base or retrieval system, without the prior written permission of the publisher.

1 2 3 4 5 6 7 8 9 0   FGR/FGR   9 0 2 1 0 9 8 7

ISBN 0-07-018546-8

The sponsoring editor for this book was Steven M. Elliot, the editing supervisor was Patricia V. Amoroso, and the production supervisor was Pamela A. Pelton.

Printed and bound by Quebecor / Fairfield.

   This book is printed on recycled, acid-free paper containing a minimum of 50% recycled, de-inked fiber.

> Information contained in this work has been obtained by The McGraw-Hill Companies, Inc. ("McGraw-Hill") from sources believed to be reliable. However, neither McGraw-Hill nor its authors guarantee the accuracy or completeness of any information published herein, and neither McGraw-Hill nor its authors shall be responsible for any errors, omissions, or damages arising out of use of this information. This work is published with the understanding that McGraw-Hill and its authors are supplying information but are not attempting to render engineering or other professional services. If such services are required, the assistance of an appropriate professional should be sought.

McGraw-Hill books are available at special quantity discounts to use as premiums and sales promotions, or for use in corporate training programs. For more information, please write to the Director of Special Sales, McGraw-Hill, 11 West 19th Street, New York, NY 10011. Or contact your local bookstore.

Contents

Preface		xi
1 Introduction to ATM Based Networks		**1**
1.1	ATM Concept	2
1.2	Control and Management Functions	5
1.3	Service Categories	9
1.4	Discussion and Bibliographic Notes	14
2 Resource Management and Traffic Control Issues		**17**
2.1	Decomposition of the Resource Management Functions	18
	2.1.1 Cell layer	20
	2.1.2 Connection layer	23
	2.1.3 Virtual network layer	28
	2.1.4 Physical network layer	31
2.2	Resource Management Architecture	32
2.3	Discussion and Bibliographic Notes	34
3 Resource Allocation to Connections		**39**
3.1	Logical "Bandwidth" Allocation to Connections	40
	3.1.1 Equivalent bandwidth allocation for heterogeneous cases	44
3.2	Models for Equivalent Bandwidth Allocation	46
	3.2.1 Single-priority system	46
	3.2.2 Multi-priority system	50
	3.2.3 Application for QoS virtual network design	58
3.3	Discussion and Bibliographic Notes	63
4 Adaptive Resource Allocation to Aggregate Traffic		**67**
4.1	Framework for Adaptive Connection Admission	68
4.2	Adaptive Aggregate Equivalent Bandwidth Estimation	70
4.3	Estimation Error Analysis	76
4.4	Connection Admission Control Analysis (QoS)	83
4.5	Trade-off Between Relaxed and Strict Source Policing	87
4.6	Discussion and Bibliographic Notes	92

5 CAC & Routing Strategies — 95
- 5.1 Problem Formulation — 96
  - 5.1.1 CAC & routing classification — 98
  - 5.1.2 Reward maximization as an objective function — 102
- 5.2 Reward Maximization Approach — 103
  - 5.2.1 Optimal solution* — 104
  - 5.2.2 Strategy based on link shadow price concept — 105
  - 5.2.3 Link shadow price evaluation — 108
  - 5.2.4 Average shadow price as a sensitivity measure* — 115
  - 5.2.5 Important features of MDPD model — 118
- 5.3 Implementation Issues — 128
  - 5.3.1 Centralized implementation — 128
  - 5.3.2 Distributed implementation — 130
  - 5.3.3 Hierarchical implementation — 131
- 5.4 Discussion and Bibliographic Notes — 133

6 Multi-point Connections — 139
- 6.1 Classification and Main Issues — 140
- 6.2 Heuristics for Minimum Cost Tree Design — 144
  - 6.2.1 Heuristics for directed cost networks — 144
  - 6.2.2 Heuristics based on minimum spanning tree — 146
- 6.3 CAC & Routing Strategies for Multi-point Connections — 149
  - 6.3.1 Strategy based on link shadow prices — MDPD — 150
  - 6.3.2 Strategy based on least loaded path routing — LLP — 150
  - 6.3.3 Strategy based on the inverse of residual link capacity — IRCF — 150
  - 6.3.4 Numerical study — 151
- 6.4 Discussion and Bibliographic Notes — 155

7 Network Performance Models — 159
- 7.1 Decomposition Methods — 160
  - 7.1.1 Decomposition into link problems — 161
  - 7.1.2 Decomposition into path problems — 162
  - 7.1.3 General features of the decomposition techniques — 164
- 7.2 Performance Model for Reward Maximization Routing — 169
  - 7.2.1 Link loading and performance functions* — 170
  - 7.2.2 Macro-state approximation* — 173
  - 7.2.3 Numerical study — 175
- 7.3 Discussion and Bibliographic Notes — 179

8 Optimal Operating Point — Fairness vs. Efficiency — 185
- 8.1 Game Theoretic Framework — 186
  - 8.1.1 Fairness concepts — 186
  - 8.1.2 The Pareto boundary for multi-service systems* — 190
- 8.2 Analysis of a Single-Link System — 192
  - 8.2.1 Simplified CAC policies — 202

8.3 General Case 208
8.4 Discussion and Bibliographic Notes 209

9 Virtual Networks as a Tool for Resource Management   213
9.1 Virtual Network Definition 214
9.2 Virtual Network Applications 215
9.3 Relation Between Virtual Network and Virtual Path Concepts ... 217
9.4 Resource Allocation to QoS Virtual Networks 221
9.5 "Bandwidth" Enforcement in Virtual Private Networks 225
9.6 Virtual Network Design 225
    9.6.1 Optimization procedure for reward maximization routing* . 227
    9.6.2 Solution analysis* 230
    9.6.3 Numerical study 232
9.7 Virtual Network Update 233
9.8 Backup Virtual Network 236
9.9 Discussion and Bibliographic Notes 237

10 Physical Resource Allocation to ATM Networks   241
10.1 ATM Network Layer Versus Transport Layer 242
    10.1.1 Dedicated transport layer 242
    10.1.2 ATM over the digital hierarchy (PDH, SDH, SONET) ... 243
10.2 Dimensioning of ATM Networks 245
    10.2.1 Continuous link bandwidth* 246
    10.2.2 Link bandwidth modularity 253
    10.2.3 Multi-hour case* 254
10.3 Adaptation to Traffic Changes 255
10.4 Survivability Issues 256
    10.4.1 Failure scenarios 256
    10.4.2 Restoration mechanisms 257
    10.4.3 Connection and GoS restoration objectives 259
    10.4.4 Design of self-healing networks* 261
10.5 Discussion and Bibliographic Notes 264

A Kalman Filter   269
A.1 Linear System Model 270
    A.1.1 Observability and controllability 272
    A.1.2 Covariance matrix 273
A.2 Discrete Kalman Filter 274
A.3 Discussion and Bibliographic Notes 276

B Markov Decision Theory   279
B.1 Markov Processes 279
B.2 Markov Decision Processes 283
    B.2.1 Discrete-time Markov decision models 283
    B.2.2 Continuous-time Markov decision models 288
B.3 Discussion and Bibliographic Notes 291

C	**Cooperative Game Theory**	**293**
	C.1 Basic Concepts	293
	C.1.1 Players	293
	C.1.2 Extensive form	293
	C.1.3 Strategic form	295
	C.1.4 Coalitional form	296
	C.2 Cooperative vs. Non-cooperative Games	297
	C.3 Different Formulations of Cooperative Games	298
	C.4 A Class of Arbitrated Solutions	299
	C.4.1 Extension to the general case	300
	C.5 Discussion and Bibliographic Notes	301
D	**Notation**	**303**
	Index	**309**

Preface

RESOURCE management in *Asynchronous Transfer Mode* (ATM) based networks is a complex issue due to a broad range of services, traffic characteristics, time scales, and performance constraints, which are integrated into one system. On top of that, the fast advancement in technology results in rapid changes of the main bottlenecks. The purpose of this book is to present a framework and models which would help one to cope with this complexity. This objective is threefold. Firstly, to present a framework for resource management synthesis and analysis which is driven by traffic source characteristics and requirements rather than by network transport technology. This approach ensures that the described solutions are less affected by changes in technology since services and traffic source characteristics are less time dependent. Secondly, to introduce an economic factor directly into the real-time resource control algorithms. In this way the objectives of network planning, dimensioning, and control can be strongly integrated. This feature simplifies interaction among different parts of the resource management system. It also provides higher efficiency and greater adaptiveness. Thirdly, the final goal of this work is to show how mathematical theories can lead to very practical algorithms. In particular we present applications of estimation theory, Markov decision theory, and game theory. These theories provide a solid basis for synthesis and analysis of a wide range of algorithms, including the traditional ones.

The book consists of ten chapters, which are distinct but at the same time build on the preceding ones to constitute a logical chain in resource management synthesis and analysis. We start from the basic concepts and important features of the ATM based networks believed to be the most likely solution for the global integrated multi-service broadband network (Chapter 1). A general discussion of the main resource management issues and associated traffic control topics is given in Chapter 2. The presented framework employs decomposition techniques, such as virtual networks and a layered structure of traffic entities, which facilitate coping with the enormous complexity of the problem. Although the focus is on ATM based networks, many of the issues treated in this book cover a much wider range of transmission technologies, including *Synchronous Transfer Mode* (STM).

A logical allocation of resources to connections is described in Chapter 3. This issue is critical since it enables the separation of algorithms dealing with the packet and connection layers. A framework for adaptive resource allocation to connections, based on measurements and derived from estimation theory, is presented in Chapter 4. The importance of this subject is underlined by the fact that new broadband services can have non-stationary and unpredictable characteristics. Chapter 5 is devoted to *connection admission control* (CAC) and routing problems. Besides defining the spectrum of possible algorithms and their implementations, we also present a generic formulation of the CAC & routing problem based on a reward maximization objective. This concept allows one to introduce economic consideration directly into the traffic control level. The framework is derived from Markov decision theory and has a very clear and simple physical interpretation. Multi-point connection applications are steadily increasing and one can expect that they will constitute a significant share of traffic in high speed networks. The main issues associated with CAC & routing of multi-point connections are discussed in Chapter 6.

The network performance models on the connection layer, presented in Chapter 7, are crucial for efficient design and operation of multi-service broadband networks. Some of these models are used in Chapter 8 to illustrate the concept of an optimal network operating point on the connection layer. Here the subject of fairness vs. efficiency is addressed in the context of game theory, which provides an excellent foundation for analysis and synthesis of different approaches from an economical perspective.

In Chapter 9 we present a coherent framework for resource management within the ATM network, which takes full advantage of the virtual network concept. This concept is used for virtual separation of the network resources in order to provide the required service guarantees on the connection and packet layer for different services and user groups. In particular we discuss the issues of virtual network design and its adaptation to changes in traffic conditions. In the described model the cost of bandwidth and switching is taken into account in a unified approach which covers virtual network dimensioning, planning, control, and even tariff allocation. The subject of resource management on the network physical layer is covered in the final part of the book (Chapter 10). In particular we concentrate on the allocation of the resources from the transport layer to the ATM layer where the transport layer can use either dedicated transmission facilities or digital hierarchy systems (SONET, SDH). We also discuss network survivability issues. Several approaches for self-healing ATM network design are described, including the application of backup virtual networks.

The book is addressed to a large audience including network designers, engineers, students, researchers, and scientists interested in resource management problems in multi-service broadband networks. It is intended for readers with different levels of familiarity with networking problems and advancement in mathematical modeling. In particular it contains introductory information about basic concepts and classifications illustrated with many figures (Chapters 1, 2, 9, 10, and the first parts of most of the other chapters). The presentation of the main issues focuses on applications rather than on theory and is intended to provide an informal and clear picture of the central problems accompanied by many numerical results which give a sense of the considered algorithms and systems behavior. To make the book self-contained and as accessible as possible, an informal introduction to Kalman filter theory, Markov decision theory, and game theory is given in the respective appendices for readers interested in the most advanced parts. These parts are identified with stars and can be omitted without losing continuity.

The work is the result of my research activities, which started at Warsaw University of Technology and continue at INRS-Telecommunications (Montreal). These include participation in several projects for large industrial centers: LM Ericsson (Sweden), CNET (France), and Nortel (previously BNR; Canada) as well as for the Canadian Institute for Telecommunications Research (CITR). During this time I cooperated with some remarkable personalities who had a very positive influence on my career. It all started with Józef Lubacz who, as a chief of the teletraffic group at Warsaw University of Technology, provided an excellent atmosphere, environment, and encouragement for work on new and difficult problems. Then at CNET I met Jim Roberts, whose deep insight into network problems helped me to focus on challenging problems. Michal Pióro introduced me to the network routing issues during our work on the project for LM Ericsson. Finally, I would like to mention Lorne G. Mason, without whom this book would have never been written and who contributed directly to the material included. His deep insight into network problems and creation of a free and inspiring atmosphere made my stay at INRS-Telecommunications the most productive period in my career. I also wish to acknowledge several people with whom I cooperated on several projects and who also contributed directly to the material contained herein. These include researchers Ke-Qiang Liao (currently Professor at UQAH), Ming Jia, Boris Shukhman, Jean Choquette, and Nicole Tétreault; and students Jisheng Zhang, Olivier Montanuy, Marek Juda, Jose Mignault, Juan M. Ramos, Anna Wielosz and Alexandre Nadeau. Lastly I am greatful to my colleagues Martin Mierzwicki, Lorne Mason, and Michal Pióro, for screening the manuscript.

ATM Network Resource Management

Chapter 1

Introduction to ATM Based Networks

THE DOMAIN of telecommunication networks is undergoing a significant transformation in several planes. The main driving force behind these changes is a feedback between the new technologies and the demand for new services. In particular, recent developments in microelectronics and photonics enable realization of communications with gigabit/sec speeds while the terabit/sec range is on the horizon. At the same time, developments in the areas of computer hardware, software, and image processing have led to the creation of applications which are ready to use these high speeds. Caught in the middle of these developments were the network operators with monopolistic attitudes and very long network investment return cycles (in the range of 20 years). The deregulation of the telecommunication market did not solve the problem automatically since the economic constraints do not justify creation of diverse networks oriented towards different services and based on different technologies. Moreover, the task of designing and implementing a high speed multi-service network is so daunting that a single organization, even a very powerful one, is not willing to take such a risk. These restrictions forced the industry and operators into a concerted effort aimed at the creation of standards for a *broadband integrated service digital network* (B-ISDN). The main focus of this endeavor was to design a transport layer and protocols which would provide a framework for integration of all services including those not foreseen at present. As a result of this effort, a *fast packet switched* (FPS) network based on the *asynchronous transfer mode* (ATM) multiplexing is considered the most likely candidate for the global integrated multi-service broadband network. Here transport integration is achieved by packetization of all transmitted information streams into a uniform format – the ATM cell. The basic concepts of ATM based networks are presented in Section 1.1. Due to the integration of many services within one system, ATM based networks have to perform many different control and management functions. In Section 1.2 we classify these functions and provide a short description of the basic ones. In order to facilitate the complex task of ATM network resource management it is convenient to divide the potential services into several categories. The classifications proposed by the ATM Forum and ITU-T are described in Section 1.3.

1.1 ATM Concept

The main concept of transport layer integration based on ATM is illustrated in Figure 1.1 for a one link network example. Information transmitted from the user is first divided into fixed length (53 octets) packets called ATM cells. The ATM cell header includes a tag which can be translated into the destination user's address. These cells are sent to the network via access links. At the network entry all cells are multiplexed into one link. Since the link can only carry the cells sequentially, the multiplexer (MUX) has a buffer which can introduce a delay in the cell transmission time. At the destination end of the link, the cells are directed into the proper access links by means of the demultiplexer (DMUX) and the tag included in the cell header. This system has three important features which are not present in *Synchronous Transfer Mode* (STM) used in digital circuit-switched networks. The first one is that the source information rate is not constrained physically by the system except for the maximum link speed. This allows integration of sources with very different and variable information rates and seamless evolution of more effective source coding algorithms over time. The second important feature is that when source A is not generating cells, another source (B or C) can use link bandwidth previously occupied by source A. This effect gives rise to the so-called statistical multiplexing gain (to be discussed in Chapter 3) which can provide increased bandwidth utilization when compared with systems where a fixed bandwidth is physically allocated to each connection (e.g., circuit switched systems). Finally, the uniform cell format simplifies broadband switch architectures since the bandwidth required by a connection does not influence the switching algorithm.

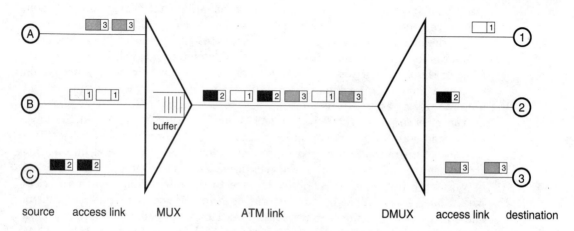

Figure 1.1. *Asynchronous Transfer Mode concept.*

1.1 ATM Concept

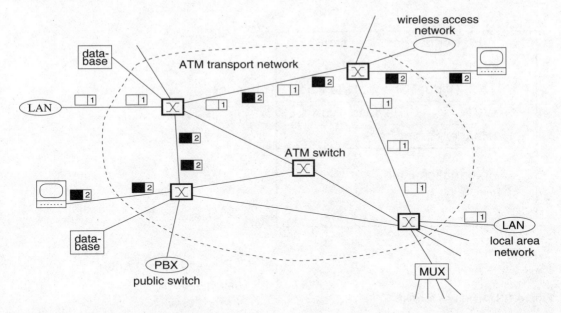

Figure 1.2. *ATM based network.*

The ATM concept is very convenient for a general switched network illustrated in Figure 1.2. Here the ATM switching nodes are connected by means of ATM links. The users are connected to the ATM switches either directly or via multiplexers in order to increase access link utilization. Cells are routed via the network using the tag included in the cell header. All cells from a particular source are sent into the network with the same tag which defines a path through the network and the destination user. These two functions of the tag are reflected in the ATM cell header format, shown in Figure 1.3. The tag has two sections associated with *virtual path* (VP) and *virtual channel* (VC) concepts: a *virtual path identifier* (VPI) and a *virtual channel identifier* (VCI). Due to the limited size of the identifier fields it may be necessary that the identifiers be reused geographically. This is realized in ATM switches, as illustrated in Figure 1.4. The basic function of the switch is to transfer each cell from the switch input port to the output port defined by the identifiers. An additional function is to replace identifiers in the cell header, if required, according to the translation table.

A particular VPI (or VCI) defines a *virtual path link* (VPL) (or a *virtual channel link* (VCL)) which corresponds to a transport of all ATM cells with common VPI (or VCI) between a point where the VPI (or VCI) is assigned and the point where that value is translated or terminated. A *virtual path connection* (VPC) (or *virtual channel connection* (VCC)) is defined by a concatenation of VP links (or VC links) and can have different VPIs (or VCIs) on different ATM links constituting the VP or VC. As illustrated in Figure 1.4 a VCC used for communication between the source and destination may consist of several concatenated VPCs and each VPC can itself carry several VCCs. In this figure the VCLs and VPLs are indicated

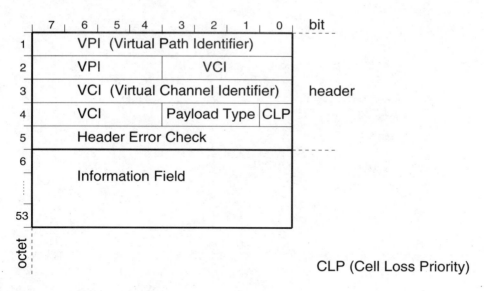

Figure 1.3. *ATM cell format.*

by continuous lines and the dotted lines indicate switched connections between the pairs of VCLs or VPLs. These connections form concatenations of VCLs and VPLs resulting in VCCs and VPCs.

Since in many cases it is sufficient to switch the cell based on VPI only, some switches analyze and translate only the VPI. These two options are referred to as the VC switch and the VP switch, respectively, as shown in Figure 1.4. Sometimes the VP switch is also referred to as the cross-connect switch. Note that in some instances many cells from different switch input ports can be addressed to the same output port at the same time. In order to avoid cell losses in such situations, buffers have to be introduced. In general the buffers can be implemented in the switch input ports, switch fabric, and switch output ports. The most efficient, from the resource utilization viewpoint, is the option where all buffering is done at the switch output ports (Hluchyj and Karol, 1988), as shown in Figure 1.4. In some switches, due to some technological constraints or other considerations, one can encounter architectures with input port buffers or a combination of input, output, and switch fabric buffers. Nevertheless, to simplify presentation, in the following considerations we assume that switches have output port buffers. In general this assumption does not limit application of the presented methods since in most cases the principles they are based on are not influenced by the switch buffering option. Note that the switch buffers can introduce a variable delay in cell transmission.

It should be emphasized that although the general structure of the ATM based network is similar to traditional packet switched networks, the problems and solutions are quite different. This is due mainly to two factors. Firstly, the speed of ATM links is significantly higher (e.g., 150 Mbps, 600 Mbps, 2400 Mbps). This feature has sweeping consequences not only for the applied technology but also

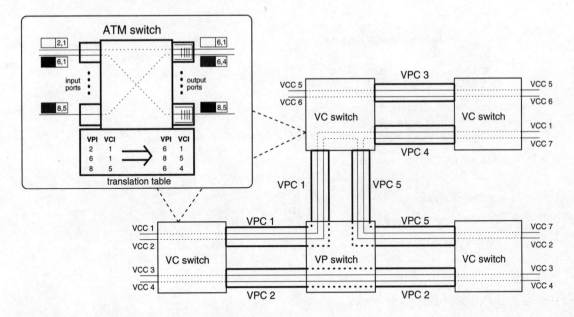

Figure 1.4. *ATM switch and example of its application.*

for traffic control due to the high bandwidth-delay product. Secondly, the available broad bandwidth can be used by a large spectrum of services, including the broadband services, which have very diverse and sometimes very stringent quality of service requirements.

1.2 Control and Management Functions

In order to consider how different types of services may use the ATM transport network and how this network is controlled and managed, it is convenient to use the protocol reference model (CCITT Rec.I.321, 1991) illustrated in Figure 1.5. The user plane provides a communication protocol for each of the services. The main function of the *ATM adaptation layer* (AAL) is segmentation of the service information format into the ATM cell format and reassemble back into the service information format. These functions are performed at the user premises. The control plane provides a protocol for signaling between different management objects located in different physical locations (network nodes, user premises). Since the signaling uses the same ATM transport layer as users, the functions of the AAL layer in this plane are similar to the ones in the user plane. The management plane is responsible for managing all user and control layers. In particular the management plane is involved in the connection setup procedure. The management plane interfaces with all layers of the control and user planes. It provides instructions to these layers and analyzes the replies.

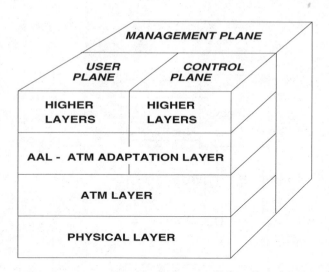

Figure 1.5. *Protocol reference model.*

Some information concerning the management and control functions is included in the cell header (Figure 1.3). The *payload type* (PT) field (three bits) is used to indicate whether the cell belongs to the user or control plane. In the case of a user data cell, this field may also be used to inform the user whether congestion is experienced on the connection path. In the case of a control cell, the PT field also specifies whether it is an *operation and maintenance* (OAM) cell or a *resource management* (RM) cell. The information contained in the *cell loss priority* (CLP) field is used when congestion occurs within the network. If CLP=1, the cell has low priority and can be discarded. The CLP bit can be set by the user to distinguish between important and less important information. Also, as will be described later, this bit can be changed from CLP=0 to CLP=1 at the user-network interface for the cells not conforming to the connection traffic contract between the user and the network.

One of the critical objectives of the management plane is to ensure that all services are served within the required *quality of service* (QoS) constraints. There are three main performance measures describing QoS for a particular connection: *cell loss rate* (CLR), *cell transfer delay* (CTD) (average or percentile), and *cell delay variation* (CDV). The fixed portion of the cell delay is mainly defined by the link propagation delay which is proportional to the link length. Additional delay may be introduced by the buffers in the switching nodes and multiplexers in the periods when the aggregate link cell arrival rate is higher than the link speed. Obviously this delay can vary between subsequent cells of the same connection. Moreover, some cells may be lost if the buffer capacity is exhausted. To provide some kind of QoS guarantees, the performance management can apply several control functions which are discussed in the next chapter. Here we focus on the control functions

1.2 Control and Management Functions

which act directly on the cells in the network in order to influence and protect the QoS of the accepted connections.

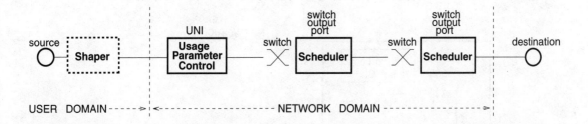

Figure 1.6. *Performance control functions acting on the cells.*

The two basic categories of these QoS control functions are indicated in Figure 1.6. The first one is located at the *user network interface* (UNI) and is called *usage parameter control* (UPC). The general objective of this function is to ensure that the connection cell stream entering the network conforms to a traffic contract approved at the connection setup time. In particular UPC controls the connection's *peak cell rate* (PCR) which defines the maximum cell rate of the connection. Additionally UPC can police some statistical parameters of the variable bit rate connections such as average cell rate and cell burst size. UPC can be based on many different algorithms, but the most attractive one, due to its simplicity and efficiency, is a policing mechanism called the *leaky bucket*. This mechanism, illustrated in Figure 1.7, consists of a pseudo-buffer, with capacity for K cells, which is served with a certain *leak cell rate* (LCR) smaller or equal to the peak cell rate. Each arriving cell increases the pseudo-buffer occupancy by one. Whenever a cell arrives when the pseudo-buffer is full, the cell is declared non-conforming and is either rejected or accepted with low priority (CLP=1). The leak rate is an upper bound on the connection admissible average cell rate. The leak rate and bucket capacity also define an upper bound on the admissible *cell burst size* (CBS) which is transmitted at the peak cell rate: $CBS_{max} = 1 + K \cdot PCR/(PCR - LCR)$.

Observe that one leaky bucket cannot police parameters associated with different time scales. For example, to control the peak rate, the leaky bucket should have LCR=PCR and a small pseudo-buffer to allow small time scale fluctuations of cell phases caused by possible variable cell delay. On the other hand, surveillance of the average cell rate requires LCR close to the average cell rate and a large pseudo-buffer to accommodate cell bursts associated with large time scale fluctuations. As a result the UPC mechanism can consist of two or more parallel leaky buckets, each of them designed for policing different parameters. In such a case it should be provided that the cells discarded or tagged by one leaky bucket do not increase the pseudo-buffer occupancies of the other leaky buckets.

There is an optional control function associated with the UPC mechanism which is referred to as *shaping*. It also acts directly on the cells in order to influence and protect the QoS of the accepted connections but is located at the user premises

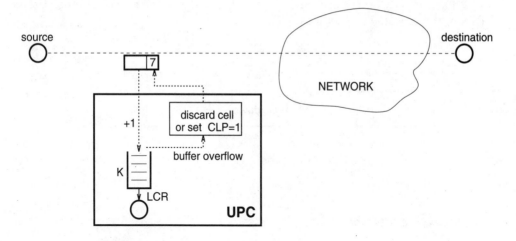

Figure 1.7. *Leaky bucket scheme.*

and is under the user control; see Figure 1.6. The objective of the shaper is to form the outgoing cell stream so that it conforms to a traffic contract approved at the connection setup time. There are three main motivations for introducing the shaping function. Firstly, the UPC algorithm discards or tags the non-conforming cells without taking into account the importance of their contents. The shaper can do the same job in a much more intelligent way by removing or tagging cells of smaller or non-critical value. Secondly, the information from the shaper can be used to influence, if possible, the cell generation process so that the bandwidth reserved for the connection is well utilized. Finally, if the information can be delayed, the parameters of the original cell stream, which can be either unpredictable or not well suited for statistical multiplexing, can be modified in order to make it more attractive to the network operator and thus reduce the connection cost.

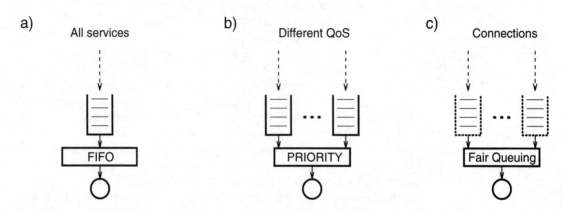

Figure 1.8. *Scheduling categories.*

TABLE 1.1 Examples of services on the application layer.

application	peak rate (bps)	peak rate/ average rate	burst duration (sec)	mean holding time (sec)
telephony	64 K	1		100
document retrieval	64 K	200	0.25	300
data on demand	64 K	200	0.04	30
video telephony	10 M	5	1.00	100
color fax	2 M	1		3
video retrieval	10 M	5	10.00	180
file transfer	2 M	1		1

The second category of QoS control functions illustrated in Figure 1.6 is located at the ATM switch output ports and concerns scheduling of the cell service by the ATM links. Three basic scheduling categories, illustrated in Figure 1.8, can be considered for ATM links. The simplest one employs a *first-in-first-out* (FIFO) buffer (Figure 1.8a). The lower priority for cells tagged with CLP=1 can be achieved by discarding these cells when the buffer state exceeds a certain threshold. In this option all services are provided with the most stringent QoS requirements. Since some of the services can have the QoS requirements relaxed, the network utilization can be increased by implementing a multi-priority system where services with distinct QoS requirements are allocated to different priorities. This option can be realized by several FIFO buffers operating with non-preemptive priority, as shown in Figure 1.8b. The third scheduling category (Figure 1.8c) is based on a fair queuing concept which is an application of the *general processor sharing* (GPS) algorithm, e.g. (Demers, Keshav, and Shenker, 1989; Parekh and Gallager, 1992; Roberts, 1994). In this case the scheduler assures that each connection is guaranteed a *minimum cell rate* (MCR) and a buffer memory for K cells. By proper selection of these parameters, in relation to the connection contracts, the scheduler can provide that, when the connection cell process conforms to the contract, the required QoS is guaranteed regardless of the behavior of the other connections. This obvious advantage is balanced by high complexity of the hardware and management algorithm implementation.

1.3 Service Categories

Due to minimal constraints on the connection's information rate and capability of ensuring very stringent QoS constraints, ATM based networks can be used for virtually all telecommunication services. A sample of these services is listed in Table 1.1 together with basic parameters defined in the RACE project BLNT (Gallassi, Rigolio, and Verri, 1990). Some of the QoS requirements for different service groups are approximately quantified in Figure 1.9.

Note that since all of these services are transformed into uniform ATM cells

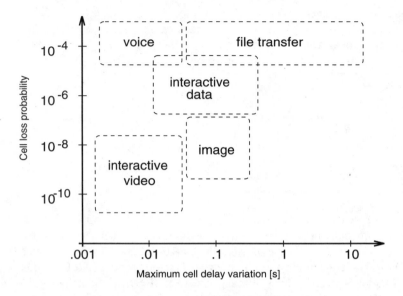

Figure 1.9. *QoS requirements for different type of services.*

by the AAL layer, the ATM layer could treat them equally by providing the most stringent QoS requirements for all connections. Nevertheless, in order to increase the utilization of network resources it is convenient to define several distinct service categories which can be treated differently on the ATM layer. In general these categories are a function of the required QoS guarantees, cell rate distribution class, and ability of the source to control its information rate during the connection. It should be emphasized that in general there is no unique correspondence between the services on the application layer and a service category on the ATM layer. For example, a data connection can use either a service category that requires information rate control by the source or a service category without this requirement which provides guaranteed delivery time but consumes more resources and therefore is more expensive. In the following we describe two definitions of ATM layer service categories described in (ATM Forum, 1995) and (ITU-T Rec.I.371, 1995), respectively.

Figure 1.10 presents a classification of the ATM layer service categories according to the ATM Forum specification. The *constant bit rate* (CBR) service category is intended for real-time applications which require tight constraints on CLR, CDV, and CTD. The traffic contract is defined by PCR, which corresponds to constant availability of a fixed quantity of bandwidth, and by *cell delay variation tolerance* (CDVT). The CDVT parameter defines the maximum cell delay variation for a stream entering the UNI which should not cause cell rejection by the UPC. Examples of typical applications are CBR video and audio connections.

The real-time *variable bit rate* (VBR) service category is similar to the CBR category except that the traffic contract is defined by the *sustainable cell rate* (SCR) and *burst tolerance* (BT) in addition to PCR/CDTV. The sustainable cell

1.3 Service Categories

	CBR	real-time VBR	non-real-time VBR	ABR	UBR
CLR Cell Loss Rate	specified	specified	specified	specified	unspecified
CTD Cell Transfer Delay	specified	specified	specified	unspecified	unspecified
CDV Cell Delay Variation	specified	specified	unspecified	unspecified	unspecified
Traffic Descriptors (Contract)	PCR/CDVT	PCR/CDVT SCR/BT	PCR/CDVT SCR/BT	PCR/CDVT MCR/ACR	PCR/CDVT
Flow Control	no	no	no	yes	no

Figure 1.10. *ATM layer service categories — ATM Forum.*

rate and burst tolerance define a continuous state leaky bucket algorithm and they are related to the previously defined discrete leaky bucket parameters as follows: SCR=LCR, BT=K/LCR. The declaration of the parameters imposing limits on some cell process characteristics allows the network to reserve, on average, a bandwidth smaller than the peak cell rate. This can be achieved due to the statistical multiplexing of VBR connections. Examples of typical applications are VBR video and voice connections. The specification of the non-real-time VBR service category differs from the real-time case only by omitting the cell delay variation constraint.

The *available bit rate* (ABR) service category is intended for non-real-time applications which can control, on demand, their transmission rate in a certain range (e.g., file transfer). The traffic contract is defined by PCR/CDVT and MCR. Based on this contract the network guarantees bandwidth at least equal to MCR. When the network has sufficient spare bandwidth, the connection is allowed to increase its cell rate up to the *allowed cell rate* (ACR); MCR $\leq$ ACR $\leq$ PCR. The value of ACR is updated periodically by a flow control algorithm in the transit ATM switches and then delivered to the traffic source by means of the *resource management* (RM) cells (more details of this scheme are given in Section 2.1.1). The contract does not specify CTD and CDV constraints, but it is expected that the network does not excessively delay the admitted cells.

The last service category, *unspecified bit rate* (UBR), is intended for non-real-time applications which do not have tight constraints on the cell delay and cell delay variations. According to the current recommendations, the network does not provide any QoS guarantees and does not reserve any network resources for this service. In particular the PCR/CDVT parameters specified in the contract are not used for bandwidth reservation purposes (but optionally can be used for the call admission control or policing purpose).

Most of the service categories defined in (ITU-T Rec.I.371, 1995) are similar to the ones defined by the ATM Forum. These categories are classified in Figure 1.11. From this figure it is easy to ascertain that the *deterministic bit rate* (DBR), *sta-*

	DBR	SBR	ABT	ABR
CLR Cell Loss Rate	specified			
CTD Cell Transfer Delay	specified			unspecified
CDV Cell Delay Variation	specified			unspecified
Traffic Descriptors (Contract)	PCR/CDVT	PCR/CDVT SCR/BT		PCR/CDVT MCR/ACR
Flow Control	no			yes

Figure 1.11. *ATM layer service categories — ITU-T.*

tistical bit rate (SBR), and ABR categories correspond to the CBR, VBR, and ABR categories in the ATM Forum classification. The new *ATM block transfer* (ABT) category has some features of the DBR, SBR, and ABR categories. It is intended for data transfer applications where the information can be transmitted in blocks of cells. The connection traffic contract is defined, in a manner similar to the SBR (VBR) service category, by PCR/CDVT and SCR/BT. Besides this contract each block admission is negotiated with the network for what can be interpreted as a traffic contract for block transfer with PCR' (PCR' $\leq$ PCR). If there is sufficient free capacity in the network, the block is transmitted with guaranteed fixed rate PCR' which is similar to the DBR category. Two options of ABT service are defined: *ABT with delayed transmission* (ABT/DT) and *ABT with immediate transmission* (ABT/IT). In the first case the source sends a block after receiving confirmation of the request acceptance from the network. In the case of ABT/IT the block is sent together with the request and if the resources are not available in the network the block is discarded and the source has to repeat the procedure. The block admission negotiation procedure is similar to the rate negotiation in the ABR category and is also based on the RM cells. The probability of block rejection is a function of the requested sustainable cell rate (more details on this scheme are given in Section 2.1.1).

From the resource management and traffic control viewpoint it is convenient to define two service macro-categories. The first one covers services which cannot use cell flow control mechanisms and henceforth is referred to as a *non-controllable traffic parameters* (NCTP) service macro-category. This macro-category consists of the CBR and VBR services. The second macro-category includes services which can use cell flow control mechanisms and henceforth is referred to as *controllable traffic parameters* (CTP) service macro-category. This macro-category consists of the ABR services. Also, the ABT services can be included in this category since the average rate of the ABT connections is a function of the network load. Nevertheless, from the accepted block viewpoint the ABT service is more similar to the DBR service. For this reason in Figure 1.11 the ABT service is classified as the one with and without flow control capabilities. Note that the UBR category is not included in the macro-categories. This follows from the fact that UBR services have the lowest priority and do not require any resource management and traffic control actions.

Observe that the ATM standard assumes that before a cell stream can be sent from the origin to the destination node, a connection (defined by VPIs and VCIs) has to be established across the network. This feature precludes introduction of connectionless services (these are services which send packets with the destination address without prior connection establishment) directly on the ATM layer. A question arises how connectionless services from the application layer should be handled. In general one can consider two possibilities. In the first one a connectionless packet is divided into ATM cells (at the AAL layer) which are sent to the destination node by means of a pre-established semi-permanent connection. In the second option an overlay packet switched network can be created by means of *connectionless servers*. This concept is illustrated in Figure 1.12. Here the connectionless servers are located in some of the ATM nodes and are connected to at least one output and one input port of the switch. The connectionless servers are connected among themselves by semi-permanent connections on the ATM layer. A connectionless packet arriving at the UNI of the ATM network is sent to the nearest connectionless server (after splitting into ATM cells, with appropriate VPI and VCI, at the AAL layer). Upon arrival at the connectionless server the packet is reassembled (at the AAL layer). Then, based on the destination address included in the packet and possibly the traffic state information, the routing algorithm chooses the next connectionless server to which the packet should be sent (after division into the ATM cells at the AAL layer). Obviously the second option can provide better resource utilization.

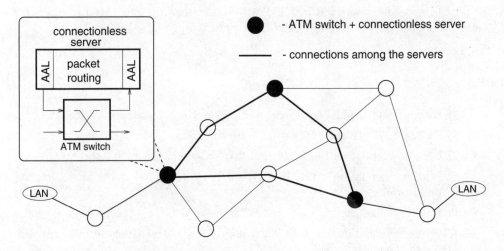

Figure 1.12. *An architecture for connectionless services.*

1.4 Discussion and Bibliographic Notes

In this chapter we presented an introduction to the basic concepts and functions of ATM based networks. The objective was to present the issues on a level required by subsequent parts of the book. The reader interested in a more detailed introduction to broadband networks can refer to several books: e.g. (De Prycker, 1991; Acampora, 1994; Chen and Liu, 1995). Also, important information can be found in the documents of standardization and recommendation bodies, principally ITU-T (ITU TelecommunicationStandarization Sector, formerly CCITT) and ATM Forum, although in general these documents require a lot of time to digest.

Service categories on the ATM layer were discussed in the last part of this chapter. As will be shown in the ensuing chapters, introduction of these service categories is very important from the resource utilization and management perspective. In the presentation we focused on the generic features of the ATM layer service categories defined by the ITU and ATM Forum. The detailed definitions of these categories can be found in (ITU-T Rec.I.371, 1995; ATM Forum, 1995).

In Section 1.3 we also gave samples of basic parameters describing the services on the application layer and the QoS requirements for such services on the ATM layer. The particular numbers should be treated only as representative since their values depend strongly on the coding techniques and the required quality of the received information (e.g., image resolution). A thorough study of these issues can be found in an extensive literature, e.g. (Anagnostou, et al., 1991; Woodruff and Kositpaiboon, 1990; Nomura, Fujii, and Ohta, 1989). More information on connectionless service over ATM networks can be found in (Vickers and Suda, 1994).

There are several possible ATM switch architectures. Although this issue is interesting and complex, it is not covered in this chapter due to the fact that from the network resource management viewpoint, in most cases, the switch can be treated as a transparent unit except for the cell buffering taking place in the switch ports. The reader interested in ATM switch architectures is referred to extensive reviews in (Acampora, 1994; Tobagi, 1990; J.S.A.C., 1991).

References

ATM Forum. 1993. ATM User-Network Interface Specification. Prentice Hall.

ATM Forum. 1995. Traffic Management Specification. 95-0013R2.

CCITT Recommendation I.311. 1991. B-ISDN General Network Aspects. Geneva.

CCITT Recommendation I.321. 1991. B-ISDN Protocol Reference Model and Its Application. Geneva.

CCITT Recommendation I.361. 1991. B-ISDN ATM Layer Specification. Geneva.

ITU-T Recommendation I.371. 1995. Traffic Control and Congestion Control in B-ISDN. Geneva.

J.S.A.C. 1991. *IEEE Journal on Selected Areas in Communication,* 9(8):1157–1360.

References

Acampora, A.S. 1994. *An Introduction to Broadband Networks.* New York: Plenum Press.

Anagnostou, M.E., et al. 1991. Quality of service requirements in ATM based B-ISDNs. *Computer Communications*, 14(4):197–204.

Chen, T.M., and Liu, S.S. 1995. *ATM Switching Systems.* Artech House.

Demers, A., Keshav, S., and Shenker, S. 1989. Analysis and simulation of a fair queuing algorithm. *Proceedings of ACM SIGCOM'89*, pp. 1–12.

De Prycker, M. 1991. *Asynchronous Transfer Mode — Solution for Broadband ISDN.* Ellis Horwood Limited.

Gallassi, G., Rigolio, G., and Verri, L. 1990. Resource management and dimensioning in ATM networks. *IEEE Network Magazine*, 5:8–17.

Hluchyj, M.G., and Karol, M.J. 1988. Queueing in high-performance packet switching. *IEEE Journal on Selected Areas in Communication,* 6(9):1587–1597.

Nomura, M., Fujii, T., and Ohta, N. 1989. Basic characteristics of variable rate video coding in ATM environment. *IEEE Journal on Selected Areas in Communication,* 7(5):752–760.

Parekh, A., and Gallager, R. 1992. A generalized processor sharing approach to flow control in integrated services networks: The single node case. *Proceedings of IEEE INFOCOM'92*, pp. 915–924. IEEE Computer Society Press.

Roberts, J.W. 1994. Virtual spacing for flexible traffic control. *International Journal of Communication Systems (John Wiley & Sons)*, 7:307–318.

Tobagi, F.A. 1990. Fast packet switch architectures for broadband integrated services digital networks. *Proceedings of the IEEE*, 78(1):133–167.

Woodruff, G., and Kositpaiboon, R. 1990. Multimedia traffic management principles for guaranteed ATM network performance. *IEEE Journal on Selected Areas in Communication*, 8(3):437–446.

Vickers, B.J., and Suda, T. 1994. Connectionless service for public ATM networks. *IEEE Communications Magazine*, (8):34–42.

Chapter

2

Resource Management and Traffic Control Issues

THE RESOURCE management and traffic control issues can be divided into four classes. The first concerns the QoS characteristics. In particular the *resource management and traffic control* (RM&TC) algorithms should ensure that the QoS requirements are fulfilled for all service categories. The second class deals with the problem of fairness which is interpreted as fair access to network resources for all users within the established priority and economic constraints (on the cell and connection layers). Efficiency issues constitute the third class, the objective of which is maximization of the network resources' utilization and minimization of their cost. The last class of issues relates to the network survivability problem. In general RM&TC algorithms should provide self-healing capabilities in face of both network component failures and unexpected traffic patterns.

The integration of all services into one uniform transport layer is seen as a major advantage of the ATM standard. Nevertheless, this integration also creates several new problems. In particular a broad range of services, traffic characteristic, time scales, and performance constraints, which are integrated into one transport system, causes the RM&TC issues to become very complex and difficult. One can compare this situation to traffic control on a highway where supersonic jet traffic is integrated with the automobile traffic. That is why, in many cases, the objective of resource management and traffic control tools in ATM based networks is to provide some kind of virtual separation between different connections or services.

The intent of this book is to present a framework and models which would help to overcome the problems and complexity of RM&TC in multi-service broadband networks. In this chapter we discuss the main issues of resource management and traffic control. At the same time, we present a general framework for RM&TC. The framework is based on decomposition of the RM&TC functions. The decomposition is done by means of virtual networks, which allows one to customize functions for different services and user groups, and by means of layered traffic entities (cell, connection, virtual network, and physical network layers) associated with different time scales. This decomposition provides a vehicle for coping with the problem's enormous complexity. In addition, the decomposition approach provides that only the cell layer RM&TC algorithms are transport technology dependent. The higher layer RM&TC issues are driven rather by the service characteristics and requirements. As a result most of the RM&TC models and algorithms presented in this book are less affected by the aging of technology and are also applicable to other types of networks such as circuit-switched and wireless. This chapter also gives a summarized introduction to the contents and interrelation of the ensuing chapters.

2.1 Decomposition of the Resource Management Functions

In this section we describe a framework which provides a coherent and effective structure for dealing with RM&TC complexities. The central concept of this framework is to decompose the RM&TC functions into relatively independent components which are easy to manage. There are two main vehicles which can be used to achieve this decomposition. The first is the concept of virtual networks. A *virtual network* (VN) is defined by a set of nodes and *virtual network links* (VNL) connecting the nodes. A VNL defines a path in the physical network which connects two nodes belonging to VN. Some resources can be allocated to VNs. In particular buffers and bandwidth can be allocated to virtual network links. Virtual networks can be used to separate RM&TC functions for services with very different traffic characteristics and QoS requirements (e.g., CTP vs. NCTP services) or to customize RM&TC functions for a group of users (e.g., virtual private networks, multi-point connections). More information on this central concept can be found in Section 2.1.3 and Chapter 9.

The second decomposition tool takes advantage of the very different time scales existing in the considered system. This feature allows us to decompose the RM&TC functions according to the layered traffic entities associated with different time scales. The four chosen layers are cell layer, connection layer, virtual network layer, and physical network layer. The main components of the RM&TC layered structure and their inter-relationship are depicted in Figure 2.1.

In the remainder of this section we discuss the main RM&TC issues associated with each of the layers. At the same time we formulate a framework for RM&TC whose general architecture is described in Section 2.2. There are five principles on which the framework is based:

1. In order to facilitate the cooperation among the RM&TC algorithms on different layers these algorithms should have a simple interface based on a common notion which can reflect in a simple way the allocation of the basic network resources such as buffers and bandwidth. In this book we will use for this purpose the concept of *qualified bandwidth* defined later on.

2. All available information should be used in an optimal way. In particular it

2.1 Decomposition of the Resource Management Functions

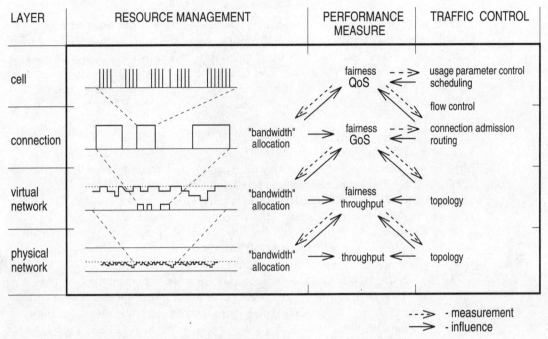

Figure 2.1. *Decomposition of the RM&TC functions.*

means that the algorithms should use, if possible, explicit measurements of the relevant network states and source traffic parameter declarations.

3. The algorithms should not be restrictive in the sense that even an undeclared traffic increase should not be rejected if there are available resources.

4. An economic factor should be introduced directly into the RM&TC algorithms. This enables the integration of the objectives of network control, planning, and dimensioning.

5. The algorithms for different service categories should be based on the same database structure and similar signaling protocol in order to reduce complexity and cost of RM&TC.

The notion of *qualified bandwidth* is defined as the bandwidth required to carry a connection with given traffic characteristics of the cell process so that the required QoS constraints are met. Thus the concept of *qualified bandwidth* also includes buffer allocation, the scheduling mechanism, and the flow control algorithms since these elements influence the QoS. To simplify the presentation in the following we will use the name "bandwidth" as the synonym of *qualified bandwidth*. Note that the term "bandwidth" management corresponds to management of the basic network resources, bandwidth and buffers, which is the focus of this book.

2.1.1 Cell layer

The RM&TC functions associated with the cell layer are the ones which act on (or are influenced by) individual cells or the cell state (e.g., switch output buffer occupancy). These functions can be grouped into four categories: usage parameter control (enforcement of the connection traffic contract), scheduling (priorities and fairness), cell flow control, and measurements. The usage parameter control and scheduling functions as well as their implementations were previously described in Section 1.2.

The flow control function concerns only CTP (controllable traffic parameters) services and is also known as reactive traffic control or congestion control. The objective of the flow control function is to adapt the connection cell rate to the network traffic condition. There are two distinct parts to this function. The first is measurement of the traffic conditions and in particular recognition of the congestion states in the switch output ports. The second is adaptation of the connection rate so that congestion is avoided and "bandwidth" is well utilized. There are three basic approaches to the cell flow control: rate-based schemes, credit-based schemes (e.g., window mechanisms), and fast reservation protocols.

In the rate-based schemes the network state (e.g., buffer occupancy) is measured directly at the switch output ports. Using this information the connection rate is adjusted at the source. In particular the source can be informed about the network state by a single-bit binary feedback (requires one bit in the ATM cell header) or by an explicit rate scheme using resource management cells. The explicit rate schemes have several advantages. They are faster and the recommended rates can be easily policed at the network entry (using information from the RM cells).

In the credit-based schemes the information about the network state is measured indirectly via cell delay on the network path (or link). A typical example is a window mechanism where the source has a certain number of credits called the window size. Each data unit transmitted by the source (e.g., cell) takes one credit which is returned to the source by the data unit receiver once the data unit has reached the destination. In the case of large delays in the network the source may run out of the credits and the transmission is suspended (also the window size can be reduced for some time). A credit-based scheme can be implemented on user-to-user or link-by-link basis.

In the case of fast reservation protocols it is assumed that the source has to reserve the "bandwidth" for transmission of a cell burst at a given rate. In one possible implementation a resource management cell is sent along the connection path. In each of the transit nodes the required "bandwidth" is reserved, if available. Once the reservation is confirmed, the source starts the transmission. The reserved "bandwidth" is released once the burst is sent.

It can be easily noticed that the last category of RM&TC functions on the cell layer, measurements, is present in each of the previously described function categories (usage parameters control, scheduling, flow control). Besides these applications, measurements of the cell process parameters (delay, losses, rate, etc.) in the switch output ports can be used by the connection layer to adapt some procedures such as routing, connection admission, and "bandwidth" allocation.

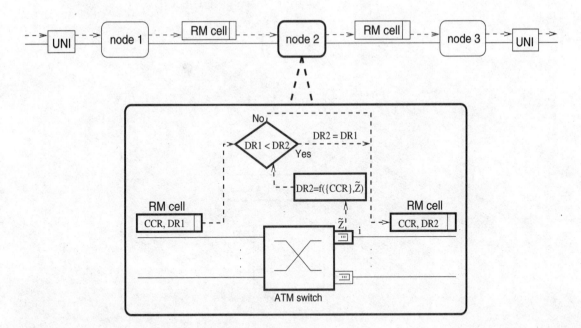

Figure 2.2. *Explicit rate flow control for ABR services.*

Explicit rate-based flow control

The presented scope of the RM&TC functions on the cell layer indicates that there is a large number of important and interesting issues in this area. Nevertheless, since the focus of the book is on "bandwidth" management, in the following we elaborate only on mechanisms which involve explicit "bandwidth" management algorithms within the network. The most important cases of such mechanisms are the rate-based flow control algorithms which use the resource management cells. These algorithms correspond to ABR and ABT service categories defined by the ATM Forum and ITU (see Section 1.3). The basic idea of the ABR protocol is illustrated in Figure 2.2. During connection, at least every 100 ms, the source sends to the destination a resource management cell with information about the *current cell rate* (CCR). At each node the RM&TC algorithm evaluates the *desired rate* (DR) for each VC based on the declared CCRs (received from RM cells) and measurements of the cell process parameters, $\tilde{Z}$, in the switch output port used by VCs. The desired rates are inserted into the passing RM cells which are sent back to the source by the destination node. If the RM cell is lost, the source automatically reduces the rate. The desired rate can be evaluated by an iterative algorithm with the objective of providing high "bandwidth" utilization and a fair share of "bandwidth" to all users.

The principle of the fast reservation protocol for ABT services is illustrated in Figure 2.3 (observe the similarity with the ABR protocol). In this case the source sends to the destination a resource management cell with a request for reservation of

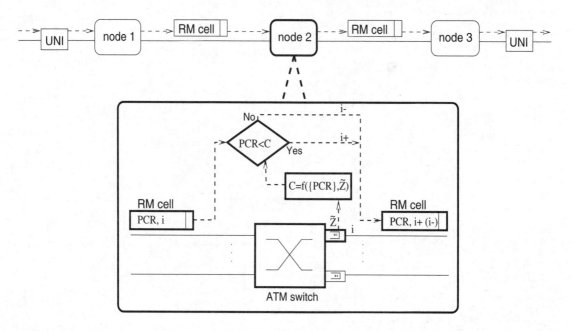

Figure 2.3. *Fast reservation protocol for ABT services.*

the "bandwidth" corresponding to the required peak cell rate (PCR) for cell block transmission. At each node the RM&TC algorithm evaluates the residual capacity (C) for each outgoing link based on previous reservations (received from RM cells) and possible measurements of the cell process parameters, $\tilde{Z}$, in the switch output ports. The decision (i+ or i–) is inserted into the RM cell which is sent back to the source by the destination node. If the decision is negative, the source suspends the block transmission which was started at the time of sending the request in the case of ABT/IT protocol (see Section 1.3). If the decision is positive, the source starts transmission of the block in the case of ABT/DT protocol. Both versions of the protocol can repeat the request after a certain time if at the moment there is not sufficient resources available. Note that the QoS service for ABT services is characterized by two different sets of parameters. The first one consist of cell characteristics (CLR, CDV, CTD) within the accepted cell block. The second comprises cell block rejection probability or/and average cell block transmission delay. The described algorithm ensures that the required constraints for the first set are met. To provide certain guarantees for the second set, the number of accepted connections has to be limited. This problem is equivalent to connection admission control based on logical "bandwidth" allocation, and will be described in the next section.

Note that although both described mechanisms, ABR and ABT, can take into account measurements on the cell layer, the outcome of the algorithm is an adaptive "bandwidth" allocation to a connection. That is why in Figure 2.1 there is no "bandwidth" allocation function on the cell layer. This feature indicates that the flow control function is involved in both the cell layer (measurement) and the connection layer ("bandwidth" allocation). The involvement of the flow control function in adaptive "bandwidth" allocation to connections will become apparent in Chapter 4 where an adaptive "bandwidth" allocation scheme for VBR (or SBR) services, which uses basically the same protocol as ABT and ABR schemes, is discussed.

It should be mentioned that the discussed flow control mechanisms for CTP services are relatively easy to implement for point-to-point connections. The situation is much more complex when multi-casting connections are considered. The main difficulty results from the fact that the signaling protocols used in flow control mechanisms are not well suited for multi-point connections.

2.1.2 Connection layer

The main RM&TC functions operating within the connection layer are: logical "bandwidth" allocation to connections, connection admission control, routing, and measurements. The objective of these functions is to deliver high "bandwidth" utilization and fair access for all users (within the considered VN) providing that the QoS constraints on the cell layer are met. The principal *grade of service* (GoS) metric on the connection layer is connection rejection probability for a given connection class. A connection class can be defined by the traffic characteristics (e.g., "bandwidth" requirement, mean holding time), origin-destination node pair, QoS requirements, and GoS requirements. The issue of fairness deals with GoS distribution among the connection classes.

Logical "bandwidth" allocation to connections. The crucial element of the resource management on the connection layer is the logical allocation of the link "bandwidth" to the connections (virtual channel connections). The importance of this allocation lies in the fact that it allows us to separate traffic control algorithms on the connection and higher layers from the cell layer. This feature can simplify significantly the RM&TC functions. Moreover, observe that the "bandwidth" allocation to virtual channels makes the system similar to the circuit-switched environment on the connection layer. This property allows us to apply the well developed resource and traffic management techniques from the circuit-switched to ATM environment. Thus the two, historically almost separated, RM&TC worlds, circuit-switched and packet-switched, can be integrated by using the best elements from each of them.

The issue of logical "bandwidth" allocation to connections is quite complex due to several requirements. In particular the "bandwidth" allocation should provide high resource utilization and QoS guarantees at the same time. Moreover, the QoS constraints can be different for different services (virtual networks) so the "bandwidth" allocation can also be a function of the QoS constraints. Another problem arises when the connection parameters are not well known in advance. In this case an adaptive "bandwidth" allocation based on measurements would be the most effective approach. All these issues are addressed in Chapters 3 and 4. In Section 3.1

we define and discuss different approaches to logical "bandwidth" allocation which can vary from minimum rate allocation via equivalent bandwidth allocation to peak rate allocation. The equivalent bandwidth approach is most attractive for NCTP services since it can take into account statistical multiplexing of VBR (or SBR) sources. Several models and approximations for equivalent bandwidth allocation are presented in Section 3.2. They include an approach for "bandwidth" allocation in multi-priority systems which can serve as a basis for resource management in virtual networks associated with different QoS requirements. In general the models for equivalent bandwidth allocation assume that the source parameters conform to the source declarations or that the source parameters are limited by parameters of a policing mechanism (UPC). Since some sources may not know their parameters in advance, they will tend to declare parameters with a safety margin. Moreover, the UPC cannot adequately control some of the statistical parameters of the connection (e.g., average rate) so again some safety margin has to be applied. These features, together with the possible non-stationary behavior of the sources, can cause network resources underutilization. To avoid these drawbacks, adaptive "bandwidth" allocation based on source declarations and measurements of some cell layer characteristics can be applied. In Section 4.2 we describe such an adaptive approach for NCTP services. The signaling protocol and database structure of this algorithm is almost identical to the ones recommended by ATM Forum and ITU for CTP services. Besides reducing the complexity and cost of RM&TC algorithms, the integration of the two algorithms provides a potential for further increases in resource utilization. The key element of this adaptive "bandwidth" allocation for NCTP services is an estimation procedure which uses measurements and an optimization framework based on a linear Kalman filter (a short introduction to the Kalman filter technique is given in Appendix A).

For further considerations it is convenient to introduce the concepts of well behaved and *worst case* sources. The well behaved source conforms to the declared source process parameters. The notion of *worst case* source is associated with the UPC mechanism parameters. Namely, due to the fact that the UPC mechanism cannot ideally control statistical parameters declared by the source, a malicious source can generate traffic which passes through the UPC mechanism but uses more network resources than the well behaved source for which the UPC mechanism was designed. The *worst case* source corresponds to the malicious user who consumes most of the network resources for given UPC parameters. In general this definition is not unique. In this book we define the *worst case* source as the one which requires maximum equivalent bandwidth allocation in a given network state.

Connection admission control. The *connection admission control* (CAC) function has two distinct aspects. The first one deals with the question of whether a new connection can be admitted on a particular link or path so that the QoS constraints are met. Henceforth this function will be referred to as CAC^l_{QoS} or CAC^p_{QoS} for link and path problems, respectively. Observe that, once the link "bandwidth" required by the new connection is evaluated, the CAC^l_{QoS} function is trivial since it is sufficient to check whether the residual (free) "bandwidth" on the considered link is larger or equal to the demanded "bandwidth." This also simplifies the CAC^p_{QoS}

function since in most cases it can be decomposed into a set of CAC^l_{QoS} problems. The algorithms for CAC_{QoS}, based on logical "bandwidth" allocation, are discussed in Chapters 3 and 4.

The second aspect of connection admission control deals with the question of whether the connection should be admitted on a link or path which ensures that the QoS constraints are met (i.e., there is sufficient residual "bandwidth"). This function is related to the GoS characteristics (connection layer performance metrics) and is henceforth referred to as CAC^l_{GoS} or CAC^p_{GoS} for link and path problem, respectively. There might be several reasons for rejecting the connection, even though it could be accepted from the resource availability viewpoint. For example, in some circumstances rejection of a multi-link connection can make room for accepting several connections on each of the links composing the path. Thus the network utilization can be increased. In the case of a mixture of wide-band and narrow-band connections, the rejection of narrow-band connections in some states can provide fair access to resources for wide-band connections which would encounter higher loss probability otherwise. In summary the objective of CAC_{GoS} is to provide fair and efficient access to the network resources by changing priorities for different connection classes when compared with the policy of accepting connections whenever possible. The CAC_{GoS} issues are analyzed extensively in Chapters 5, 6, and 8.

Decomposition of the CAC problem into a set of link problems (CAC^l_{QoS}, CAC^l_{GoS}) requires efficient protocols for communication between the CAC manager and link resource managers. The traditional approach is based on meta-signaling, B-ISUP standard (ITU-T Rec. Q.2761-Q.2764, 1993), which provides a general protocol for communication between the network control units. This generality requires significant processing capacity and can cause noticeable delays in the CAC procedure; see, e.g. (Veeraraghavan et al., 1995). In this book we advocate an alternative approach which is based on resource management cells. In this case the protocol is faster, simpler, and does not require additional processing power since it is consistent with the signaling protocol for ABR and ABT services (see Figures 2.2 and 2.3). This approach is described in Section 4.1 and its ramification on resource management on the virtual network layer are discussed in Section 9.3.

Connection routing. The objective of connection routing is to find an optimal path to carry the connection. The optimality criterion is not straightforward due to several potential metrics which should be considered. The metrics can be grouped into two categories: QoS metrics and resource cost metrics. The QoS metrics were already defined (CLR, CTD, and CDV) and in general each connection class requires that the QoS metrics do not exceed predefined constraints. The resource cost metrics are the ones which take into account the cost of the resources engaged in the connection. This cost can be static or dynamic (e.g., dependent on network traffic parameters or a state) and can be subject to constraints. In general the problem of routing path optimization can be formulated as a multi-criterion minimization problem with constraints. Nevertheless, in broadband networks the issue of efficient resource utilization is dominant and in this book we focus on minimization of the "bandwidth" cost subject to the QoS and cost constraints. The issues and algorithms associated with this approach are discussed in Chapter 5 for point-to-point connections and in Chapter 6 for multi-point connections.

The concept of virtual paths is closely related to the issue of routing. In this book we distinguish two types of VPs from a connection routing viewpoint. If some connections are carried on two or more concatenated virtual paths, we name these paths *node-to-node virtual paths* (VP_{nn}). On the other hand the *end-to-end virtual path* (VP_{ee}) denotes a virtual path used by VC connections carried only on this path. These options are illustrated in Figure 2.4. The implementation of routing algorithms can be significantly simplified by utilization of the end-to-end virtual path connections assigned to alternative paths considered for routing recommendation. In this case establishment of a new VC connection does not require any changes in the routing tables of the transient nodes. Otherwise the routing tables between concatenated node-to-node VPCs have to be updated. The issue of potential "bandwidth" allocation to VPC is discussed together with the virtual network layer issues.

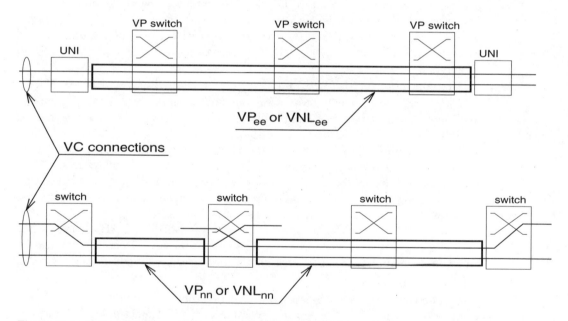

Figure 2.4. *Examples of end-to-end and node-to-node VPs or VNLs.*

CAC & routing. In general the issues of connection admission control and routing are closely connected. In particular, "bandwidth" cost constraints in the connection routing problem correspond to a CAC_{GoS}^p problem aiming at providing high network throughput and/or priority for certain connection classes. This feature is shown formally in Chapter 5 where a model for CAC & routing is derived from Markov decision theory (a short introduction to Markov decision theory is given in Appendix B). In this case each connection is characterized by a reward parameter, and the objective of CAC & routing is to maximize the reward from the accepted connections. The CAC & routing decision is based on state-dependent link shadow prices interpreted as a predicted price for seizing link "bandwidth" by the connection. The path with a minimal shadow price cost (sum of link shadow prices over the links constituting the path) is recommended by routing functions. The CAC^p_{GoS} function compares the value of the new connection reward parameter with the recommended path cost. If the gain is positive, the connection is accepted; otherwise the connection is rejected. There are several important features of the reward maximization approach. Firstly, the GoS of each connection class can be simply controlled by varying the values of connection reward parameters (the higher the reward parameter, the smaller connection, the loss probability). Secondly, the algorithm is based on measurements of connection flows in the network so it can automatically adapt to varying traffic conditions. Also, the algorithm can easily incorporate economic factors by relating the real tariff or resources cost to the reward parameters and reward from the network. Finally the approach provides a general framework for studying, constructing, and optimizing other CAC & routing strategies. All these features are discussed in Sections 5.1.2 and 5.2 for point-to-point connections and in Section 6.3 for multi-point connections. It is also shown that in case of simplified CAC & routing strategies the CAC^p_{GoS} function can be separated from the routing function in order to reduce complexity. In this case the CAC^p_{GoS} function is usually decomposed into a set of link CAC^l_{GoS} functions (in other words the CAC^p_{GoS} decision is done independently by each link constituting the chosen path). Several CAC^l_{GoS} models are discussed in Chapter 8.

Performance models on connection layer. Network performance models on the connection layer are required in many phases of network development and operation. In particular they can serve for comparison of different design options, optimization of routing algorithms, network resource allocation, network survivability tests, identification of unexpected problems, etc. Unfortunately exact performance models are not practical for any network, except the smallest ones, due to the enormous cardinality of the problem. In Chapter 7 we describe an approximate methodology based on decomposition of the network problem into a set of link or path problems which can be solved by means of repeated substitution. Models based on this methodology can be applied for networks with different CAC & routing strategies. In Section 7.2 we extend this methodology to networks with CAC & routing based on the reward maximization principle. In this case the model evaluates the network performance and optimizes the routing parameters at the same time.

Fairness issue. Many of the CAC & routing strategies provide tools for achieving practically an arbitrary operating point defined as the GoS distribution among the connection classes. In particular this is the case for the reward maximization approach which can control almost continuously the GoS distribution. The question arises: Which operating point is both fair and efficient? The natural notion of fairness is GoS equalization but it can be shown that this approach might lead to low efficiency. On the other hand, the traditional single criterion formulations, involving global throughput maximization or minimum average delay, yield efficient solutions but do not provide any guarantees of fairness. Thus there is a need for a framework which can cope with these dual aspects. This issue is addressed in Chapter 8 where it is shown that co-operative game theory provides an excellent and precise framework for efficiency and fairness trade-off considerations (a short introduction to co-operative game theory is given in Appendix C). In this case the efficiency is provided by the Pareto optimality requirement while fairness is achieved by satisfying a set of fairness axioms.

Application of co-operative game theory to the synthesis and analysis of fair-efficient CAC & routing algorithms is presented in two steps. First, in Section 8.2, we study a one link case which corresponds to the decomposed CAC. In particular, the arbitration schemes from game theory are compared with two traditional CAC objectives: traffic maximization and loss equalization. Moreover, it is analyzed whether simplified policies (co-ordinate convex policies, "bandwidth" reservation policies, and dynamic "bandwidth" reservation policies) can be used to achieve fairness objectives of the arbitration schemes. In Section 8.3 we extend study to the general network case.

Measurements. Measurements on the connection layer constitute an important part of the network resource management system. There are two main applications of measurements. Firstly, they can be used to adapt CAC & routing algorithms to changing traffic patterns or failures. The objective of these measurements is to provide high resource utilization, fairness, robustness, and reliability. The measured parameters can be GoS metrics and/or some parameters of connection processes (e.g., connection flow averages). For example, the reward maximization algorithm, described in Chapter 5, applies measurements of carried connection flows to adapt the "bandwidth" cost function. Also the measured GoS distribution among the connection classes can be used to provide fairness by adapting the reward parameters. Secondly, measurements on the connection layer are necessary for resource management on the higher layers. In particular the GoS metrics and/or connection flow distribution within a virtual (or physical) network can be used to adapt resource allocation and topology of this network.

2.1.3 Virtual network layer

Since the notion of a virtual network in ATM based networks is not standardized we start from a definition which will be applied throughout this book. A *virtual network* (VN) is defined by a set of the network nodes and a set of *virtual network links* (VNL) connecting the nodes. The virtual network link defines a path (consisting of one or more physical links) between two VN nodes and is referred to by a *virtual network link identifier* (VNLI). Several virtual networks can co-exist in a physical network. They can constitute independent entities, but in some cases a virtual network can be nested in another virtual network. Examples of virtual networks are shown in Figure 2.5. As in the case of virtual paths, we define two types of VNLs: node-to-node and end-to-end (see Figure 2.4). A set of resources can be allocated to the virtual network. In particular this set can include "bandwidth"

2.1 Decomposition of the Resource Management Functions

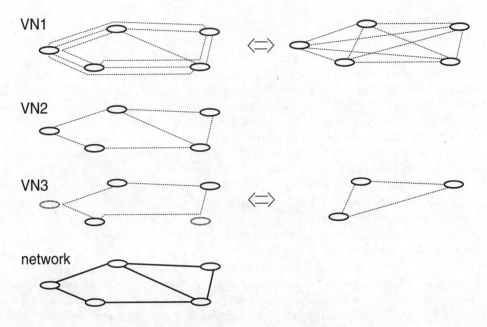

Figure 2.5. *Examples of virtual network configurations.*

and a set of resource management objects. If "bandwidth" is allocated to the VNL, it is managed by a VNL resource manager. In particular if a connection (VCC or VPC or nested VNL) requires allocation of the VNL "bandwidth," it has to ask the VNL manager which decides whether this "bandwidth" can be allocated.

There are two main motivations for using the virtual network concept. The first is separation of management functions in order to make them customized to particular needs of some services and user groups (does not necessarily require resource allocation). The second is virtual separation of "bandwidth" in order to simplify the resource management functions and provide QoS and GoS guarantees for some services and user groups. In general the VN applications can be divided into three classes: service, user and management oriented.

Service oriented virtual networks are created to separate management functions specialized for different services (e.g., NCTP vs. CTP services) and/or to simplify the QoS management (each QoS class is served by a separate virtual network). Allocation of "bandwidth" to service oriented VNs aims at providing sufficient GoS and fairness for different services. Moreover, "bandwidth" allocation to QoS virtual networks can increase "bandwidth" utilization and simplify "bandwidth" allocation to connections as shown in Section 3.2.2. In the service oriented VNs the set of VN nodes includes, in general, all nodes to which the users are connected.

User oriented virtual networks are created for some group users who have specific requirements (e.g., guaranteed throughput, customized control algorithms, "bandwidth" management under the user control, increased security and reliability, group tariff, etc). The two most likely applications are private networks and multi-point connections. Note that in most cases the set of VN nodes in a user oriented VN will include only a subset of all nodes.

Management oriented virtual networks are created to facilitate some of the management functions (not associated with any particular service or user group). The first application is related to fault management and is called a backup virtual network. The "bandwidth" allocated to a backup VN should provide that in case of a link or node failure, all (or a given fraction of) connections affected by a failure can be restored in the backup VN. The second application is aimed at simplification of the "bandwidth" reservation procedure during connection setup. In particular, if all connections are routed via end-to-end VNLs, the connection admission procedure has to ask only the VNL resource manager at the connection origin node for the required "bandwidth" (no need for "bandwidth" reservation in the transit nodes).

In the following we focus on application of the virtual network concept for resource management. There are three generic issues associated with this subject: virtual network design, virtual network adaptation to environment changes, and design of the backup virtual network.

The objective of the VN design is to optimize VN topology and "bandwidth" allocation based on the predicted traffic demand and performance constraints (GoS, fairness criteria). The issue of VN design is discussed in Chapter 9. In particular in Section 9.6 a general model for design of VN topology and "bandwidth" allocation is presented. The model applies the reward maximization routing which is optimized at the same time. This formulation allows one to incorporate economic interrelation between link resource cost, connection setup cost, routing, and network dimensioning.

Once the VN is set up the initial design may be not optimal, due to either discrepancies between the real and predicted traffic patterns or changes in traffic patterns and cost factors. Thus an adaptive procedure is required to adjust the VN "bandwidth" allocation and topology. This procedure should be based on measurements of parameters relevant to the GoS, fairness, and resource cost. In Section 9.7 we discuss this issue including an adaptive model consistent with the model for VN design presented in Section 9.6.

One of the important issues of resource management is protection against link and node failures. The objective is to ensure that all (or a given fraction of) connections affected by the failure can be restored on the non-affected paths (one failure at a time is assumed). A backup virtual network is a very effective tool for this kind of fault management. The backup network is defined by distribution of restoration virtual paths and virtual channels for existing connections. The main design issue is to optimize this distribution so that the resources allocated to the backup VNLs are minimized. Note that due to the assumption of only one element failure at a time, the same "bandwidth" of a particular VNL can be used by different restoration virtual paths. In Sections 9.8 and 10.4 we discuss approaches and models for designing the backup VN.

Besides the mentioned generic problems, to take full advantage of the virtual network concept several application oriented problems have to be resolved. In particular, the resource allocation to service oriented virtual networks is a function of the scheduling and flow control procedures on the cell layers. These issues are discussed in Section 9.4 including application of models for equivalent bandwidth allocation presented in Section 3.2.2. Another application oriented problem is related to "bandwidth" allocation enforcement in virtual private networks. While in other virtual networks this function can be realized logically by the connection admission algorithm and UPC mechanism, in virtual private networks the connection admission procedure can be under the user's responsibility. Possible solutions to this problem are discussed in Section 9.5.

Before we move to the next layer it is important to clarify the relation between the VNL and VPC concepts since in the literature one can encounter many works dealing with "bandwidth" allocation to virtual paths. First note that there is different motivation for creation of the two entities. The VNL concept is created for resource management purposes. The VPC concept is created to simplify routing and switching of the cells throughout the network. For some particular resource cost function and constraints the optimization process of routing and "bandwidth" allocation can result in VNLI=VPI, which corresponds to "bandwidth" allocation to virtual paths. Thus the resource allocation to virtual paths is a restricted case of the virtual network problem. This restriction has several disadvantages which are discussed in more details in Section 9.3.

2.1.4 Physical network layer

From the resource management viewpoint the physical network can be seen as a set of nodes and transmission links connecting the nodes. Thus the generic objectives of the RM&TC algorithms on this layer are similar to the ones for virtual networks. The first issue is to create network connection topology and to allocate "bandwidth" to each link for a given traffic demand and node configuration constraints. The design should ensure that the network overall throughput can support traffic demand under required performance and reliability constraints. The second issue is to adapt the network topology and resource allocation based on the measurements of performance indicators from the lower layers. The third issue is to provide self-healing capabilities in case of equipment failures.

While these objectives are similar to the objectives on the virtual network layer, realization of these functions can be quite different. This follows from the fact that the transport layer, which provides transmission facilities for the ATM layer, can be realized in many ways characterized by different constraints and functionality. In particular, usually the "bandwidth" can be allocated to ATM links only in certain modules. The time needed to allocate additional modules can vary from seconds (if the transport layer is based on a digital hierarchy system with cross-connect capabilities) to weeks (if new installation is required). Moreover, the transport layer can have its own survivability mechanisms which influence the failure protection mechanisms on the ATM network layer. We discuss important features of different transport layer options in Section 10.1, including digital hierarchy systems (PDH, SDH, SONET).

The initial allocation of resources from the transport layer to the ATM network is described in Section 10.2. We consider several options including multi-hour traffic design. Adaptation to traffic changes is described in Section 10.3. Depending on the underlying transport layer two options are considered: fast adaptation and slow adaptation. The survivability issues are treated in Section 10.4. First we discuss possible failure scenarios and restoration mechanisms. Then the objectives of the connection and GoS restoration are described followed by generic formulations for self healing network design.

2.2 Resource Management Architecture

As described in the previous section the basic resource management objects can be distributed along the different time scales (layered structure) and different types of services and users (virtual network applications). The third dimension of the problem is added by the fact that these objects can be implemented in different locations and in different manners (e.g., centralized vs. decentralized). This additional dimension is illustrated in Figure 2.6 where a general architecture of the resource management objects is presented. In the following we briefly describe the main functions of the objects from Figure 2.6 and their interrelations. To facilitate the presentation we divide the description into three parts associated with layers where explicit resource allocation is performed.

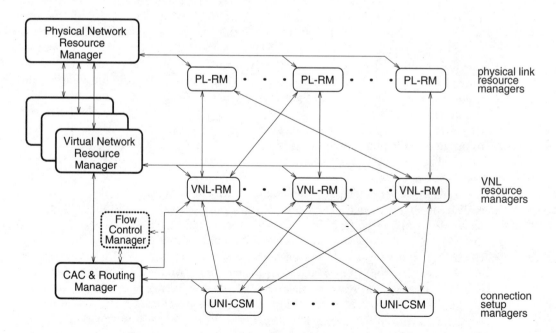

Figure 2.6. *Resource management architecture.*

Connection layer

Connection admission control and routing manager (CAC&RM): The main function of this object is to provide the UNI-CSM objects with CAC & routing information which would facilitate establishment of the connections within VN. In particular CAC&RM is responsible for evaluation of functions used to translate the connection declared parameters into the "bandwidth" which should be reserved for the connections. Moreover, CAC&RM should design a set of alternative paths for each OD pair (these paths can be implemented in the form of VPCs). Another important CAC&RM function is to provide the VN resource manager with connection class performance metrics which can be used to adapt the VN topology and its resource allocation. The CAC&RM manager can be implemented in centralized or decentralized (over UNI-CSM and VNL-RM) fashion.

User-network interface connection setup manager (UNI-CSM): The UNI-CSM object is responsible for routing and "bandwidth" reservation for connections originating at this interface. Based on the routing recommendation from CAC&RM, the UNI-CSM asks the resource managers of VNLs constituting the chosen path to reserve the required "bandwidth." If there is not enough resources on this path, another path can be tried or the connection can be rejected.

Flow control manager (FCM): The flow control manager object exists only in the virtual networks for CTP services such as ABR and ABT. Its function is to supervise the flow control algorithms from stability and efficiency viewpoints. This function requires communication with VNL-RMs, where the ABR and ABT algorithms make decisions concerning "bandwidth" allocation to connections.

Virtual network layer

Virtual network resource manager (VN-RM): The main function of the VN-RM object is to update the VN topology and resource allocation based on the performance measures from the connection layer. When there is a need to increase the "bandwidth" allocation, the VN-RM sends the demand to the network resource manager. If there is not enough resources, the network resource manager can realize only a part of the demand or reject the demand. Virtual network resource manager can be implemented in centralized or decentralized (over VNL-RM) manner.

Virtual network link resource manager (VNL-RM): The main function of the VNL-RM object is link connection admission control. In particular VNL-RM decides whether a connection demand (VCC or VPC or nested VNL) should be granted the requested "bandwidth." Several fixed or state dependent policies can be considered (e.g. complete sharing, coordinate convex, "bandwidth" reservation, dynamic "bandwidth" reservation, shadow price). Optionally, the VNL-RM can ask directly the link resource managers (L-RM) to increase the "bandwidth" allocation to the VNL.

Network layer

Physical network resource manager (PN-RM): There are two main functions of the PN-RM object. The first is to update the network topology and resource allocation based on the performance measures from the VN layer. The second is to allocate resources to virtual networks. The second function has to take into account fairness criteria when there is not enough resources to satisfy all VN demands. Physical network resource manager can be implemented in a centralized or decentralized (over PN-RM) manner.

Physical link resource manager (PL-RM): The PL-RM object cooperates with the PN-RM object. Its function is to control admission of new VNs and to update resource allocation of the existing VNs. As in the case of VNL-RM, several different policies can be applied which should take into account fairness and efficiency criteria.

2.3 Discussion and Bibliographic Notes

The objective of this chapter was to present the main issues of resource management and traffic control in broadband networks with focus on the areas described in this book. A framework for resource management was formulated based on two decomposition techniques which allow the decomposition of complex issues into manageable entities. The decomposition is done by applying the concepts of virtual networks and layered traffic entities associated with different time scales. In general the concept of virtual networks is not new and was discussed in the literature mainly in the context of virtual private networks, e.g. (Mason et al., 1990; Walters and Ahmed, 1992; Crocetti et al., 1994). Also "bandwidth" allocation to virtual paths, e.g. (Saito, 1994), can be treated as a particular case of management and service oriented VNs. In this book we generalize the virtual network concept so it can be treated as a fundamental tool enabling simplification and increased efficiency of network resource management. This coherent approach based on clear definitions is described in more details in Chapter 9.

The concept of layered traffic entities associated with different time scales was introduced in (Filipiak, 1989) and (Hui, 1990). In this book we modified two components of this structure. The virtual path layer is replaced by the virtual network layer. This modification follows from the fact that, as shown in Chapter 9, application of virtual paths for resource management has several drawbacks. These drawbacks can be avoided if the functions of virtual paths and virtual networks are separated. In this case the virtual paths are used for the purpose of routing and switching simplification, which is consistent with the original motivation for their introduction, while the virtual networks facilitate the complex resource management functions. We also removed from the layer hierarchy the burst layer. To be more precise this layer is integrated with the connection layer. This integration is justified by the fact that, from the resource management viewpoint, allocation of resources to a cell burst is equivalent to allocation of resources to a connection. Thus the resource allocation to bursts of the same connection can be treated as an adaptive allocation of resources to this connection.

The term *resource management* is very broad and may have many interpretations. In this book we focus on the algorithms which deal explicitly with the allocation of basic resources to the layered traffic entities. The basic resources correspond to the "bandwidth," which is defined as a combination of bandwidth and buffers linked by the scheduling mechanism and QoS requirements of the traffic using this bandwidth. Nevertheless the requirements for the basic resources determine requirements for other resources such as interfaces, switches, processing capacity, etc. Thus the material presented in this book covers indirectly a much wider resource domain.

Many of the traffic control mechanisms are coupled strongly with the resource management issues. Therefore, although the book focuses on resource management algorithms, it also includes a description of many traffic control issues. For the same reason some of the traffic control issues are not discussed in detail in this book. This is the case with the traffic control mechanisms associated with the cell layer (UPC, scheduling, flow control) which are not directly involved in the network explicit "bandwidth" management algorithms. The reader interested in these traffic control algorithms is referred to a substantial literature on these issues; policing — e.g. (Rosenberg and Lague, 1994; Mason, Pelletier, and Lapoint, 1996); scheduling — e.g. (Demers, Keshav, and Shenker, 1989; Parekh and Gallager, 1992; Golestani, 1994; Hun and Kesidis, 1996); flow control — e.g. (Jain, 1995; Mitra, 1992; Liao and Mason, 1994; Letourneau and Mason, 1996). There is also abundant literature dealing with performance models on the cell layer which consider the ATM network and switch characteristics, e.g. (Kesidis, 1996; Onvural, 1994).

In Section 2.2 we described a schematic representation of resource management architecture in terms of objects which perform specific functions and exchange required information among themselves. In a sense what is shown in Figure 2.6 is only a tip of an iceberg. The issue is how all the resource management algorithms should be implemented in the network operating system. There is a lot of activities in the areas of general network management systems with the objective to create a flexible, scalable, and portable framework which would provide universal objects, interfaces, and information architecture for all network needs. The main standardization activities are going on at ITU-T, *Telecommunications Network Management* (TNM) specification (CCITT Rec.M.3010, 1992), *Telecommunications Information Network Architecture* (TINA) Consortium (De la Funete, Pavón, and Moreno, 1995), and ATM Forum (ATM Forum, 1994). The generality of these efforts involves a risk of not fulfilling the needs of resource management objectives which often have strong time constraints. An interesting idea to incorporate the requirements of the network resource management into the TINA framework is presented in (Woodruff, 1996) under the name *Network Resource Management Information Model*.

Although in this book we focus on the fast packet switched networks based on the ATM standard, many parts of the included material are also relevant to other transfer technologies based on the circuit-switched or packet-switched concepts. This is for two reasons. First, although the ATM standard is based on the packet-switched concept, at the same time the information transfer is based on fixed virtual channels. Second, the presented material is driven by service and user characteristics rather than by technology. As a result many of the presented solutions can be applied to other environments. In particular the CAC & routing algorithms on the connection layer described in Chapter 5 are also applicable to circuit-switched multi-rate networks (Dziong and Mason, 1994). The adaptive "bandwidth" allocation to connections based on the Kalman filter (Chapter 4) can be extended to CDMA based wireless access networks (Dziong, Jia, Mermelstain, 1996). The concept of virtual networks (Chapter 9) can be useful in almost every multi-class-service and multi-class-user network. Also, allocation of physical resources to ATM networks and providing self-healing capabilities (Chapter 10) has many common elements with circuit switched and packet switched networks, especially in the case of using a digital hierarchy as the transmission facilities.

References

ATM Forum. 1994. M4 Interface Requirements and Logical MIB: ATM Network Element View. Version 1.

ATM Forum. 1995. Traffic Management Specification. atmf 95-0013R2.

CCITT Recommendation M.3200. 1992. TMN Management Services Overview. Geneva.

CCITT Recommendation M.3010. 1992. Principles for a Telecommunications Management Network. Geneva.

CCITT Recommendation M.3020. 1992. TMN Interface Specification Methodology. Geneva.

ITU-T Recommendation Q.2761-Q.2764. 1993. B-ISDN User Part. Draft. Geneva.

ITU-T Recommendation I.371. 1995. Traffic Control and Congestion Control in B-ISDN. Geneva.

Crocetti, P., Fratta, L., Gallassi, G., and Gerla, M. 1994. ATM Virtual Private Networks: Alternatives and performance comparisons. In *Proceedings of IEEE ICC'94*, pp. 608–612. IEEE Computer Society Press.

De la Funete, L.A., Pavón, J., and Moreno, J.C. 1995. The TINA-C approach to TMN. In *Proceedings of ISS'95*, Vol. 2, pp. 227–231. Berlin: VDE-VERLAG GMBH.

Demers, A., Keshav, S., and Shenker, S. 1989. Analysis and simulation of a fair queuing algorithm. *ACM SIGCOM'89*.

Dupuy, F., Nilsson, G., and Inoue, Y. 1995. The TINA Consortium: Towards Networking Telecommunications Information Services. In *Proceedings of ISS'95*, Vol. 2, pp. 207–211. Berlin: VDE-VERLAG GMBH.

Dziong, Z., and Mason, L.G. 1994. Call admission and routing in multi-service loss networks. *IEEE Transactions on Communications* 42(2/3/4):2011–2022.

Dziong, Z., Xiong, Y., and Mason, L.G. 1996. Virtual network concept and its applications for resource management in ATM based networks. In *Proceedings of Broadband Communications'96, An International IFIP-IEEE Conference on Broadband Communications*, pp. 223–234. Chapman & Hall.

Dziong, Z., Jia, M., and Mermelstain, P. 1996. Adaptive traffic admission for integrated services in wireless access networks (CDMA). *IEEE Journal on Selected Areas in Communication,* 14(9):1737–1747.

References

Filipiak, J. 1989. M-Architecture: A structural model of traffic management and control in broadband ISDNs. *IEEE Communications Magazine*, 27(5):25–31.

Golestani, S.J. 1994. A self-clocked fair queueing scheme for broadband applications. In *Proceedings of IEEE INFOCOM'94*, pp. 636–646. IEEE Computer Society Press.

Hui, J.Y. 1990. *Switching and Traffic Theory for Integrated Broadband Networks*. Kluwer Academic Publisher.

Hung, A., and Kesidis, G. 1996. Bandwidth scheduling for wide-area ATM networks using virtual finishing times. *IEEE/ACM Transactions on Networking*, 4(1):49–54.

Jain, R. 1995. Congestion control and traffic management in ATM networks: Recent advances and a survey. Invited submission to *Computer Networks and ISDN Systems*.

Kesidis, G. 1996. *ATM Network Performance*. Kluwer Academic Publisher.

Létourneau, E., and Mason, L.G. 1996. Integration strategies for flow control mechanisms in ATM networks. In *Proceedings of The 10th ITC Specialist's Seminar on "CONTROL IN COMMUNICATIONS,"* pp. 177–188. Lund, Sweden, September.

Liao, K.-Q., and Mason, L.G. 1994. A congestion control framework for broadband ISDN using selective window control. In *Proceedings of Broadband Communications'94, An International IFIP Conference on Broadband Communications*.

Mason, L.G., Dziong, Z., Liao, K.-Q., and Tetreault, N. 1990. Control architectures and procedures for B-ISDN. In *Proceedings of the 7th ITC Specialist Seminar*, Morristown, USA.

Mason, L.G., Pelletier, A., and Lapoint, J. 1996. Towards optimal policing in ATM networks. *Computer Communication*, 19(3):194–204.

Mitra, D. 1992. Asymptotically optimal design of congestion control for high speed data networks. *IEEE Transactions on Communication*, 40(2):301–311.

Onvural, R.O. 1994. *Asynchronous Transfer Mode Networks — Performance Issues*. Artech House.

Parekh, A., and Gallager, R. 1992. A generalized processor sharing approach to flow control in integrated services networks: The single node case. In *Proceedings of IEEE INFOCOM'92*, pp. 915–924. IEEE Computer Society Press.

Rosenberg, C., and Lague, B. 1994. A unified heuristic framework for source policing in ATM networks. *IEEE/ACM Transactions on Networking*, 2(4):387–397.

Saito, H. 1994. *Teletraffic Technologies in ATM Networks*. Artech House.

Walters, M.S., and Ahmed, N. 1992. Broadband virtual private networks and their evolution. In *Proceedings of ISS'92*.

Woodruff, G. 1996. Network resource management information model. Canadian Institute for Telecommunications Research (CITR) report, rev.0.

Veeraraghavan, M., La Porta, T.F., and Lai, W.S. 1995. An alternative approach to call/connection control in broadband switching systems. In *Proceedings of the 1st IEEE International Workshop on Broadband Switching Systems*, pp. 319–332. Poznan, Poland.

Chapter

3

Resource Allocation to Connections

THE ISSUE of resource allocation to connections is critical from a resource management and traffic control viewpoint. This follows from the fact that resource allocation to connections in ATM based networks provides a tool for taking advantage of two traffic control domains which historically were almost totally separated: packet switched networks and circuit-switched networks. This bridging is exemplified by the fact that on one hand resource allocation to connections is tightly connected with all traffic control mechanisms at the cell level which influence the QoS. On the other hand it provides for straightforward cooperation with resource management and traffic control mechanisms on the connection (CAC & routing function) and higher layers (virtual and physical networks) influencing the network resource utilization and GoS of connection classes.

Besides being important, the issue of resource allocation to connections is also difficult due to the lack of physical separation between resources used by different connections in ATM based networks. This feature dictates that, in general, only logical "bandwidth" allocation can be applied to connections. By logical allocation, which could be also called virtual allocation, we mean that although we associate a certain amount of "bandwidth" with each connection, there is no physical mechanism which could ensure this allocation. Nevertheless a logical "bandwidth" allocation to connections can significantly simplify the CAC_{QoS}, as will be shown in this chapter. There are several possible options for logical "bandwidth" allocation in fast packet switched networks. They range from minimum rate allocation via equivalent bandwidth allocation (also called effective bandwidth allocation) to peak rate allocation, and are discussed in Section 3.1. For further considerations we have chosen the equivalent bandwidth approach, initially proposed in (Turner, 1987), due to its potential effectiveness resulting from taking into account statistical multiplexing of variable-bit-rate sources at the cell layer.

In Section 3.2 we present several coherent models and approximations for evaluation of equivalent bandwidth allocation to connections on several types of ATM links including systems with priorities. The models are constructed under very general assumptions and require only a cell layer performance model for QoS evaluation in a non-priority system with homogeneous sources. This is achieved by a decomposition of the priority system into a set of non-priority systems and by applying appropriate bounds. It is also shown how the multi-priority models can be applied for resource allocation to the service oriented virtual networks where each virtual network is associated with particular QoS constraints.

3.1 Logical "Bandwidth" Allocation to Connections

The phrase "logical "bandwidth" allocation to connections" is used to underline the fact that in fast packet switched networks all connections on a particular link share the same link "bandwidth." Thus the physical "bandwidth" allocation to connections does not exist in this environment in a strict sense. While the sharing principle has several advantages (e.g., statistical multiplexing on the cell layer), the concept of logical "bandwidth" allocation was created to take advantage of simple and well known techniques developed for circuit-switched networks such as CAC & routing mechanisms. In the following we define several possible concepts for logical "bandwidth" allocation in networks with statistical QoS guarantees. To simplify the presentation we start with a consideration of an ATM link (multiplexer) with FIFO buffer and bandwidth, L, which is carrying connections with the same QoS constraints. A connection source of class j ($j \in J$, where J denotes a set of connection classes) generates a stationary cell process characterized by the set of source parameters, $\mathbf{h}_j$. Let d_j denote the logical "bandwidth" allocated to a class j connection.

In general logical "bandwidth" allocation can be realized in many ways. Nevertheless, in all cases the common objective is to provide that when all the link "bandwidth" is allocated to the accepted connections the quality of service constraints are not violated for any connection class:

$$\sum_{j \in J} x_j d_j = L \quad \Longrightarrow \quad \psi_j \leq \psi^c, \quad j \in J \quad (3.1)$$

where x_j is the number of accepted connections from class j and ψ_j, ψ^c denote the vectors of connection quality of service indicators and their constraints, respectively. If the condition (3.1) is fulfilled, the connection admission control (CAC$_{\text{QoS}}$) can be based on a simple verification whether the residual link capacity, $C = L - \sum_{j \in J} x_j d_j$, is sufficient to accept a new connection from class i:

$$d_i \leq L - \sum_{j \in J} x_j d_j \quad (3.2)$$

Note that the optimal logical "bandwidth" allocation is achieved when the quality of service indicators are tight to their constraints:

$$\sum_{j \in J} x_j d_j = L \quad \Longrightarrow \quad \{\psi_j : j \in J\} \models \psi^c \quad (3.3)$$

3.1 Logical "Bandwidth" Allocation to Connections

where $\models$ indicates that at least one of the performance indicators is tight to the constraint (while others are smaller than or equal to the constraints). In general the logical "bandwidth" allocation can be a function of source parameters, link bandwidth, buffer length, QoS constraints, and the accepted connection mixture. However, it should be underlined that, from the implementation viewpoint, it is desirable to have d_j independent of the current connection state in order to simplify the connection admission algorithm (CAC_{QoS}).

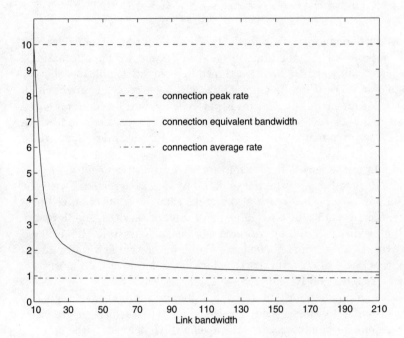

Figure 3.1. *Connection equivalent bandwidth vs. link bandwidth.*

To illustrate basic alternatives for logical "bandwidth" allocation let us consider a homogeneous case where all connections have the same source parameters. In this case it follows from condition (3.3) that the optimal "bandwidth" allocation is given by

$$d_j = \frac{L}{x_{\max}} \qquad (3.4)$$

where $x_{\max}$ is the number of connections for which $\psi_j \models \psi^c$. An example of the optimal logical "bandwidth" allocation to connections is shown in Figure 3.1 for an on-off source class as a function of the link bandwidth. The on-off source is defined by its cell peak rate (PCR), $P = 10$, average cell burst length, $\tilde{B} = 200$, and average silence length, $\tilde{S} = 2000$ (unless stated otherwise we assume exponential distributions for the burst and silence lengths). The properties of the function are general. In particular it can be shown that for variable-bit-rate connections with the peak rate, P_j, larger than the average rate, m_j, we have

$$m_j \leq d_j(L) \leq P_j \qquad (3.5)$$

$$\lim_{L\to\infty}[d_j(L)] = m_j \qquad (3.6)$$

Moreover, if for a given number of connections the QoS is a continuous monotonic function of L around the considered values of L (which is the case for typical ATM link models), we can assume that for any type of source traffic parameters the curve is monotonic and convex. This feature is analogous to monotonicity and concavity of trunk utilization vs. link capacity for a given connection loss probability in the Erlang model for circuit-switched loss systems, see e.g. (Girard, 1990).

Equivalent bandwidth allocation. Henceforth the term equivalent bandwidth will be used for the optimal logical "bandwidth" allocation and its approximations. The optimal logical "bandwidth" allocation shown in Figure 3.1 indicates that by taking advantage of statistical multiplexing at the cell layer one can provide high resource utilization. This attractive feature is partly traded-off by some potential weaknesses connected with performance guarantees and the equivalent bandwidth evaluation. There are three main sources of criticism of the equivalent bandwidth approach.

The first arises from the fact that the equivalent bandwidth concept works optimally only if the declared connection parameters, $\mathbf{h}_j^d$, are in agreement with the actual connection parameters, $\mathbf{h}_j$, which also implies that the connection cell process is stationary. These conditions are seldom met for many services, and in reality the equivalent bandwidth required by a source can be different from the declared value and moreover it can be time dependent. We describe this feature by introducing the notion of declaration error, $\mathbf{c}_j(t)$, which defines the actual parameters of source j at time t as $\mathbf{h}_j(t) = \mathbf{h}_j + \mathbf{c}_j(t)$.

To cope with the unpredictability of source parameters and malicious users one can incorporate a source policing mechanism (see Chapter 1) which ensures that the connection parameters cannot exceed certain values, $\mathbf{h}_j^p$, defined by the policing mechanism parameters. In this case the equivalent bandwidth allocation based on the policing algorithm parameters will guarantee the required QoS. The potential weakness of this approach is that there is no perfect mechanism which could police statistical parameters of the source; therefore the equivalent bandwidth derived from the UPC parameters will be higher than the one based on source declarations. Moreover, since in many cases the source cannot predict accurately its statistical parameters, this approach can force sources to declare overestimated parameters which will also reduce the network resource utilization.

Finally, evaluation of the optimal equivalent bandwidth allocation to connections may require complex models, especially in the case of heterogeneous connection classes and systems with priorities. Also, as it will be shown in the second part of this section, the optimal logical "bandwidth" allocation to a connection can depend on the link state. These features, combined with the source parameter estimation problem, suggest that the gain from statistical multiplexing of connections is difficult to attain.

The remainder of this chapter is focused on presentation of concepts and models which are designed to overcome the above mentioned potential weaknesses of the equivalent bandwidth approach. In the second part of this section it is shown how one can eliminate dependency of equivalent bandwidth allocation on the link state without sacrificing high resource utilization. Simple and general models for evaluation of equivalent bandwidth for multi-priority systems with heterogeneous connection classes are presented in Section 3.2. The issues of source cell process non-stationarity and unpredictability are addressed in Chapter 4.

Peak rate allocation. One of the simplest schemes for logical "bandwidth" allocation is to use connection peak rate as the "bandwidth" allocated to a connection, $d_j = P_j$. This allocation determines an upper bound for all optimal "bandwidth" allocations as shown in Figure 3.1. There are several advantages to this option. In particular the cell peak rate can be easily enforced by a policing mechanism at the user-network interface, so the performance guarantee can be easily achieved. Moreover, the peak rate allocation does not depend on the link parameters nor on the link state; thus the implementation of this option is very simple from both the user and network viewpoints. The obvious weakness of this approach is that the network resources can be significantly underutilized when the variable-bit-rate sources constitute a significant share of the traffic (see Figure 3.1).

Minimum rate allocation. The minimum rate allocation is suitable for controllable data services (CTP). In this case the network allocates to the connection "bandwidth" corresponding to the requested minimum cell rate. The traffic exceeding the minimum cell rate can be still carried in the network if, at the moment, spare capacity exists. In general, the resources required by the excess traffic are not guaranteed. Thus the excess traffic should be controlled by the source using some feedback information concerning the network state as described in Chapters 1 and 2. The main advantage of this scheme is its potential for high resource utilization. This is achieved at the expense of not providing any guarantees for the excess traffic.

Multi-link path connections. In the presented concepts of logical "bandwidth" allocation it is assumed that the connection traffic parameters at the entry to the link multiplexer are the same as the ones generated by the source (or the output of the UPC). Nevertheless, in the case of connections using paths with many links and/or multiplexers, one can argue that the upstream link stages or switches can alter the cell connection distribution offered to the next link. This is caused by the delays introduced by buffers and by the interactions with other connections in the upstream multiplexers and switches. Taking into account these effects, if at all possible, is a daunting task. Fortunately, in general, the QoS requirements for the non-controllable services are very stringent. This feature causes that the cell process alterations for these services are very small and are limited to small time scale fluctuations which can be dealt with by introducing certain simple modifications to the performance models used for evaluation of "bandwidth" allocation. In particular, for a constant-bit-rate source the performance model can assume a Poissonian cell arrival distribution to protect the system against small time scale fluctuations see, e.g. (Liao and Mason, 1990). As for the controllable services, the connection cell rates can be adaptively adjusted based on the QoS measurements, so the alteration of the cell distribution process is not critical. These arguments

TABLE 3.1 Description of tested examples.

Example	j	P	$\tilde{B}$	$\tilde{S}$	L	$\mathcal{D}^c$	$\mathcal{D}^c_h$	$\mathcal{D}^c_l$
1	1	10	200	2000	50	10		
	2	constant bit rate						
2	1	10	100	300	50	10		
	2	1	100	300				
3	1	10	100	300	200	10		
	2	1	100	300				
4	1	2	220	440	100		2	500
	2	10	2000	20000				
5	1	2	22	44	100		1	500
	2	1	50	150				

indicate that although all models presented in this chapter are constructed for systems with single-link connections, they can be easily extended to multi-link path connections. This feature allows one to apply these schemes for the $\text{CAC}^l_{\text{QoS}}$ and $\text{CAC}^p_{\text{QoS}}$ functions (see Section 2.1.2).

3.1.1 Equivalent bandwidth allocation for heterogeneous cases

In this section[1] we consider an ATM multiplexer which serves several connection classes with different traffic characterization. Let the vector $\mathbf{x} = [x_j]$ denote the multiplexer state where x_j, $j \in J$ denotes the number of class j connections carried on the multiplexer (class j connections are characterized by the same cell process distribution). For the sake of presentation simplicity we assume that the number of connections, x_j, is a real number and its functions (described later) are continuous (except for some indicated cases), which should be treated as an interpolation of the discrete case. To consider the equivalent bandwidth allocation in the heterogeneous case it is convenient to construct an admissible region in the domain of connection class normalized throughputs

$$U_j = \frac{x_j m_j}{L}, \quad j \in J \tag{3.7}$$

The admissible region is limited by a boundary which separates the states below the boundary, where the QoS constraints are met for all connection classes, from those above the boundary, where the QoS constraints are not met for at least one

---

[1]Portions of Sections 3.1.1 and 3.2 are reprinted from (Dziong et al., 1990; Dziong, Liao, and Mason, 1993) with kind permission from Elsevier Science – NL, Sara Burgerhartstraat 25, 1055 KV Amsterdam, The Netherlands.

3.1 Logical "Bandwidth" Allocation to Connections

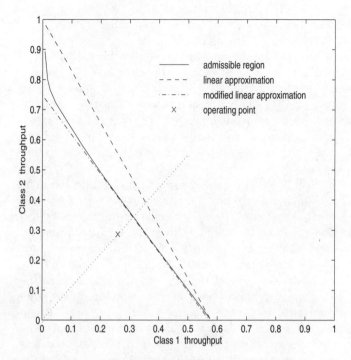

Figure 3.2. *Admissible regions (Example 1).*

connection class. Observe that if the boundary was linear (hyper-plane) the equivalent bandwidth allocation for each class would be state independent and equal to the allocation in the homogeneous case. Unfortunately this is not the case and in general the boundary is convex. To give a simple explanation of this feature let us consider a link which is offered two connection classes. The first is characterized by a variable bit rate and the second has a constant bit rate. Note that under stringent QoS constraints the contribution of constant-bit-rate connections in the statistical gain of the other connection class is negligible and in practice we can assume that the connections from the variable-bit-rate class see a link with the capacity $L(x_2) = L - x_2 m_2$. This permits evaluation of the equivalent bandwidth for the first class using the homogeneous model for each state of constant-bit-rate connections. An admissible region for this model is presented in Figure 3.2 (continuous line) for Example 1 from Table 3.1. Observe that the equivalent bandwidth function for the variable-bit-rate connection class is presented in Figure 3.1. This example indicates that the convexity of the admissible region boundary is caused by differences between the statistical gains of different connection classes, which is quantified by difference between normalized equivalent bandwidth functions for homogeneous cases, $d_j(L)/P_j$.

The convexity of the admissible region implies that any equivalent bandwidth allocation that provides maximum link utilization is state dependent. It is obvious that this feature is not attractive from an implementation point of view. For this reason in the following we discuss two approximations for equivalent bandwidth allocation which result in state independent "bandwidth" allocation.

In the first one the equivalent bandwidth allocated in the heterogeneous case is the same as in the homogeneous case. Henceforth we refer to this model as the linear approximation. The graphical interpretation of this approach for the case of two connection classes is given in Figure 3.2. The important advantage of the linear approximation is its simplicity. The weak side of this approach is that in some states (the area between exact and linear boundary) the QoS constraints are not met. Thus the linear approximation is reasonable for the cases where the gap between the exact and approximated boundaries is small (e.g., small peak to link rate ratio).

An alternative approximation is based on the construction of a linear surface (hyper-plane) tangent to the admissible region. This linear surface is treated as a boundary of a new admissible region. Then the fixed equivalent bandwidth allocation for each connection class can be evaluated from the maximum number of connections $x_j^{\max}$ in the new admissible region ($d_j = L/x_j^{\max}$). Henceforth this approach is called the modified linear approximation. One issue left to be clarified is the choice of the tangent point. In general the objective is to maximize the average throughput. While the exact solution would require knowledge of the state distribution and complex analytical tools, one can use a simplified approach in which the throughput is maximized for the mixture of connections proportional to the predicted or measured average state (operating point). This approach is illustrated in Figure 3.2. The advantage of this approach is that it provides the required QoS and state independent "bandwidth" allocation. The potential link underutilization in states far from the predicted operating point can be avoided by updating the tangent point based on measurements.

3.2 Models for Equivalent Bandwidth Allocation

In the first part of this section we present a very general and simple model for an approximate evaluation of the admissible region boundary. This model can be used to calculate the modified linear approximation described in the previous section. In the second part we will show how the equivalent bandwidth allocation concepts and models for single priority systems can be extended to multi-priority systems.

3.2.1 Single-priority system

We start by introducing two parameter characterization of the statistical properties of a connection background process. Consider an ATM link in a state **x** located on the admissible region boundary. The class j connection background process is defined as a superposition of cell processes of all connections carried on the link except the one under consideration. The two mentioned parameters are background softness, S_j, and background sparseness, $\mathcal{H}_j$. The background softness is defined

3.2 Models for Equivalent Bandwidth Allocation

as the difference between the sum of peak rates over all background connections and the sum of equivalent bandwidths allocated for these connections:

$$S_j = \sum_{i \in J \setminus \{j\}} x_i P_i - \sum_{i \in J \setminus \{j\}} x_i d_i \qquad (3.8)$$

The background sparseness is defined as the difference between the equivalent bandwidth required to carry all background connections in the ATM link under consideration and the average rate of these connections:

$$\mathcal{H}_j = \sum_{i \in J \setminus \{j\}} x_i d_i - \sum_{i \in J \setminus \{j\}} x_i m_i \qquad (3.9)$$

The main idea behind this two parameter description is that it comprises information about the performance of the system under consideration (by means of the equivalent bandwidth values). This is an important feature since the first two or three cell process moments provide only weak information relative to QoS, especially in the case of non-renewal processes (Holtzman, 1990).

Let us consider for awhile a homogeneous case where all connections belong to the same class. As shown in the previous section (Figure 3.1) the function $d_j(L)$ is monotonic. Thus by varying the multiplexer capacity one can obtain, for each connection class, unique relations between the equivalent bandwidth and background softness, $d_j = f_j^{\mathcal{S}}(\mathcal{S}_j)$, and between the equivalent bandwidth and background sparseness, $d_j = f_j^{\mathcal{H}}(\mathcal{H}_j)$. In general these functions can be evaluated from an analytical model or by measurements in a test bed multiplexer.

Non-linear approximation. The key assumption in the non-linear approximation is that, for each point on the boundary of the admissible region, the background softness and sparseness of each connection class (characterized by d_j) should be at least as large as the background softness and sparseness (respectively) in the homogeneous case for class j resulting in the same equivalent bandwidth allocation, d_j. Additionally it is assumed that the equivalent bandwidth evaluated from the linear approximation, d'_j, is a lower bound for d_j. This can be stated as follows:

$$d_j = \max\{f_j^{\mathcal{S}}(\mathcal{S}_j), f_j^{\mathcal{H}}(\mathcal{H}_j), d'_j\}, \quad j \in J \qquad (3.10)$$

This set of implicit equations can be used to determine the equivalent bandwidth allocation for each type of connection by means of repeated substitution with initial values evaluated from the linear approximation.

Numerical examples

To verify accuracy of the non-linear approximation one needs a model for evaluation of the equivalent bandwidth, d_j, and the functions $f_j^{\mathcal{S}}(\mathcal{S}_j)$, $f_j^{\mathcal{H}}(\mathcal{H}_j)$, for the homogeneous case. In the following we present numerical results for two different analytical models. The first one is a modified fluid approximation (Liao and Mason, 1990) with average cell delay, $\mathcal{D}$, as the QoS measure and infinite buffer. This model has good accuracy over a wide range of source parameters. A sample of the

TABLE 3.2 Accuracy of linear and non-linear approximations ($\mathcal{D}^c = 10$).

		case			linear approximation		non-linear approximation	
j	P	$\tilde{B}$	$\tilde{S}$	L	x	$\mathcal{D}$	x	$\mathcal{D}$
1	10	200	2000	50	1	75.30 ($\pm$ 12.8)	1	10.61 ($\pm$ 3.18)
2	1	22	44		124	22.06 ($\pm$ 3.84)	105	2.34 ($\pm$ 0.49)
1	10	200	2000	50	12	28.74 ($\pm$ 3.72)	12	6.21 ($\pm$ 1.55)
2	1	22	44		62	12.55 ($\pm$ 1.65)	43	2.45 ($\pm$ 0.60)
1	10	100	300	50	1	21.01 ($\pm$ 1.05)	1	11.93 ($\pm$ 0.54)
2	10	10	30		14	12.17 ($\pm$ 0.46)	13	7.13 ($\pm$ 0.20)
1	10	100	300	50	6	14.14 ($\pm$ 1.13)	6	8.11 ($\pm$ 0.78)
2	10	10	30		7	8.83 ($\pm$ 0.66)	6	4.86 ($\pm$ 0.41)
1	10	200	2000	100	1	60.28 ($\pm$ 10.3)	1	10.44 ($\pm$ 2.58)
2	1	22	44		271	24.44 ($\pm$ 3.79)	251	2.82 ($\pm$ 0.43)

results is presented in Table 3.2. Each case represents a mixture of two connection classes characterized by different parameters (peak rate, P_j, average burst length, $\tilde{B}_j$, average silence length, $\tilde{S}_j$). The number of connections from the first class, x_1, is fixed while the number of connections from the second class, x_2, is evaluated by the non-linear approximation to provide average delay of each connection class within the constraint, $\mathcal{D}^c = 10$ (expressed in cell service time units). Simulation is used to check the actual QoS (95% confidence intervals are indicated in the brackets). The design based on the linear approximation is also included in Table 3.2. The results show that the non-linear approximation provides that the QoS of the class with the worst performance is close to the constraint. As for the linear approximation, the performance is significantly worse than the design constraint.

The second analytical model, used as the input for the non-linear approximation, is *Modulated Markov Poisson Process* (MMPP) (Lucantoni, 1989; Fischer, 1989). Since the MMPP model exhibits non-negligible error over a certain range of parameters, we also use the MMPP model (version for heterogeneous connections) as a reference model. The rationale behind this approach is that although both the heterogeneous and homogeneous versions of the MMPP model have some errors, these errors will have the same magnitude so the relative difference between the tested models for the admissible boundary evaluation will be close to the case with exact performance models. This argument is also supported by the fact that the linear and non-linear approximations are not attached to any specific performance model. A sample of the results is presented in Figure 3.3 for Examples 2 and 3 from Table 3.1. Besides the three models already indicated (linear approximation, non-linear approximation, and MMPP) the results include a model in which statistical multiplexing is allowed only between the connections belonging to the same class (separation between the classes). For realistic cell rate processes this approach can be treated as a lower bound for the admissible region since it does not take into

3.2 Models for Equivalent Bandwidth Allocation

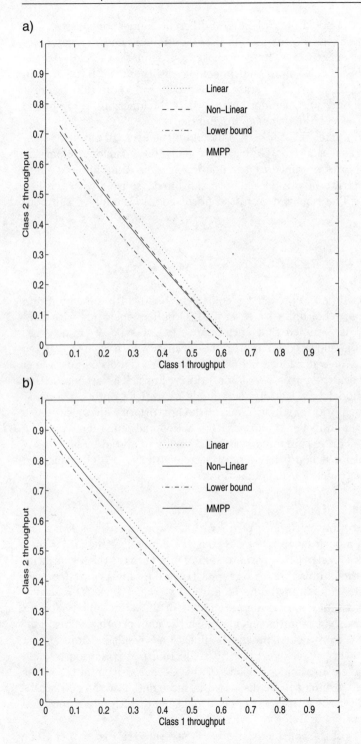

Figure 3.3. *Admissible regions: (a) Example 2; (b) Example 3.*

account the gain from statistical multiplexing of different connection classes. The results confirm that the linear approximation is too optimistic and in some cases it can introduce significant errors. The non-linear approximation is close to the reference model. The results also show that in some cases there is a large drop in link utilization when we move from the state $x_1 = 0$ to $x_1 = 1$ (note that the first connection class is more bursty). To underline this effect the functions in Figure 3.3 are plotted for $x_1 \geq 1$ (except the linear approximation).

While the presented numerical studies are based on the average cell delay constraint, the results given in (Monteiro, Gerla, and Fratta, 1991) include comparison of the admissible boundaries obtained by the non-linear approximation and simulation model for the mixture of variable-bit-rate and bursty sources with cell loss probability constraint. The reported results confirm good accuracy of the non-linear approximation.

3.2.2 Multi-priority system

In this section we describe an approach for equivalent bandwidth allocation and buffer dimensioning in an ATM link with several FIFO buffers operating with non-preemptive priority. It is assumed that each connection or type of traffic (e.g., connectionless services) is allocated an appropriate priority which is characterized by its QoS constraints. The central idea behind the approach is a decomposition of the considered priority scheme into a set of non-priority links. The decomposition procedure employs a model based on upper bounds for cell loss probability and average cell delay for each priority traffic. Each of the non-priority links serves as a means of evaluation of the equivalent bandwidth allocation and buffer dimensioning related to one priority. This decomposed system is readily analyzed by employing the methods developed for a non-priority multiplexer with one FIFO queue and with the same QoS constraints for all connections.

Decomposition model

For the sake of presentation clarity we begin with a two priority system. The central idea of the decomposition technique is that the admissible region for high priority connections, H, is constructed without taking into account the low priority connections. Then another admissible region is constructed for low priority connections but this time in the domain including high priority connections. This second admissible region, T, is a function of some parameters of the admissible region for high priority connections. Note that in this approach the high priority connections are characterized by two equivalent bandwidth allocations (resulting from H and T). This means that to accept a high priority connection two conditions should be checked. Important features and efficiency of this decomposition approach as well as some possible extensions will be discussed in more detail at the end of this section. A basic model for the evaluation of equivalent bandwidth allocations and the related buffer dimensioning follows.

The general structure of the considered system is presented in Figure 3.4a. The link has two FIFO buffers with capacities K_l, K_h, where indexes l, h, indicate low

3.2 Models for Equivalent Bandwidth Allocation

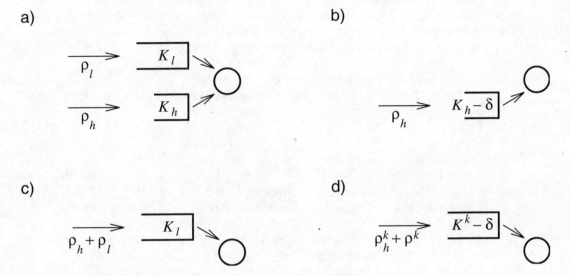

Figure 3.4. *Structures of the two-priority and modified systems.*

and high priority, respectively. The average cell rates offered to the system are denoted as ρ_l, ρ_h. Each connection can be allocated to one of the two priorities characterized by different QoS constraints: cell loss probability, $\mathcal{B}_l^c$, $\mathcal{B}_h^c$ and/or average cell delay, $\mathcal{D}_l^c$, $\mathcal{D}_h^c$. In other words, all cells from a given connection are entering one buffer (K_l or K_h) and the objective of the connection admission control (CAC$_{\text{QoS}}$) is to provide that the QoS constraints are met for each connection in progress. We consider two types of server: synchronous and asynchronous. In the synchronous case, time is slotted with cell service time period and a cell to be served has to wait for the beginning of the slot. In the asynchronous case the cell can be served immediately on condition the server is free. Throughout this section it is assumed that a model for the buffer dimensioning and equivalent bandwidth allocation for the link with one FIFO buffer is available. To simplify notation the waiting time is expressed in cell service time units.

High priority traffic*
Consider a modified system (Figure 3.4b) where the low priority traffic is neglected and the buffer capacity is given by $K_h' = K_h - \delta$, where $\delta = 0$ in the case of a synchronous server and $\delta = 1$ in the case of an asynchronous server. Let $\mathcal{B}_{j,h}$, $\mathcal{D}_{j,h}$ denote the QoS for class j high priority connection in the original system. Assuming that the state of high priority connections in the modified system is the same as in the original system, it can be shown (Dziong, Liao, and Mason, 1993) that the following relations hold:

$$\mathcal{B}_{j,h}' \geq \mathcal{B}_{j,h} \qquad (3.11)$$

$$\mathcal{D}_{j,h}' \geq \mathcal{D}_{j,h} - \delta \qquad (3.12)$$

where the prime indicates the parameters in the modified system. In fact for $\delta = 0$ these relations become equalities since in the synchronous system the low priority cells are not influencing the high priority cells. These results show that the QoS values in the non-priority system can be used to evaluate upper bounds for the QoS values for high priority connections in the original system.

Let us apply a model for the buffer dimensioning and "bandwidth" allocation in the modified, non-priority system under the following QoS constraints:

$$\mathcal{B}_h^{c'} = \mathcal{B}_h^c \tag{3.13}$$
$$\mathcal{D}_h^{c'} = \mathcal{D}_h^c - \delta \tag{3.14}$$

where it is assumed that $\mathcal{D}_h^c > 1$ for $\delta = 1$. This procedure gives the buffer capacity K_h' and the admissible boundary, H, defined in the domain of high priority connection states, $\mathbf{x}_h$. Based on Equations (3.11 and 3.12) it is easy to show that if the buffer capacity, $K_h = K_h' + \delta$, and equivalent bandwidth allocations determined by H are used in the original system for the high priority connections, the grade of service constraints are met for all classes of high priority connections

$$\mathcal{B}_{j,h} \leq \mathcal{B}_h^c \tag{3.15}$$
$$\mathcal{D}_{j,h} \leq \mathcal{D}_h^c \tag{3.16}$$

for any state within the admissible region. Observe that this result is valid regardless of the level of low priority traffic. In other words the evaluated buffer dimension and equivalent bandwidth allocation for high priority connections are independent of the analogous values for low priority connections.

Low priority traffic*

Consider a modified system where both the low and high priority connections are offered to the low priority buffer K_l (K_h is not used), as illustrated in Figure 3.4c. Let $\mathcal{B}_l$, $\mathcal{D}_l$ denote the cell loss probability and the average cell delay for low priority cells, respectively, and $\mathcal{B}$, $\mathcal{D}$ denote the average cell loss probability and average delay for both the high and low priority cells in the original system. Let us also define the total offered cell rate ρ, and offered cell rates of high and low priority traffic, ρ_h, ρ_l as

$$\rho = \frac{\sum_{j \in J} x_j m_j}{L}, \quad \rho_h = \frac{\sum_{j \in J} x_{j,h} m_j}{L}, \quad \rho_l = \frac{\sum_{j \in J} x_{j,l} m_j}{L} \tag{3.17}$$

where $x_{j,h}$, $x_{j,l}$ denote numbers of class j connections allocated to high and low priority, respectively. It is clear that $\rho = \rho_l + \rho_h$. In the following we assume that the difference between the average cell delay in the considered systems with finite buffers and the same systems with infinite buffers is negligible for the states within the admissible region (henceforth this assumption will be refered to as **A.3.1**). This assumption is reasonable in the ATM environment since the cell loss constraints are very stringent.

Assuming that the state of connections of both types in the modified system is the same as in the original system, it can be shown (Dziong, Liao, and Mason,

3.2 Models for Equivalent Bandwidth Allocation

1993) that

$$\rho \mathcal{B}'' \geq \rho_l \mathcal{B}_l \qquad (3.18)$$

and under assumption **A.3.1**

$$\mathcal{D}'' = \mathcal{D} \qquad (3.19)$$

where the double prime indicates parameters in the modified system. Note that in the original system we have

$$\rho_l \mathcal{D}_l + \rho_h \mathcal{D}_h = \rho \mathcal{D} \qquad (3.20)$$

Thus by using Equation (3.19) we arrive at the following relation for average delay of the low priority cells:

$$\mathcal{D}_l = \frac{\rho}{\rho_l} \mathcal{D}'' - \frac{\rho_h}{\rho_l} \mathcal{D}_h \qquad (3.21)$$

Based on this relation and Equation (3.18) the following QoS constraints for the modified system can be constructed:

$$\mathcal{B}^{c''} = \left[\frac{\rho_l}{\rho} \right]_{\min} \mathcal{B}_l^c \qquad (3.22)$$

$$\mathcal{D}^{c''} = \left[\frac{\rho_l}{\rho} \mathcal{D}_l^c + \frac{\rho_h}{\rho} \mathcal{D}_h \right]_{\min} \qquad (3.23)$$

where $[\ ]_{\min}$ denotes the minimum value over $\mathbf{x} \in X^b$ and X^b denotes the set of states for which both the state of high priority traffic and the state of low priority traffic are located on the boundaries of the corresponding admissible regions: H and T. Let us assume for the moment that the constraints (3.22) and (3.23) can be evaluated. These constraints can be used in the model for the buffer dimensioning and equivalent bandwidth allocation in the modified non-priority system. Note that the resulting equivalent bandwidth allocation defined by the admissible boundary T concerns both high and low priority connections. If these results are applied in the original system together with previously evaluated K_h and H, it is easy to show that the following relations are valid:

$$\mathcal{B}_l \leq \mathcal{B}_l^c \qquad (3.24)$$
$$\mathcal{D}_l \leq \mathcal{D}_l^c \qquad (3.25)$$

Up to now we assumed that the values of $\mathcal{B}^{c''}$ and $\mathcal{D}^{c''}$ are known, but in fact they are functions of the final result, thus making the algorithm implicit. To find the solution one can apply an iterative algorithm with initial values given by

$$\mathcal{B}^{c''} = \mathcal{B}_l^c \qquad (3.26)$$
$$\mathcal{D}^{c''} = \mathcal{D}_l^c \qquad (3.27)$$

Note that in many practical applications the QoS constraints for high priority traffic are several orders of magnitude smaller than the ones for low priority traffic. Thus the upper bound

$$\mathcal{D}_l \leq \frac{\rho}{\rho_l}\mathcal{D}''\tag{3.28}$$

achieved by neglecting the second term in Equation (3.21) can be used to construct a simpler constraint for the modified system

$$\mathcal{D}^{c''} = \left[\frac{\rho_l}{\rho}\right]_{\min}\mathcal{D}_l^c\tag{3.29}$$

An alternative upper bound for the average delay of low priority cells can be developed under assumption (henceforth refered to as **A.3.2**) that the average delays in the original system can be approximated by an $M/D/1$ system with infinite buffers and an asynchronous server. In other words we assume that the cell superposition process is close to Poissonian for each priority. In this case the following relation can be developed (Dziong, Liao, and Mason, 1993) from the solution of $M/D/1$ system:

$$\mathcal{D}'' = \mathcal{D}_l(1-\rho_h)\tag{3.30}$$

Thus to evaluate equivalent bandwidth allocation in the modified system we can use the following constraint for the average delay

$$\mathcal{D}^{c''} = \mathcal{D}_l^c(1-[\rho_h]_{\max})\tag{3.31}$$

where $[\ ]_{\max}$ denote maximum value over $\mathbf{x} \in X^b$. The advantage of this constraint is that all its input parameters can be evaluated a priori.

The proposed algorithm provides that the QoS constraints are met for the superposition of low priority connections. But if we assume (**A.3.3**) that for any link state on the admissible boundaries, $\mathbf{x} \in X^b$, the proportion of cell losses for different connection classes with low priority is similar in both original and modified systems,

$$\frac{\mathcal{B}_{j,l}}{\mathcal{B}_{i,l}} \simeq \frac{\mathcal{B}''_{j,l}}{\mathcal{B}''_{i,l}}, \quad i,j \in J\tag{3.32}$$

the following relations are also valid for each connection class:

$$\mathcal{B}_{j,l} \leq \mathcal{B}_l^c\tag{3.33}$$

Under an analogous assumption for average delays we have

$$\mathcal{D}_{j,l} \leq \mathcal{D}_l^c\tag{3.34}$$

General case*

The approach for the system with two priorities can be easily extended to the general case with N priorities. Let $i = 1, ..., N$ denote the priority index where $i = 1$ and $i = N$ correspond to the lowest and highest priority, respectively. To evaluate buffer dimension and equivalent bandwidth allocation for a particular class k it is convenient to introduce an aggregated description of all connections with higher priority, ρ_h^k, $\mathcal{B}_h^k$, $\mathcal{D}_h^k$, defined as follows:

$$\rho_h^k = \sum_{i>k} \rho_i \tag{3.35}$$

$$\mathcal{B}_h^k = \frac{\sum_{i>k} \mathcal{B}_i \rho_i}{\rho_h^k} \tag{3.36}$$

$$\mathcal{D}_h^k = \frac{\sum_{i>k} \mathcal{D}_i \rho_i}{\rho_h^k} \tag{3.37}$$

Consider a modified system where all connections with priority $i \geq k$ are offered to the kth buffer (the other connections are neglected) and the buffer capacity is given by $K^{k'''} = K^k - \delta$, as illustrated in Figure 3.4d. Observe that from the kth priority traffic viewpoint the system is analogous to the one for low priority connections described in the previous section with the exception that the buffer capacity is reduced in the case of the asynchronous server to take into account the influence of the connections with the lower priority as was done for the bounds evaluation for high priority connections. Thus based on Equations (3.11), (3.12), (3.18), and (3.21) the following bounds hold:

$$\rho^k \mathcal{B}^k \leq (\rho_h^k + \rho^k) \mathcal{B}''' \tag{3.38}$$

$$\mathcal{D}^k - \delta \leq \frac{(\rho_h^k + \rho^k)}{\rho^k} \mathcal{D}''' - \frac{\rho_h^k}{\rho^k}(\mathcal{D}_h^k - \delta) \tag{3.39}$$

In case the original system can be approximated by an $M/D/1$ system with infinite buffers and an asynchronous server (analogous to assumption **A.3.2**) it follows from Equations (3.12) and (3.30) that

$$\mathcal{D}''' \geq (\mathcal{D}^k - \delta)(1 - \rho_h^k) \tag{3.40}$$

Thus to evaluate the buffer capacity and equivalent bandwidth allocation for priority k connections we can solve the modified system with constraints given by

$$\mathcal{B}^{c'''} = \left[\frac{\rho^k}{\rho_h^k + \rho^k}\right]_{\min} \mathcal{B}^{k,c} \tag{3.41}$$

$$\mathcal{D}^{c'''} = \left[\frac{\rho^k}{\rho_h^k + \rho^k}(\mathcal{D}^{k,c} - \delta) + \frac{\rho_h}{\rho_h^k + \rho^k}(\mathcal{D}_h^k - \delta)\right]_{\min} \tag{3.42}$$

or alternatively

$$\mathcal{D}^{c'''} = (\mathcal{D}^{k,c} - \delta)(1 - [\rho_h^k]_{\max}) \tag{3.43}$$

For the system with an asynchronous server and $k = 1$ we have $\delta = 0$ since there is no traffic with lower priority.

Numerical examples

The following presents a discussion of the efficiency of the decomposition approach for buffer dimensioning and equivalent bandwidth allocation in a two priority system. To simplify this presentation we consider the same homogeneous source classes for high and low priority traffic so both admissible regions can be presented in the same figure (in general the two source classes can be characterized by different parameters for different priorities). Since the presentation is oriented towards the relative differences between admissible regions we used the linear approximation with only the average delay constraint (infinite buffers are assumed) as the non-priority model. The connections are modeled as on-off sources. The modified fluid approximation (Liao and Mason, 1990) is used as the cell layer performance model for homogeneous connections in the linear approximation for both the synchronous and asynchronous systems.

The examples of admissible regions for high priority connections, H, and for superposition of high and low priority connections, T, are shown in Figure 3.5 for Examples 4 and 5 from Table 3.1. The indexes s, a correspond to synchronous and asynchronous servers, respectively. Different admissible regions for total state of connections correspond to different bounds for QoS constraints: T_1 — basic approach (3.23), T_2 — high priority factor neglected (3.29), T_3 — $M/D/1$ approximation (3.31), T_0 — initial region in the iterative procedure (3.27). The state of high priority connections carried by the multiplexer cannot exceed the boundary H and the state of all connections carried by it cannot exceed the boundary T, in order to meet QoS constraints. These boundaries can be translated into equivalent bandwidth allocations for high priority connections, $d_{j,h}$, and for all connections, d_j, respectively. As previously indicated in the case of a new low priority connection the connection admission control (CAC$_{QoS}$) has to consider one condition

$$d_j \leq L - \sum_{i \in J} x_i d_i \qquad (3.44)$$

while in the case of a new high priority connection two conditions

$$d_{j,h} \leq L - \sum_{i \in J_h} x_{i,h} d_{i,h} \qquad (3.45)$$

$$d_j \leq L - \sum_{i \in J} x_i d_i \qquad (3.46)$$

have to be checked.

Observe that in the case of a synchronous server the modified system (Figure 3.4b) used to evaluate the boundary H^s is exact since we have equality in Equations (3.11) and (3.12) for $\delta = 0$. To consider the tightness of bounds (3.11) and (3.12) in the case of an asynchronous server ($\delta = 1$) let us define two factors

$$\eta_b = \frac{1}{\mathcal{D}_h^c} \qquad (3.47)$$

$$\eta_d = \frac{1}{K_h} \qquad (3.48)$$

3.2 Models for Equivalent Bandwidth Allocation

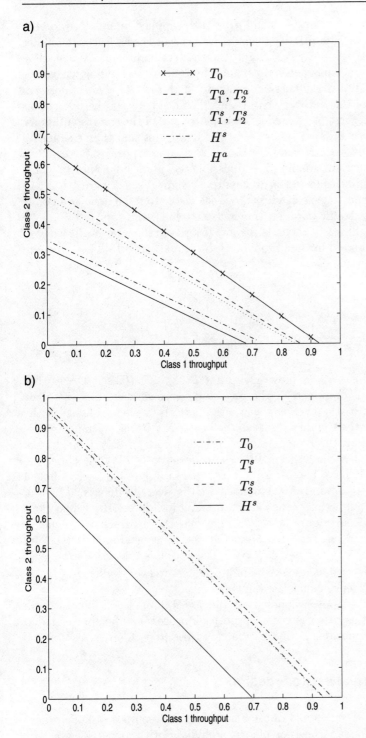

Figure 3.5. *Admissible regions: (a) Example 4; (b) Example 5.*

It can be easily shown that for $\eta_b \to 0$ and $\eta_d \to 0$ the applied bounds provide the maximum throughput of high priority traffic. In other words, the larger the buffer capacity and the larger the delay constraint, the tighter the bounds. When these values are small there is a gap between the bounds and the corresponding real values causing some reduction of the admissible region for high priority connections. An indication of the size of this gap is given by the difference between the boundaries H^s and H^a. The former can be treated as an upper bound of the exact solution for the system with an asynchronous server, H_e^a. It is important that from a practical point of view the gap between H^a and H_e^a can be treated as a positive factor. This follows from the fact that the smaller the admissible region for high priority traffic the larger the total admissible region (compare T_1^s and T_1^a in Figure 3.5a). In fact, as discussed later, in some cases it can be reasonable to reduce the admissible region for high priority traffic in order to increase the link utilization.

To evaluate the tightness of bounds for low priority traffic, we begin with an analysis of the upper bound for loss probability $\mathcal{B}_l$, Equation (3.18). Let η denote the ratio of average delays

$$\eta = \frac{\mathcal{D}_h}{\mathcal{D}_l} \tag{3.49}$$

It can be easily shown that for $\eta \to 0$ we have

$$\mathcal{B}'' = \frac{\rho_l}{\rho} \mathcal{B}_l \tag{3.50}$$

Note that in practical cases it is most likely that $\mathcal{D}_h^c \ll \mathcal{D}_l^c$. This indicates that the admissible region T evaluated by the proposed approach can be close to the maximum one. As far as the average delay of low priority cells is concerned, the basic approach applies the equality (3.21) so the constraint $\mathcal{D}^{c''}$ is optimal. In case the approach is based on the bound where the high priority factor is neglected, Equation (3.28), it is clear that the smaller the ratio $\eta^c = \mathcal{D}_h^c/\mathcal{D}_l^c$, the closer the result is to the optimal one. This feature is illustrated in Figure 3.6 for Example 4 where the gap between T_1 and T_2 (expressed as the average difference in link utilization) is given as a function of the delay constraint for low priority connections ($\mathcal{D}_l^c$) for the example from Figure 3.5a. Observe that the behavior of the gap is opposite to the predicted one for $\mathcal{D}_l^c < 50$ in the case of asynchronous server and for $\mathcal{D}_l^c < 144$ in the case of synchronous server. The reason is that in this range we have $\mathcal{D}^{c''} < \mathcal{D}_h^c$ so H is used as a lower bound for T_2. Moreover, we have $T_1^s = H^s$ for $\mathcal{D}_l^c < 64$ so the difference equals zero in this range.

The efficiency of the approach based on the $M/D/1$ model is illustrated in Figure 3.5b. Since in this case the cell superposition process can be approximated by the Poissonian distribution, T_1 and T_3 are very close to each other.

3.2.3 Application for QoS virtual network design

So far we have discussed the quality of the bounds in the decomposed model where the admissible region for superposition of low and high priority connections, T, and the resulting equivalent bandwidth allocations are independent of the state of high

3.2 Models for Equivalent Bandwidth Allocation

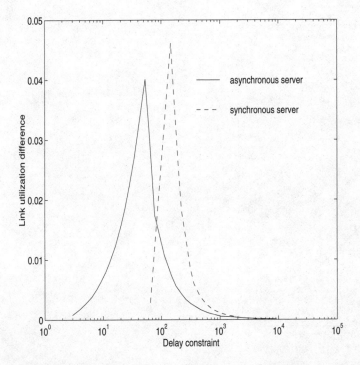

Figure 3.6. *Gap between T_1 and T_2 versus constraint for low priority connections (Example 4).*

priority connections. While this independence makes the model very simple, it is clear that for some states of high priority connections, with $\rho_h < [\rho_h]_{\max}$, more low priority connections could be accepted without compromising the QoS constraints. In particular the boundary T_0 can be interpreted as a limiting admissible boundary for $\rho_h \to 0$. In the following we show that the increased resource utilization can be achieved while retaining the fixed (state independent) equivalent bandwidth allocations.

To illustrate this issue let us consider a cross section (for $U_2 = 0$) of the optimal admissible region in the domain of link utilizations by high and low priority connection. Such a cross section, for the synchronous server, is presented in Figure 3.7a (continuous line) where $U_{1,h}$ and $U_{1,l}$ denote, respectively, the normalized throughputs of the high and low priority connections from the first class. The optimal admissible region is obtained by varying the maximum allowed throughput of high priority connections, $[\rho_h]_{\max}$, in the decomposition model. Note that the cross section boundary forms a corner where its linear part is determined by the maximum throughput of high priority connections. The other slanted part determines the trade-off between throughputs of low and high priority connection classes. These parts are henceforth called the north and northeast, respectively. From now on we will use the term "corner" to denote a point in each cross section of the admissible boundary (representing the relation between a pair of low and high priority

60   *Resource Allocation to Connections*

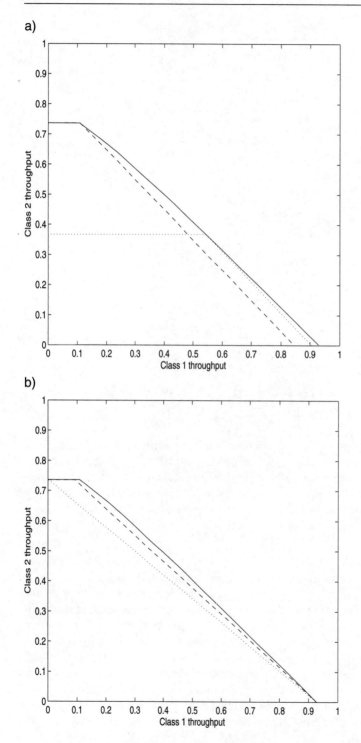

Figure 3.7. *Cross-sections of the admissible regions (Example 4).*

throughputs, $U_{i,h}$, $U_{j,l}$, when all other throughputs are set to zero) where its north and northeast sections meet.

Note that the northeast part of the cross section of the optimal boundary in Figure 3.7a is slightly concave. The fixed equivalent bandwidth allocations achieved in the decomposition model correspond to linearization of the northeast part of the boundary indicated by the dashed line in Figure 3.7a. This cross section is defined by the equation $U_{1,h} + U_{1,l} = 0.84$, which states that the maximum throughput of the superposition of low and high priority connections from the same class does not depend on the share of high priority connections, in the decomposition model. In the following we will describe two approaches which can reduce the gap between the optimal boundary and the boundary related to the basic decomposition model while maintaining the fixed equivalent bandwidth allocation feature.

The first approach is based on limiting the maximum allowed throughput of high priority connections, $[\rho_h]_{\max}$, in the decomposition model. The dotted line in Figure 3.7a represents the case where the maximum throughput of the high priority connections is reduced by half. Observe that in this case the gap between the optimal and the linear northeast parts of boundaries is significantly reduced (northeast cross section defined by $U_{1,h} + U_{1,l} = 0.91$ and $[U_{1,l}]_{\max} = 0.93$). Application of this mechanism is reasonable when the high priority connections do not constitute a majority of the traffic. In other words the maximum "bandwidth" allocated to high priority connections should provide the required performance at the connection level (GoS). This approach fits very well into the concept of service oriented virtual networks presented in Chapters 2 and 9. In this case the high priority traffic would be carried on a virtual network (VN_h) with "bandwidth" allocated to VNLs according to the GoS constraints. To simplify our discussion let us consider the case with one class of connection traffic parameters (the class index is omitted). In this case the dotted line in Figure 3.7a represents the complete admissible regions of a particular link. Let U_h^c and U_l^c denote the admissible region's corner point coordinates. We can assume that the "bandwidth" allocated to the high priority virtual network link is defined as

$$L_h = \frac{U_h^c}{U_h^c + U_l^c} \cdot L \qquad (3.51)$$

Thus the equivalent bandwidth allocation to connections within the VNL_h is given by

$$d_{h1} = \frac{L_h}{U_h^c} \qquad (3.52)$$

As far as low priority connections are concerned we can assume that the VN_h is nested in a network which handles both high and low priority connections. Although this is a physical network, it will be treated as virtual network, VN_l, in order to facilitate subsequent generalization. In this case the VNL_l connection admission algorithm (CAC_{QoS}) has to consider the total state of both the high and low priority connections. The equivalent bandwidth allocation to connections within the VNL_l is given by

$$d_l = \frac{L}{U_h^c + U_c^l} = d_{h1} \qquad (3.53)$$

Note that the equivalent bandwidth allocation to the high priority connections is the same in both VNs.

Another option is to assume creation of a separate virtual network for low priority connections. The "bandwidth" allocated to the low priority virtual network link (L_l) is given by

$$L_l = \frac{U_l^c}{U_h^c + U_l^c} \cdot L = L - L_h \qquad (3.54)$$

The equivalent bandwidth allocation within the separate VNL$_l$ is given by

$$d_{h2} = d_l = \frac{L_l}{U_l^c} \qquad (3.55)$$

Obviously the separation of the VNs results in lower resource utilization due to the lack of statistical multiplexing between high and low priority traffic at the connection leyer. On the other hand, the connection admission control (CAC_{QoS}) is based on the low priority connection state only, which simplifies the resource management. The two options can be easily expanded to a general multi-priority system. The first option will result in a sequence of nested virtual networks where the $k + 1$ priority VN is nested in the kth priority VN. In this case the CAC_{QoS} algorithm of the kth priority VN has also to consider the connection states of all higher priorities. Using the second alternative will result in a set of independent VNs.

Another approach to increasing resource utilization is based on creating a linear surface adjoined to the optimal northeast boundary from the inside of the admissible region. This linear surface is treated as a new admissible region boundary. In this case CAC_{QoS} is based on two equivalent bandwidth allocations for the high priority connections which are different from the one for a low priority connection of the same type. In the following we describe an algorithm for defining the adjoining points.

We start with a remark that in general the optimal northeast part of the admissible boundary might be neither convex nor concave. In particular it is convex in the cross sections relating connection classes with the same priority but can be concave in the northeast part of the cross section relating connection classes with different priority. The issue of convexity can be solved by applying the linear or modified linear approximation in the decomposition model. Then, the approach for generating a linear surface adjoined to the northeast part of the boundary assumes that it is concave. In this case the linear surface can be defined by the points of maximum throughput for low priority traffic classes and one corner per each high priority connection class (in such a way that all other corners are on or outside the new admissible boundary). This option is indicated by the dashed line in Figure 3.7b. The corresponding equivalent bandwidth allocations are as follows:

$$d_{1,h1} = \frac{L}{[U_{1,h}]_{\max}}, \quad d_{1,h2} = \frac{L}{U_{1,h'}}, \quad d_{1,l} = \frac{L}{[U_{1,l}]_{\max}} \qquad (3.56)$$

where $U_{1,h'}$ denotes the point on the $U_{1,h}$ axis being also a member of the linear surface containing all points of maximum throughput for low priority connections and the corner which minimizes $U_{1,h'}$ (in the example presented it is the corner from the cross section $U_{1,h}$, $U_{2,l}$). Then the connection admission conditions are given by

$$d_{j,h1} \leq L - \sum_{i \in J_h} x_{i,h} d_{i,h1} \tag{3.57}$$

$$d_{j,h2} \leq L - \sum_{i \in J} (x_{i,h} d_{i,h2} + x_{i,l} d_{i,l}) \tag{3.58}$$

for high priority connections and by

$$d_j^l \leq L - \sum_{i \in J} (x_{i,h} d_{i,h2} + x_{i,l} d_{i,l}) \tag{3.59}$$

for low priority connections. The described equivalent bandwidth allocations can be easily obtained from the decomposition model, since the only required information are the corner points and the maximum throughputs of the low priority traffic for $\rho_h = 0$. It is obvious that in general many different linear surfaces adjoined to the northeast boundary can be created. This flexibility in choosing the northeast part of the admissible boundary can be used to provide maximum throughput close to the operating point. To give an indication of the gain from application of two equivalent bandwidth allocations for high priority connections, the cross section of the admissible region with one equivalent bandwidth allocation for each type of traffic is also shown in Figure 3.7b (dotted line).

Observe that the second of the presented approaches to resource utilization increase could be also applied to the nested virtual network concept. A general discussion of resource allocation to QoS virtual networks, based on the results presented in this section, is given in Section 9.4.

3.3 Discussion and Bibliographic Notes

The literature on logical "bandwidth" allocation to connections in ATM networks is extensive. It all started with the idea of linear approximation proposed in (Turner, 1987). The validity of the equivalent bandwidth approach was proven by implementation of this concept in trial networks and products, e.g. (Guerin, Ahmadi, and Naghshineh, 1991). In this chapter we concentrated on the important features and general models which illustrate how the equivalent bandwidth allocation can be applied to provide efficient and reliable CAC_{QoS} algorithms. The cell layer performance models, which can serve for equivalent bandwidth calculation, are not discussed in this book. The interested reader can easily find them in the literature. There are three main methodologies for the performance evaluation of ATM links. The first group of models uses the *modulated Markov Poisson process* (MMPP) for cell traffic description and the matrix geometric approach for solving the performance problem, e.g. (Neuts, 1981; Heffes and Lucantoni, 1986; Fischer, 1989). The second methodology is based on fluid approximation which in general is more suitable for the ATM network environment, e.g. (Kosten, 1974; Anic, Mitra, and Sondhi, 1982; Daigle and Langford, 1986; Maglaris et al., 1988). The third approach is based on the central limit theorem and large deviation approximations, e.g. (Hui, 1990; Kelly, 1991; Elwalid and Mitra, 1993; Kesidis, Warland, and Chang, 1993). Its advantage is that it can provide an approximate explicit formulation for a modified linear approximation.

The models for equivalent bandwidth allocation presented in this chapter are generic in the sense that they are not limited to a particular type of source or a particular performance model. In fact the only performance characteristics required at the cell layer are the ones for the ATM multiplexer with homogeneous connections. Thus any reasonable analytic performance model can be applied. Alternatively these performance characteristics can be measured in test bed experiments with real sources. Besides evaluation of equivalent bandwidth allocation to connections, the model for multi-priority systems can also serve as a design for resource allocation to QoS virtual networks. This issue is discussed in a wider context in Chapter 9.

In this chapter we focused on the equivalent bandwidth allocation concept for systems where connections share link bandwidth and buffers. This approach provides statistical guarantees for QoS. Another important option is bandwidth allocation coupled with buffer allocation by means of a "fair queuing" scheduler (Demers, Keshav, and Shenker, 1989; Parekh and Gallager, 1992; Golestani, 1994; Roberts, 1994) which is an application of the general processor sharing, GPS, concept. This approach combined with a policing mechanism can provide deterministic guarantees for QoS (e.g., no cell losses). This important advantage is achieved at the expense of significant increase in the complexity of the link scheduler. As far as resource allocation efficiency is concerned, it is inferior to the optimal system with statistical guarantees since the GPS based system has to be designed for the *worst case*. Evaluation of "bandwidth" allocation in the GPS based systems can be easily done based on the link scheduler parameters.

References

Anic, D., Mitra, D., and Sondhi, M.M. 1982. Stochastic theory of a data-handling system with multiple sources. *B.S.T.J.*, 61(8)1871–1894.

Boyer, P.E., and Tranchier, D.P. 1992. A reservation principle with applications to the ATM traffic control. *Computer Networks and ISDN Systems*, 24:321–334.

Daigle, J.N., and Langford, J.D. 1986. Models for analysis of packet voice communications systems. *IEEE Journal on Selected Areas in Communication*, SAC-4(6):847–855.

Demers, A., Keshav, S., and Shenker, S. 1989. Analysis and simulation of a fair queuing algorithm. *ACM SIGCOM'89*, 1989.

Dziong, Z., Choquette, J., Liao, K.-Q., and Mason L.G. 1990. Admission control and routing in ATM networks. *Computer Networks and ISDN-Systems,* 20(1-5):189–196.

References

Dziong, Z., Liao, K-Q., and Mason, L.G. 1993. Effective bandwidth allocation and buffer dimensioning in ATM based networks with priorities. *Computer Networks and ISDN-Systems,* 25:1065–78.

Elwalid, A., and Mitra, D. 1993. Effective bandwidth of general markovian traffic sources and admission control of high speed networks. *IEEE/ACM Transactions on Networking,* 1(3):329–343.

Elwalid, A., and Mitra, D. 1995. Analysis, approximations and admission control of a multi-service multiplexing system with priorities. In *Proceedings of IEEE INFOCOM'95,* pp. 463–473. IEEE Computer Society Press.

Fischer, W. 1989. The MMPP Cookbook. Technical Report, INRS-Telecommunications.

Gallassi, G., Rigolio, G., and Fratta, L. 1989. Bandwidth assignment and bandwidth enforcement policies. In *Proceedings of IEEE GLOBECOM'89,* IEEE Computer Society Press.

Gallassi, G., Rigolio, G., and Fratta, L. 1990. Bandwidth assignment in prioritized ATM networks. In *Proceedings of IEEE GLOBECOM'90,* IEEE Computer Society Press.

Girard, A. 1990. *Routing and Dimensioning in Circuit-Switched Networks.* Addison-Wesley.

Golestani, S.J. 1994. A self-clocked fair queuing scheme for broadband applications. In *Proceedings of IEEE INFOCOM'94,* IEEE Computer Society Press.

Guerin, R., Ahmadi, H., and Naghshineh, M. 1991. Equivalent capacity and its application to bandwidth allocation in high-speed networks. *IEEE Journal on Selected Areas in Communication,* 9(7):968–981.

Heffes, H., and Lucantoni, D.M. 1986. A Markov modulated characterization of packetized voice and data traffic and related statistical multiplexerpPerformance. *IEEE Journal on Selected Areas in Communication,* SAC-4(6):856–868.

Holtzman, J.M. 1990. Coping with broadband traffic uncertainties: Statistical uncertainty, fuzziness, neural networks. In *Proceedings of IEEE GLOBECOM'90,* IEEE Computer Society Press.

Hui, J.Y. 1988. Resource allocation for broadband networks. *IEEE Journal on Selected Areas in Communication* 6(9):1598–1608.

Hui, J.Y. 1990. Switching and traffic theory for integrated broadband networks. Kluwer Academic Publisher.

Kelly, F. 1991. Effective bandwidths at multi-class queues. *Queueing Systems,* 9:5–15.

Kesidis, G., Warland, J., and Chang, C. 1993. Effective bandwidths for multiclass Markov fluids and other ATM sources. *IEEE/ACM Transactions on Networking,* 1(4):424–428.

Kosten, L. 1974. Stochastic theory of a multi-entry buffer. *Delft Progress Report,* 1:10-18.

Liao, K-Q., Dziong, Z., Mason, L.G., and Tetreault N. 1992. Effectiveness of leaky bucket policing mechanism. In *Proceedings of IEEE ICC'92,* IEEE Computer Society Press.

Liao, K.-Q., and Mason, L.G. 1990. A heuristic approach for performance analysis of ATM systems. In *Proceedings of IEEE GLOBECOM'90,* IEEE Computer Society Press.

Lucantoni, D.M. 1989. New results on the single server queue with a batch Markovian arrival process. Submitted to *Stochastic Models.*

Maglaris, B., Anastassiou, D., Sen, P., Karlson, G., and Robbins, J.D. 1988. Performance models of statistical multiplexing in packet video communications. *IEEE Transactions on Communications,* 36(7):834–844.

Monteiro, J.A.S., Gerla, M., and Fratta, L. 1991. Statistical multiplexing in ATM networks. In *Performance Evaluation* 12:157–167.

Neuts, M.F. 1981. *Matrix-Geometric Solutions in Stochastic Models: An Algorithmic Approach.* The Johns Hopkins Univ. Press.

Parekh, A., and Gallager, R. 1992. A generalized processor sharing approach to flow control in integrated services networks: The single node case. In *Proceedings of IEEE INFOCOM'92,* IEEE Computer Society Press.

COST 224. 1992. *Performance evaluation and design of multiservice networks,* ed. Roberts J.W. COST 224 final report. Commision of the European Communities.

Roberts, J.W. 1994. Virtual spacing for flexible traffic control. *International Journal of Communication Systems (John Wiley & Sons),* 7(4):307–318.

Tedijanto, T.E., and Gun, L. 1993. Effectiveness of dynamic bandwidth management mechanisms in ATM networks. In *Proceedings of IEEE INFOCOM'93,* IEEE Computer Society Press.

Turner, J.S. 1987. The challenge of multi-point communication. In *Proceedings of the 5th ITC Seminar,* Lake Como, Italy.

Turner, J.S. 1992. Managing bandwidth in ATM networks with bursty traffic. *IEEE Network,* 6(5):50–58.

Woodruff, G., and Kositpaiboon, R. 1990. Multimedia traffic management principles for guaranteed ATM network performance. *IEEE Journal on Selected Areas in Communication,* 8(3):437–446.

Chapter

4

Adaptive Resource Allocation to Aggregate Traffic

WHEN the algorithms discussed in Chapter 3 are used directly for connection admission control (CAC_{QoS}) of real-time services, they can provide QoS guarantees on the condition that the traffic generated by the sources conforms to the source traffic declarations, or on the condition that the design is done for the *worst case* traffic defined by the policing mechanism (UPC) parameters. There are two potential drawbacks of such an approach. First, it may be difficult for some sources to correctly predict their parameters in advance so the sources will tend to declare parameters with a safety margin, resulting in potential network underutilization. The second weakness is caused by the fact that the policing mechanism cannot control well the statistical parameters of the connection (e.g., average rate) so the *worst case* design has to be applied, resulting again in potential network underutilization. Moreover, the strict requirements concerning declaration of the source traffic parameters can be seen as a kind of unwanted restrictiveness from some user's point of view. In this chapter we describe a framework and models for adaptive equivalent bandwidth allocation where the drawbacks are avoided. Still, this approach requires models for evaluation of the equivalent bandwidth based on the declared parameters. Thus the models presented in Chapter 3 are also useful within this framework.

The central element of the framework is an estimation of the aggregate equivalent bandwidth required by all connections served by each of the switch output ports. In order to use all the available relevant information the estimation procedure employs source parameter declarations and direct measurement of connection superposition cell process parameters in the switch output ports. In this scheme the connection admission procedure can verify resource availability on all considered links by means of resource management cells passing through the transit nodes. One of the key advantages of the proposed framework is that the resource management and traffic control algorithm for real time and controllable data services, such as ABR or/and ABT, are based on the same database structure and signaling protocol. Besides reduction in algorithm complexity and cost this feature can also increase resource utilization by enabling coordination of both algorithms. Another important feature of the approach is that it is not restricted to any particular link or source model. In other words, any reasonable model for analytical evaluation of the connection equivalent bandwidth can be used in this framework.

The applied estimation process (Section 4.2), based on a linear Kalman filter, is decomposed into two parts. First the algorithm estimates certain parameters of the link connection superposition process. Then, based on the estimated and declared parameters, the aggregate equivalent bandwidth required by all connections on the link is estimated. The accuracy of the estimation process is analyzed in three stages (Section 4.3). First we concentrate on the evaluation of the measurement error. Then we analyze the errors of estimates provided by the Kalman filter. Finally we verify the accuracy of the model used for estimation of the aggregate equivalent bandwidth and the "bandwidth" reserved for the estimation error. The accuracy of the connection admission control (CAC_{QoS}) algorithm is analyzed under non-stationary conditions and large, non-Gaussian, declaration errors (Section 4.4).

In Section 4.5 we extend the estimation model to take into account the influence of source policing mechanisms. The numerical study of this option illustrates the trade-off between strict and relaxed source policing.

4.1 Framework for Adaptive Connection Admission

There are three principles on which the framework is based:

- All available information should be used in an optimal way. In particular it means that the algorithms should use explicit measurements of the relevant network states and the source traffic parameter declarations.

- The algorithms should not be restrictive in the sense that even an undeclared traffic increase should not be rejected if there are available resources (the magnitude and frequency of the undeclared traffic increase should rather be controlled by an appropriate tariff function).

- The algorithms for different service categories (CTP and NCTP) should be based on the same database structure and a similar signaling protocol in order to reduce complexity and cost of resource management and traffic control.

The basic concept of the framework is illustrated in Figure 4.1. In response to the new connection request the CAC algorithm at the user-network interface, UNI, sends a resource management cell (henceforth called CAC cell) to the destination UNI with information about the connection declared traffic parameters, $\mathbf{h}_j^d$, and required equivalent bandwidth, d'_j. In each node the TC&BM algorithm estimates the aggregate equivalent bandwidth, D', which should be reserved for all connections carried on each of the outgoing links. This is done based on the measurements of the superposition cell process parameters, $\tilde{Z}$, in the switch output ports and the

4.1 Framework for Adaptive Connection Admission

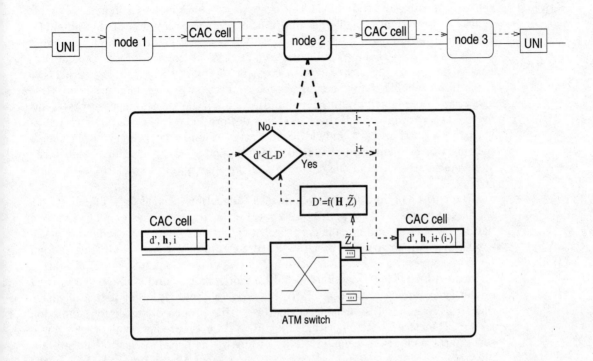

Figure 4.1. *CAC based on aggregate "bandwidth" estimation.*

aggregate declared parameters of accepted connections, **H**. When the CAC cell arrives at a node, the algorithm verifies whether the requested "bandwidth" d'_j is smaller than (or equal to) link residual capacity, $C = L - D'$. This information is inserted into the passing CAC cell and the "bandwidth" is reserved if $d'_j \leq C$. At the destination UNI the cell is returned to the origin UNI in order to reserve resources on the return path.

The central idea to this approach is a simple observation that the CAC algorithm does not need to know the precise values of the equivalent bandwidth required by each individual connection. On the contrary, it is the cell superposition process which defines the quality of service, and that is why the aggregate equivalent bandwidth is employed as the main control variable. Another advantage to this approach is that such a scheme can be much less restrictive compared to methods based on a policing mechanism. This is caused by two factors: statistical multiplexing of the "bandwidth" allocation errors and fast adaptiveness.

Statistical multiplexing of the errors reduces the amount of "bandwidth" reserved for these errors. Namely, in the case of *worst case* design (a policing mechanism enforces "bandwidth" allocation to the connection) the "bandwidth" allocated to the connection includes reservation for a possible error in the declared connection parameters. Although one can claim that the same problem exists with the aggregate "bandwidth" concept, it can be easily shown that standard deviation of the error in the estimation of the aggregate process parameters is in general significantly smaller than that of the sum of errors in the estimation of the individual connection parameters. Thus the aggregate "bandwidth" approach ensures more precise estimation of the "bandwidth" required for all connections. Fast adaptiveness is related to the fact that measurement of the cell superposition process is much more reliable than the sum of each connection process measurements (this again can be explained by statistical multiplexing of the measurement errors). This feature ensures that even if there is a large declaration error it can be quickly mitigated by adjusting the "bandwidth" allocation. Thanks to both effects the source policing can be significantly relaxed and its function changed from "bandwidth" enforcement to control of the magnitude of the aggregate "bandwidth" estimation error.

Comparison of the adaptive "bandwidth" allocation schemes for CTP and NCTP services (Figures 4.1, 2.2 and 2.3) shows that both approaches employ the same database structure and signaling protocol. In addition to significant reduction of the algorithms' complexity and cost this feature can also increase resource utilization by close cooperation of both algorithms. For example, the explicit rate allocation algorithms for CTP connections would be more efficient and stable if parameters of the aggregated cell process of NCTP services could be predicted in advance. Since this prediction is an inherent feature of the proposed connection admission for NCTP services, both algorithms may use a common prediction algorithm where some parts would be parameterized according to the needs of each of the two applications (e.g., to accommodate differences in the considered time horizon). In the remainder of this chapter we present an implementation of the proposed framework for connection admission control (CAC_{QoS}) for NCTP services. First we focus on a model for aggregate equivalent bandwidth estimation and its error analysis. Then, the adaptive CAC_{QoS} mechanism is studied. Two types of sources are considered: without and with a policing mechanism.

4.2 Adaptive Aggregate Equivalent Bandwidth Estimation

General framework[1]

Let $\hat{D}$ denote an estimate of the aggregate equivalent bandwidth. This estimate is a function of the declared parameters of accepted connections, $\mathbf{H}$, and measured parameters, $\tilde{Z}$, of the connection superposition process at the switch output port. $\hat{D}$ may be different from the actual value, D, due to declaration and estimation errors. If we assume that the "bandwidth" reserved for accepted connections, D', equals $\hat{D}$, the QoS constraints can be violated with high probability. In order to keep this probability at an acceptable level we introduce a "bandwidth", $\mathcal{R}$, reserved for the estimation error so the connection acceptance rule is based on the "bandwidth" reserved for aggregate traffic defined as

$$D' = \hat{D} + \mathcal{R} \qquad (4.1)$$

---

[1] Portions of Sections 4.2-4.5 are reprinted, with permission, from (Dziong, Juda, and Mason, 1996). ©1996 IEEE.

4.2 Adaptive Aggregate Equivalent Bandwidth Estimation

Thus the objective of the estimation process is to find $\hat{D}$ and $\mathcal{R}$ which fulfill

$$Pr\{D > \hat{D} + \mathcal{R}\} \leq \epsilon_1 \qquad (4.2)$$

where ϵ_1 is the estimation error constraint.

Note that from the connection admission control point of view we are interested in discrete points of time. Thus the system is modeled in the discrete time domain where t_k ($k = 1, 2...$) denotes the kth instant of the system state change, $W_{k-1} \to W_k$, caused either by a new connection or by a connection release. In general the system state can be defined by some parameters being a function of the cell rate superposition process, $A(t)$. Based on estimation theory, it can be shown that, by applying a recursive discrete filter, the state estimate of our system, $\hat{W}_k$, and the covariance matrix of its error, $\mathcal{P}_k$, can be evaluated as a function of the following parameters: $\mathbf{h}_k^d$, $\tilde{Z}_k$, $\hat{W}_{k-1}$ (where $\mathbf{h}_k^d$ denotes the declared parameters of the connection added or released in the transition $W_{k-1} \to W_k$). In addition, the parameters, $\mathbf{q}_k$, $\tilde{Y}_k$, defining the declaration and measurement error distributions (assumed to be Gaussian) are required. $\hat{D}_k$ and $\mathcal{R}_k$ estimation fits very well into the framework of recursive discrete filters. This suggests that the natural choice for the state description would be the aggregate equivalent bandwidth so the required estimates, $\hat{D}_k$, $\mathcal{R}_k$, would be obtained directly. While this approach is possible, in the following we describe another state description which simplifies the estimation process and ensures that it is not limited to a particular model for equivalent bandwidth evaluation.

There are two difficulties with direct estimation of $\hat{D}_k$. The relationship between the equivalent bandwidth and the parameters which can be directly measured, $\hat{D}_k = f_{\hat{D}}(\tilde{Z}_k)$, is in general non-linear. This feature implies that a non-linear filter should be applied which is typically more complex than a linear filter. In addition, although one can find in the literature several models for evaluation of the equivalent bandwidth, these algorithms are relatively complex and do not provide a universal closed form solution. These characteristics significantly increase the complexity of the problem since the function $f_{\hat{D}}$ should be inverted in the estimation algorithm.

To avoid the problems mentioned we chose as the state description a vector of the superposition cell rate process parameters which can be directly measured at the switch output port. Thus a linear filter can be applied and the inverse function problem becomes trivial. Having the estimate of the system state, $\hat{W}_k$, the estimate of aggregate equivalent bandwidth can be evaluated from a differential approach:

$$\hat{D}_k = f_{\Delta \hat{D}}(\hat{W}_k - W_k^d, D_k^d) \qquad (4.3)$$

where W^d, D^d denote the values evaluated from the declared parameters. The differential approach provides that the considered CAC_{QoS} algorithm can work with any reasonable model for evaluation of the connection equivalent bandwidth since this model is not directly used in the estimation procedure. The structure of the control system based on the differential approach is illustrated in Figure 4.2.

Now we will move on to investigate a particular version of the described scheme where the mean and variance of the instant cell rate of the connection superposition

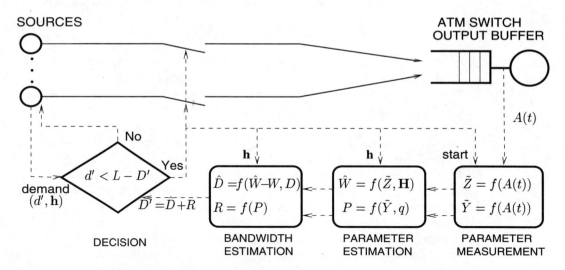

Figure 4.2. *Structure of the control system based on the differential approach.*

process are chosen as the system state description and parameters to be measured. While at first glance it might be viewed as a bold simplification, observe that the equivalent bandwidth can be evaluated from the full set of declared parameters which take into account all important source features, while the estimated parameters serve only to correct this "bandwidth" allocation. The implied assumption is that in general the sources do not change their basic features associated with the type of connection (otherwise a more sophisticated set of estimated and measured parameters would have to be chosen). The choice of the instant cell rate mean and variance is also supported by the burst scale performance analysis; see e.g. (COST, 1992). This analysis indicates that in a robust CAC_{QoS} algorithm for real-time applications the arrival link cell rate should only exceed link capacity with a very small probability, while the buffer dimensioning should take care of the cell scale congestion (small time scale variability factor). In this case the temporal characteristics of the cell rate process are not critical. Obviously, in the (unlikely) case of very large buffers the chosen parameters should be complemented by a parameter(s) characterizing the cell rate autocorrelation function.

Concerning the source declaration parameters $\mathbf{h}_k^d$, we assume that they include the predicted connection average cell rate, m_k^d, cell rate variance, v_k^d, and variances of these prediction errors, $\mathbf{q}_k = [v_{e,k}^m, v_{e,k}^v]^T$, respectively. These values can be declared directly by the source or can be estimated based on the source type, declared policing parameters (peak rate, sustained rate, and burst tolerance), and long term statistics concerning this source type behavior. Throughout the section we assume that the function for equivalent bandwidth evaluation from the declared parameters, $d_j^d = f_d(\mathbf{h}_j^d)$, is given (e.g., one of the models discussed in Chapter 3).

4.2 Adaptive Aggregate Equivalent Bandwidth Estimation

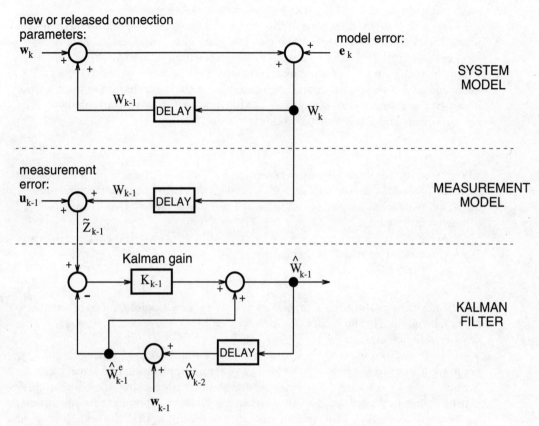

Figure 4.3. *System model and Kalman filter.*

Estimation of the cell rate mean and variance*

The objective of the estimation process is to provide the best estimate of the mean, M_k, and variance, V_k, of the instant cell rate of the connection superposition process. We assume that these two variables are independent although the proposed approach can also take into account the correlated case. The state of our system is defined as $W_k = [M_k, V_k]^T$. The dynamics of the system model is illustrated in Figure 4.3 and is described by

$$W_k = W_{k-1} + \mathbf{w}_k + \mathbf{e}_k \qquad (4.4)$$

where $\mathbf{e}_k$ denotes the model error and $\mathbf{w}_k = [\alpha_m m_k^d, \alpha_v v_k^d]^T$ denotes either the declared mean and variance of the accepted connection ($\alpha_m = \alpha_v = 1$) or the normalized declared mean and variance of the released connection ($\alpha_m = -\hat{M}_k/M_k^d$, $\alpha_v = -\hat{V}_k/V_k^d$ where index d denotes parameters evaluated from declarations). For the time being we assume that the model error, $\mathbf{e}_k = [\delta_k^m, \delta_k^v]^T$, is a Gaussian random variable with zero mean and covariance matrix, $\mathcal{Q}_k$. The interpretation of the model error distribution parameters and their evaluation based on the declaration error parameters, $\mathbf{q}_k$, will be discussed in Section 4.3.

The system model in Figure 4.3 is complemented by the measurement model which provides the estimates of the cell rate mean and variance in state k, $\tilde{Z}_k = [\bar{M}_k, \bar{V}_k]^T$. To differentiate these estimates from the final estimates, $\hat{W}_k = [\hat{M}_k, \hat{V}_k]^T$, we consider $\bar{M}_k$ and $\bar{V}_k$ as the measured mean and variance of the cell rate. The delay in the measurement model corresponds to the fact that the result of the parameter measurements is required at the time of the next state change. The measurement model is defined as follows:

$$\tilde{Z}_k = W_k + \mathbf{u}_k \qquad (4.5)$$

where $\mathbf{u}_k = [\bar{\delta}_k^m, \bar{\delta}_k^v]^T$ is the measurement error. The measurement error is assumed to be a Gaussian random variable with zero mean and known covariance matrix, $\tilde{Y}_k$. The evaluation and interpretation of the measurement error parameters are discussed in Section 4.3.

Besides the state estimate we are interested in assessing the characteristics of the estimation error which is defined as

$$\Delta_k = \hat{W}_k - W_k \qquad (4.6)$$

Since under the assumptions presented this error is a random variable with zero mean, the quality of the estimation process can be well assessed by the estimation error covariance matrix, $\mathcal{P}_k$.

The system and measurement models fit very well into the framework of linear recursive filters. In the following we demonstrate an application of a linear, discrete Kalman filter to the considered system (more information about the Kalman filter can be found in Appendix A). In general the Kalman filter provides an optimal least-squares estimate of the system state on the condition that the system is linear and the model and measurement errors are Gaussian random variables. The block diagram of the applied filter is shown in Figure 4.3. The system state estimate update is given by

$$\hat{W}_k = \hat{W}_k^e + \mathcal{K}_k[\tilde{Z}_k - \hat{W}_k^e] \qquad (4.7)$$

where $\hat{W}_k^e = \hat{W}_{k-1} + \mathbf{w}_k$ denotes the state estimate extrapolation and $\mathcal{K}_k$ is the Kalman filter gain. From the form of Equation (4.7) it is clear that the gain is a weight which determines how much confidence should be given to the measurement vs. the declaration.

To simplify further considerations let us introduce the transition matrix, Φ_k, with the following elements: $a_{11} = \frac{M_{k-1} + \alpha_m m_k}{M_{k-1}}$, $a_{22} = \frac{V_{k-1} + \alpha_v v_k}{V_{k-1}}$, and $a_{12} = a_{21} = 0$. Thus the system model can be described by

$$W_k = \Phi_k W_{k-1} + \mathbf{e}_k \qquad (4.8)$$

The Kalman gain is defined by

$$\mathcal{K}_k = \mathcal{P}_k^e[\mathcal{P}_k^e + \tilde{Y}_k]^{-1} \qquad (4.9)$$

where $\mathcal{P}_k^e$ denotes the estimate error covariance matrix extrapolation given by

$$\mathcal{P}_k^e = \Phi_{k-1}\mathcal{P}_{k-1}\Phi_{k-1}^T + \mathcal{Q}_k \qquad (4.10)$$

where $\mathcal{P}_{k-1}$ is the updated estimate error covariance matrix evaluated from

$$\mathcal{P}_{k-1} = [I - \mathcal{K}_{k-1}]\mathcal{P}^e_{k-1} \qquad (4.11)$$

Note that under the Gaussian assumption of the model and measurement errors, the estimation error distributions are also Gaussian with zero mean and variance defined by diagonal elements of the error covariance matrix, $\hat{v}^m_{e,k}$, $\hat{v}^v_{e,k}$, for cell rate mean and variance, respectively. These variances will be used to evaluate the "bandwidth" $\mathcal{R}_k$ reserved for the error in evaluation of $\hat{D}_k$ as described later in the section. Concerning the initial values we assume that the system is empty at the time $k = 0$ so $W_0 = 0$ and $\mathcal{P}_0 = 0$.

Estimation of aggregate equivalent bandwidth

To estimate the aggregate equivalent bandwidth from the differential approach defined by Equation (4.3) one needs to evaluate sensitivity of D with respect to the cell rate mean and variance around the declaration point D^d (to simplify presentation the time index is omitted in this part of the section). Note that the sensitivity function should not be complex since it is used in on-line evaluations. To fulfill this requirement we derive this function from a simple expression suggested for approximate equivalent bandwidth evaluation (COST, 1992; Gach, Mialaret, and Allard, 1992)

$$d^d = \gamma \cdot m^d + \theta \cdot v^d \qquad (4.12)$$

Based on Equation (4.12), for a particular state of accepted connections, we have

$$D^d = \gamma \cdot M^d + \theta \cdot V^d \qquad (4.13)$$

Since the values D^d, M^d, V^d are known for each link state, one can evaluate parameters γ, θ from two recent link states. Alternatively one can assume that the coefficient γ is independent of the current state and can be evaluated off-line using the link speed, QoS constraint, and average traffic mixture. Then the coefficient θ, for a given state, is given by

$$\theta = \frac{D^d - \gamma \cdot M^d}{V^d} \qquad (4.14)$$

Finally the estimated equivalent bandwidth of the connection superposition process is evaluated from

$$\hat{D} = \gamma \cdot \hat{M} + \theta \cdot \hat{V} \qquad (4.15)$$

It should be emphasized that, contrary to its appearance, Equation (4.15) serves only for evaluation of the correction to the declared equivalent bandwidth D^d implied by deviation of the estimated parameters, $\hat{M}, \hat{V}$, from the declarations, M^d, V^d, (D^d, M^d, V^d and the function for equivalent bandwidth allocation are hidden in γ and θ).

As to "bandwidth" reserved for the estimation error, under the Gaussian assumptions of the estimated mean and variance errors, the estimated "bandwidth" error, $\delta_D = \hat{D} - D$, also has Gaussian distribution. Thus under the assumption of mean and variance independence we have

$$\mathcal{R} = \mathcal{U}(\epsilon_1)\sqrt{\gamma^2 \hat{v}_e^m + \theta^2 \hat{v}_e^v} \tag{4.16}$$

where $\hat{v}_e^m$, $\hat{v}_e^v$ denote diagonal elements of the error covariance matrix, $\mathcal{P}$, and $\mathcal{U}(\epsilon_1)$ denotes a coefficient derived from the normalized Gaussian distribution which fulfills that

$$Pr\{D > \hat{D} + \mathcal{R}\} \leq \epsilon_1 \tag{4.17}$$

4.3 Estimation Error Analysis

Measurement error*

An optimal cell process measurement is complex and involves analysis of the autocorrelation function, see, e.g. (Yamada and Sumita, 1991). In this chapter we adopt an approach which is simple to implement but still enables the analysis of important features of the proposed traffic admission control model.

The first stage of the measurement process provides information concerning the instant cell rate. We assume that this estimate is given. In the second stage, mean M_k, variance V_k of the instant rate and the measurement errors are estimated. In the following we analyze a standard approach based on instant cell rate samples, s_i, taken in regular intervals in the period $(t_k, t_k + 1)$ and assumed to be independent. (A more general approach based on the autocorrelation function of the instant cell rate process would require estimation of the autocorrelation function, possibly using declarations and measurements.)

The standard estimates of the measured cell rate mean and variance are given by

$$\bar{M}_k = \frac{\sum_i s_i}{N_k} \tag{4.18}$$

$$\bar{V}_k = \frac{\sum_i (s_i - \bar{M}_k)^2}{N_k - 1} \tag{4.19}$$

where N_k denotes the number of the samples. Let us define the measurement errors for mean and variance of instant rate as

$$\bar{\delta}_k^m = \bar{M}_k - M_k \tag{4.20}$$
$$\bar{\delta}_k^v = \bar{V}_k - V_k \tag{4.21}$$

respectively. The theoretical values of these error variances are given by

$$v_{e,k}^m = \frac{V_k}{N_k} \tag{4.22}$$

$$v_{e,k}^v = \frac{V_k^4 - (V_k)^2}{N_k} \tag{4.23}$$

4.3 Estimation Error Analysis

where V_k, V_k^4 denote the variance and fourth central moment of the instant rate, respectively. Since in our framework only measured values are available we apply the following estimates:

$$\bar{v}_{e,k}^m = \frac{\bar{V}_k}{N_k} \qquad (4.24)$$

$$\bar{v}_{e,k}^v = \frac{\bar{V}_k^4 - (\bar{V}_k)^2}{N_k} \qquad (4.25)$$

where

$$\bar{V}_k^4 = \frac{\sum_i (s_i - \bar{M}_k)^4}{N_k - 1} \qquad (4.26)$$

In the following we analyze the accuracy of the approach presented by means of the connection admission process simulation at one switch output port (the details of the applied simulation model are given in Section 4.4). The selected parameters are as follows (Example 1): link capacity $L = 25$, connection arrival rate $\lambda = 200^{-1}$, mean connection holding time $\mu^{-1} = 5 \times 10^4$; source parameters: peak rate $P = 1$, fixed burst length $\tilde{B} = 50$, exponentially distributed silence length with average $\tilde{S} = 70$; simulation run-time $= 4 \times 10^6$. The measurement quality is assessed for the case where the sources conform to the declared parameters. In this example it is assumed that the sampling period equal to the average of the on-off source period, 120, provides sufficient independence of samples from a practical viewpoint. This is consistent with the results presented in (Yamada and Sumita, 1991).

The state duration is a critical factor for the quality of the measurement process; therefore the information concerning measured mean, variance, and their errors are gathered as a function of the number of samples, N. The analyzed variables related to the mean cell rate measurement are defined as follows:

- measured average error of the measured mean instant rate:

$$\bar{m}_e^m(N_k) = \frac{\sum \bar{\delta}_k^m(N_k)}{L_N} \qquad (4.27)$$

where L_N expresses the size of the state population with N_k samples.

- measured variance of the error of the measured mean instant rate:

$$\bar{V}_e^m(N_k) = \frac{\sum (\bar{\delta}_k^m(N_k))^2}{L_N} - (\bar{m}_e^m)^2 \qquad (4.28)$$

- estimated variance of the error of the measured mean instant rate:

$$\hat{V}_e^m(N_k) = \frac{\sum \bar{v}_{e,k}^m(N_k)}{L_N} \qquad (4.29)$$

Figure 4.4a presents results obtained for the mean instant rate measurement. One can notice that an underestimation appears for the short state periods. This spurious fault finds an explanation in the nature of the connection admission algorithm. When the state duration is short, the measured value is in general more

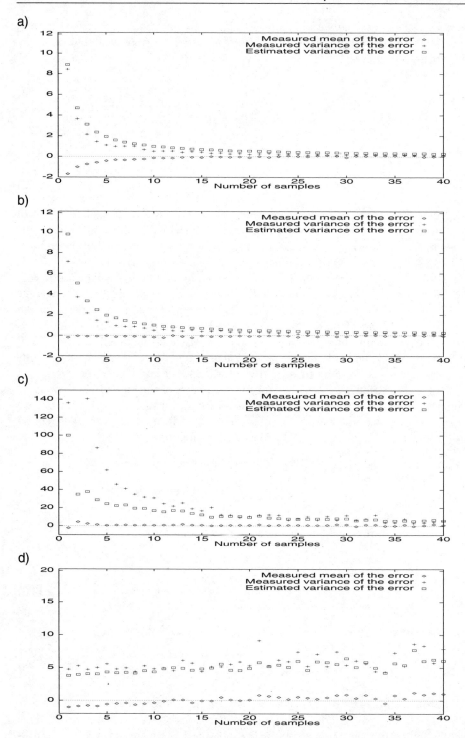

Figure 4.4. *Instant rate mean (a,b) and variance (c,d) measurement analysis (Example 1).*

likely to be significantly different from the declared one. In the figure this fact is expressed by the error variance which indeed is increased for small state durations. As a consequence, the overestimation of the mean cell rate results in an overestimation of the equivalent bandwidth. Thus a new connection is more likely to be rejected (no state change), while in the opposite case (underestimation of the mean cell rate), a new connection is more likely to be accepted. This effect causes more short duration states with an underestimated mean cell rate. The correctness of the above reasoning is illustrated in Figure 4.4b. In this case the admission decisions were made based on the declared aggregate "bandwidth" instead of the estimated one. The bias caused by the feedback is significant only when the number of rejected connections is large, as is the case in the chosen example. It can be removed by an appropriate modification of the state duration definition (introduction of state changes when the connections are rejected). Note that in both cases the variance of the error estimated from Equation (4.25) is close to the measured value.

Variables chosen to test the quality of cell rate variance measurement have been defined in an analogous way to the ones used for the cell rate mean measurement quality assessment:

- measured average error of the measured instant rate variance:

$$\bar{m}_e^v(N_k) = \frac{\sum \bar{\delta}_k^v(N_k)}{L_N} \tag{4.30}$$

- measured variance of the error of the measured instant rate variance:

$$\bar{V}_e^v(N_k) = \frac{\sum (\bar{\delta}_k^v(N_k))^2}{L_N} - (\bar{m}_e^m)^2 \tag{4.31}$$

- estimated variance of the error of the measured instant rate variance:

$$\hat{V}_e^v(N_k) = \frac{\sum \bar{v}_{e,k}^v(N_k)}{L_N} \tag{4.32}$$

Results are presented in Figure 4.4c. Discrepancy in the measured and estimated variance of the error for the small number of samples, shows that, as could be expected, formula (4.19) cannot guarantee reliable results. The problem can be alleviated by applying an approximation based on a moving window with a fixed number of samples. Since the window span can contain several states, the following approximation for the measured variance can be used

$$\bar{V}_k = \frac{V_k^d \cdot \bar{V}_k^w}{V_k^{w,d}} \tag{4.33}$$

where V_k^d denotes the declared variance in state k, $\bar{V}_k^w$ is the measured variance based on the samples from the whole window, and $V_k^{w,d}$ is the average declared variance in the window. The efficiency of the applied mechanism may be verified through the comparison of Figure 4.4c (standard procedure) and Figure 4.4d (procedure enhanced by the moving window with 40 samples). Improvement of the accuracy, particularly for the short state durations, is significant.

Mean and variance estimation error*

Let us define the estimation errors for mean and variance of the instant cell rate as

$$\hat{\delta}_k^m = \hat{M}_k - M_k \quad (4.34)$$
$$\hat{\delta}_k^v = \hat{V}_k - V_k \quad (4.35)$$

respectively. Under the Gaussian assumption of the model and measurement errors, the estimation error distributions are also Gaussian with zero mean and variance defined by the diagonal elements of the error covariance matrix, $\hat{v}_{e,k}^m$, $\hat{v}_{e,k}^v$. The error covariance matrix is a function of the measurement and model errors [see Equations (4.9) and (4.10)]. Since the assessment of the measurement error was already presented, in the following we focus on the model error.

In general the model error is a function of the declaration errors which are defined as a difference between the actual and declared connection parameters

$$\mathbf{c}_j(t) = \mathbf{h}_j - \mathbf{h}_j^d \quad (4.36)$$

where time t indicates that in general the error can be non-stationary. After transformation of the declaration error into the discrete time domain the model error is defined by

$$\mathbf{e}_k = \mathbf{c}_{j,k} + \sum_i [\mathbf{c}_{i,k} - \mathbf{c}_{i,k-1}] \quad (4.37)$$

where index j corresponds to the new or departing connection and index i corresponds to an existing connection. Observe that in general the periods between the state transitions are significantly shorter than the connection duration. This indicates that although the connection parameters might be different from the declarations, one can expect that there will be a large autocorrelation between the values in the subsequent system states. Therefore the terms under summation in Equation (4.37) will be small. Moreover, under the assumption of statistical independence of the connection instant rate processes, these terms will be positive and negative. These premises lead to the conclusion that the second term in Equation (4.37) can be neglected so the model error can be approximated by the declaration error of the new or departing connection

$$\mathbf{e}_k \cong \mathbf{c}_{j,k} \quad (4.38)$$

This approximation is exact when each connection process is stationary. Based on Equation (4.38) the covariance matrix of the model error, $\mathcal{Q}_k$, is defined by the predicted variances of declaration errors, $\mathbf{q}_k = [v_{e,k}^m, v_{e,k}^v]$, for cell rate mean and variance, respectively. These values can be evaluated based on statistical measurements of real sources. Note that in our model it is assumed that the declaration errors have Gaussian distribution with zero mean. Nevertheless, the algorithm can cope very well with distributions quite different from the Gaussian, as will be shown in Section 4.4.

To verify the accuracy of the instant rate mean and variance estimation we use Example 2 which differs from Example 1 in declaration error distribution. In this case the source declared parameters are still the same ($m^d = 0.41$, $v^d = 0.25$),

4.3 Estimation Error Analysis

TABLE 4.1 Instant rate mean estimation results (Example 2).

	$\bar{m}_e^m$	$\bar{M}$	$\bar{V}_e^m$	$\hat{V}_e^m$	$\hat{m}_e^m$
Model 1	-0.090	18.5	4.99	4.78	-0.068
Model 2	-0.091	18.5	4.69	4.79	-0.064

TABLE 4.2 Instant rate variance estimation results (Example 2).

	$\bar{m}_e^v$	$\bar{V}$	$\bar{V}_e^v$	$\hat{V}_e^v$	$\hat{m}_e^v$
Model 1	-0.053	9.27	4.39	4.09	-0.733
Model 2	-0.134	9.11	4.23	4.20	-0.018

but the actual ones are generated with the Gaussian error with zero mean and variances: $v_e^m = 0.05$ (for practical reasons the distribution is limited to the range $m \in [0.2, 0.8]$), $v_e^v = 0.01$ (in this case the distribution is limited to the range $v \in [0.1, 0.4]$). Note that due to the relation between the connection mean and variance, $v = m \cdot (P - m)$, the peak rates, P, can have different values for each connection.

A set of the following variables will be used in the sequel:

- the measured average error of the measured instant rate mean and variance (respectively):

$$\bar{m}_e^m = E[\bar{M}_k - M_k], \qquad \bar{m}_e^v = E[\bar{V}_k - V_k] \qquad (4.39)$$

- the measured variance of the error of the measured instant rate mean and variance (respectively):

$$\bar{V}_e^m = E[(\bar{M}_k - M_k)^2] - (\bar{m}_e^m)^2, \qquad \bar{V}_e^v = E[(\bar{V}_k - V_k)^2] - (\bar{m}_e^v)^2 \qquad (4.40)$$

- the estimated variance of the error of the measured instant rate mean and variance (respectively):

$$\hat{V}_e^m = E[\bar{v}_{e,k}^m], \qquad \hat{V}_e^v = E[\bar{v}_{e,k}^v] \qquad (4.41)$$

- the measured average error of the estimated instant rate mean and variance (respectively):

$$\hat{m}_e^m = E[\hat{M}_k - M_k], \qquad \hat{m}_e^v = E[\hat{V}_k - V_k] \qquad (4.42)$$

- average of the measured instant rate mean and variance (respectively):

$$\bar{M} = E[M_k], \qquad \bar{V} = E[V_k] - V_k \qquad (4.43)$$

The results corresponding to the instant rate mean and variance estimation are presented in Table 4.1 and Table 4.2, respectively. *Model 1* corresponds to the

TABLE 4.3 Accuracy of the equivalent bandwidth estimation (Example 2).

	$\hat{m}_e^m$	$\hat{m}_e^v$	$\bar{E}[\hat{D} - D]$	$\bar{E}[\hat{D}]$
Model 2	-0.020	-0.003	-0.243	22.1
Model 3	-0.054	0.019	-0.010	22.3

TABLE 4.4 Quality of the reserved equivalent bandwidth estimation (Example 3).

Imposed ϵ_1	$\hat{Pr}\{D_k > \hat{D}_k + \mathcal{R}_k\}$
10^{-1}	3.2×10^{-2}
10^{-2}	3.5×10^{-3}
10^{-3}	4.7×10^{-4}
10^{-4}	1.1×10^{-4}
10^{-5}	1.3×10^{-5}

presented improved measurement algorithms. The mean instant rate is estimated with a satisfactory accuracy, the error does not exceed 0.5% of the estimated value. However, the results of the instant rate variance estimation are less adequate. One can notice that the measurement part works appropriately, but the estimation produced a significant error. This is explained by a strong correlation between the measured values of instant rate variance, $\overline{V}_k$, and variance of the instant rate variance error, $\overline{v}_{e,k}^v$ since the same set of samples is used to evaluate both values. As a result, whenever, for statistical reasons, the value of the measured variance is underestimated it has larger weight in the Kalman filter due to the underestimated measurement error. This effect gives, on average, underestimation of the estimated variance of instant rate. This bias can be mitigated by extending the window span used to evaluate the fourth moment. The results for the extended window with 100 samples are presented in Tables 4.1 and 4.2 (*Model 2*). The quality of variance measurement is noticeably upgraded.

Equivalent bandwidth estimation error

The evaluation of the equivalent bandwidth estimation accuracy is based on Example 2. Error assessment is presented in Table 4.3 (*Model 2*), where $\bar{E}[\ ]$ denotes measured average value. Note that although the mean and variance of the instant rate are estimated correctly, the equivalent bandwidth is slightly underestimated. This is caused by the fact that the exact relation between the equivalent bandwidth and the instant rate mean and variance is slightly non-linear and non-symmetrical with respect to the central point of source parameter distribution while the applied approximation (4.15) is linear. To illustrate this effect we evaluated "optimal" coefficients, θ, γ, by integrating the exact equivalent bandwidth function. The results are presented in the Table 4.3 (*Model 3*). In this case the error is negligible.

To verify the accuracy of the "bandwidth" reserved for the estimation error, $\mathcal{R}_k$, the probability $Pr\{D_k > \hat{D}_k + \mathcal{R}_k\}$ was estimated in a simulation experiment. In order to reduce declaration error distribution deformations caused by the physical constraints, the declaration cell rate variance error was reduced: $v_{e,k}^v = 0.001$

(Example 3). The estimated probabilities $\hat{Pr}\{D_k > \hat{D}_k + \mathcal{R}_k\}$ are presented in Table 4.4 for different values of the constraint ϵ_1. Note that in all cases the estimated probabilities are very close to the constraint defined by Equation (4.2).

4.4 Connection Admission Control Analysis (QoS)

Connection admission procedure

As defined in Section 4.1, in the presented framework a new connection is accepted if
$$d'_k \leq L - D'_{k-1} \qquad (4.44)$$
Note that in this condition the estimate from state $k-1$ is treated as a prediction for state k. The implied assumption is that the aggregate connection process can be treated as stationary for the period of duration of the two states. While estimation of $D'_{k-1} = \hat{D}_{k-1} + \mathcal{R}_{k-1}$ was already presented, interpretation and evaluation of the "bandwidth" reserved for a new connection, d'_k, requires additional clarifications. In particular it is important to define more precisely the design objective of the connection admission procedure and how the quality of this procedure should be judged. The answers to these questions are not straightforward. Note that the main criterion of the CAC_{QoS} procedure is to ensure that the QoS constraints are met (on the cell layer). In our model this requirement is fulfilled when the actual "bandwidth" required by the admitted connections does not exceed the link capacity, $D_k \leq L$. Obviously strict execution of this condition might cause resource underutilization. Thus, to improve resource utilization, we allow that $D_k > L$ with a certain small probability:
$$Pr\{D_k > L\} \leq \epsilon_2 \qquad (4.45)$$
There is one drawback to this formulation. Namely, this probability depends strongly on the connection arrival process and connection "bandwidth" requirements. In particular the smaller the traffic level, the smaller $Pr\{D_k > L\}$. This feature indicates that the condition (4.45) is not convenient for the design of the connection admission algorithm.

To deal with this issue we employ another definition of the CAC_{QoS} procedure quality which is independent from the connection arrival process and connection "bandwidth" requirements. It is based on the following conditional probability
$$Pr\{D_k > L \mid d'_k = L - D'_{k-1}\} \leq \epsilon_2 \qquad (4.46)$$
In this case the quality is defined for the critical case where the residual capacity is equal to the one required by a new connection and is independent from the connection arrival process and connection "bandwidth" requirements. Observe that condition (4.46) is also fulfilled when
$$Pr\{D_k > d'_k + D'_{k-1}\} \leq \epsilon_2 \qquad (4.47)$$
The latter condition constitutes the basis for design and evaluation of the proposed CAC_{QoS} procedure.

Analogously to D'_{k-1} the "bandwidth" reserved for a new connection, d'_k, can be decomposed into two parts: the equivalent bandwidth, d^d_k, evaluated from the declared parameters and the "bandwidth," ζ_k, reserved for the declaration error. Note that the sum $\zeta_k + \mathcal{R}_{k-1}$ could be evaluated from the superposition of the d_k and D_{k-1} distributions. Nevertheless from the connection admission viewpoint it is more convenient to separate evaluation of ζ_k from evaluation of $\mathcal{R}_{k-1}$. The main reason behind this approach is that in this case the processing of the CAC control cell in a transit ATM node is limited to a simple comparison of two numbers (otherwise some more complex calculation would have to be performed).

Based on the Gaussian assumption, the "bandwidth" reserved for the declaration error could be evaluated from

$$\zeta_k = \mathcal{U}(\epsilon_3)\sqrt{\gamma^2 v_e^m + \theta^2 v_e^v} \tag{4.48}$$

where the value of the constraint ϵ_3 can be chosen between $\epsilon_3 = \epsilon_2 - \epsilon_1$ (conservative) and $\epsilon_3 = \epsilon_2/\epsilon_1$ (optimistic). Another possibility is to apply an approach where the parameter ζ_k is defined by the connection peak rate, P_k:

$$\zeta_k = P_k - d^d_k \tag{4.49}$$

This approach is simpler and safer since the peak rate defines the upper boundary for the error. In the following we use the second approach.

Numerical study

We start from a description of a simulation model used for assessment of the proposed algorithms. To avoid excessive complexity we simplified the model as much as possible to concentrate on the main issues. In particular only the instant cell rate layer is modeled and the link buffer has zero length. It should be stressed that the bufferless case was chosen only to simplify the exact equivalent bandwidth evaluation. This choice does not restrict the analyzed CAC model applications nor limits generality of the results. This follows from the fact that the CAC algorithm employs the concept of equivalent bandwidth which separates the issue of CAC adaptiveness from buffer dimensioning and performance measurement on the cell layer. Moreover, the bufferless case exemplifies the burst scale layer model whose performance is critical to a robust CAC_{QoS} algorithm for real-time services.

The requests for connections of particular class are generated with intensity λ (Poissonian distribution) and mean holding time μ^{-1} (exponential distribution). The connections are of on-off type and are described by the peak rate, P, average burst length, $\tilde{B}$, (with programmable distribution) and the average silence length, $\tilde{S}$ (exponential distribution).

The QoS constraint (cell loss probability, $\mathcal{B}^c$) is set to a relatively high value $\mathcal{B}^c = 10^{-2}$ in order to achieve a reliable estimate of the cell loss probability distribution under non-stationary traffic conditions. The estimation error constraint is set to the same value, $\epsilon_1 = 10^{-2}$.

The declaration error can be generated in many ways. The results presented in the previous section are based on the Gaussian model for this error distribution (Examples 2 and 3), which conforms with the Kalman filter assumptions. In this section we introduce another error model which is significantly different from that of the Kalman filter. Namely, the error generator has two cyclic states ("on" and "off") with the same period T. When the generator is in the "on" state all connections accepted in this state are generated with the burst length larger than the declared one ($\tilde{B}' = \tilde{B} + \Delta \tilde{B}$). In the "off" state all new connections are generated with the declared burst length (for entire duration of the connection). Note that in this case the error has no zero mean. The reason for this model is to evaluate the adaptation scheme under more stressing conditions.

The binomial distribution is used to evaluate both the equivalent bandwidth from the source declarations, $d_k^d = f_d(\mathbf{h}_k^d)$, and the exact aggregate equivalent bandwidth, D_k, based on the actual connection parameters (to assess the accuracy of the adaptation scheme).

The performance of the CAC mechanism under the deterministic and non-stationary error is illustrated in two examples. The first one (Example 4) is defined by: (1) link and connection parameters: $L = 25$, $\lambda = 200^{-1}$, $\mu^{-1} = 5 \cdot 10^4$; (2) source parameters: $P = 1$, $\tilde{B} = 50$, $\tilde{S} = 70$, $v_e^m = 0.05$, $v_e^v = 0.002$; and (3) error generator parameters: $T = 2 \cdot 10^5$, $\tilde{B}' = 100$

A sample of the equivalent bandwidth allocation dynamics, during a part of the simulation run, is presented in Figure 4.5. In Figure 4.5a the total number of connections and the number of connections with modified burst length are given. During the "on" period of the error generator the parameters of almost all connections are modified, while at the end of the "off" period almost all connections have the declared parameters.

The trajectory of the actual aggregate equivalent bandwidth, D_k, and the difference between the estimated and actual aggregate equivalent bandwidth, $\hat{D}_k - D_k$, are presented in Figure 4.5b. The estimated aggregate equivalent bandwidth tracks well the actual value. It can be noticed that the highest underestimation of the equivalent bandwidth occurs during the "on" period of the error generator. Nevertheless the "bandwidth" reserved for estimation error ensures that the actual aggregate equivalent bandwidth does not exceed the link capacity. This result shows that the proposed scheme is robust since during the error generation period the actual declaration error is not only on average twice higher than the declared one but also is always positive. This robustness results from two factors. First, at the moment of a new connection admission there is provision for the equivalent bandwidth equal to its peak rate, so no matter how malicious the source could be, the QoS is kept under the constraint during the new state. Although in the following states this provision disappears (the connection is included in the aggregate equivalent bandwidth), the system has time to correct the "bandwidth" allocation by means of the measurement process incorporated in the estimation of the aggregate equivalent bandwidth.

The efficiency of the proposed adaptive algorithm is underlined in Figure 4.5c, where the estimated aggregate equivalent bandwidth allocation, $\hat{D}_k$, and the declared aggregate equivalent bandwidth, D_k^d are plotted. The large gap between the two trajectories, in the periods when the number of modified connections is significant, indicates that the connection admission scheme based solely on the declared parameters would allow too many connections and the actual aggregate equiva-

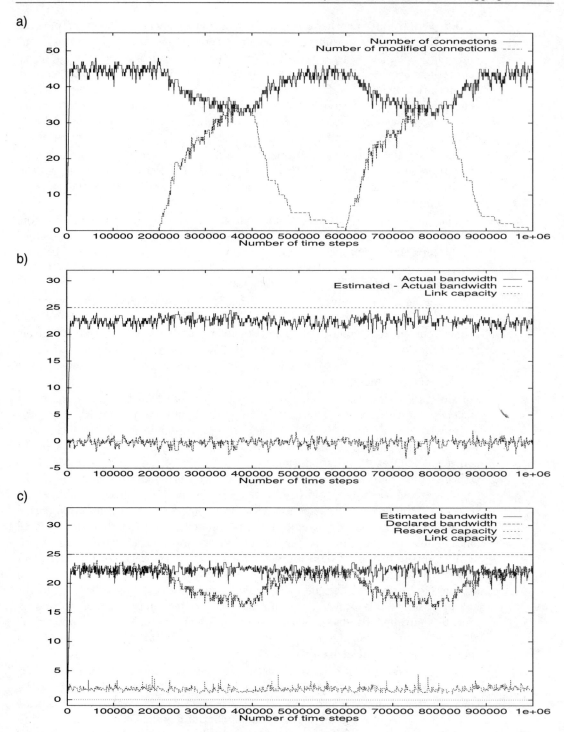

Figure 4.5. *Trajectories of the system variables vs. simulation time (Example 4).*

lent bandwidth would significantly exceed the allocated capacity. Additionally the "bandwidth" reserved for the estimation error $\mathcal{R}_k$ is also depicted in Figure 4.5c.

The second example (Example 5) has the same type of error as Example 4 (deterministic and non-stationary) but includes two connection classes and is defined by: (1) link and connection parameters: $L = 50$, $\lambda_1 = 1000^{-1}$, $\mu_1^{-1} = 5 \times 10^4$, $\lambda_2 = 500^{-1}$, $\mu_2^{-1} = 6 \times 10^4$; (2) source parameters: $P_1 = 1$, $\tilde{B}_1 = 50$, $\tilde{S}_1 = 70$, $v_{e,1}^m = 0.1$, $v_{e,1}^v = 0.1$, $P_2 = 4$, $\tilde{B}_2 = 20$, $\tilde{S}_2 = 80$, $v_{e,2}^m = 0.2$, $v_{e,2}^v = 0.2$; and (3) error generator parameters: $T = 10^5$, $\tilde{B}_1' = 150$, $\tilde{B}_2' = 60$.

A sample of the system dynamics, during a part of the simulation run, is presented in Figure 4.6. In Figure 4.6a the trajectories of connection numbers are given. The trajectories of the estimated aggregate equivalent bandwidth, $\hat{D}_k$, and the declared aggregate equivalent bandwidth, D_k^d, are presented in Figure 4.6b. The trajectory of the cell loss probability (estimated in the time windows used for the cell variance measurements) is given in Figure 4.6c. The results show that the algorithm is also robust in the multi-class environment.

The example with correlated sources (Example 6) has the same system and source parameters as Example 3 except for the variance of the declared variance error which is increased to $v_{e,k}^v = 0.1$. The correlated sources are generated every 10^5 time units starting at time 5×10^4 and the number of correlated sources is limited to $n = 10$. The system trajectories are presented in Figure 4.7. The results show that the variance estimation process reacts a little bit too slowly to the sudden variance increase caused by the correlated sources (Figure 4.7b). This feature results in a slight overpassing of the QoS constraint when the correlation is introduced (Figure 4.7c). The slower reaction, compared to the case with the deterministic error case (Example 4), is due to the larger variance measurement error. Obviously, this drawback can be avoided by introducing a very large variance of the declared variance error, v_e^v, so a bigger weight is allocated to the variance measurement in the estimation process. Nevertheless this approach would result in an increased "bandwidth" reservation for the estimation error at the times when sources are not correlated.

4.5 Trade-off Between Relaxed and Strict Source Policing

In this section we consider the case where the source parameters are controlled by means of the policing mechanism. Observe that, in general, introduction of the policing mechanism reduces the magnitude of the source declaration error, seen at the switch output ports. Thus, there is a potential to further increase the link resource utilization without compromising the QoS performance. In the following we will extend the adaptive algorithm presented in the previous sections to take advantage of the policing mechanism. We will also give numerical examples to illustrate the accuracy of this approach and to discuss the advantages and disadvantages of the source parameter enforcement.

We start with the assumption that the source declaration errors obey a Gaussian distribution with zero mean. When there is no policing mechanism, this assumption

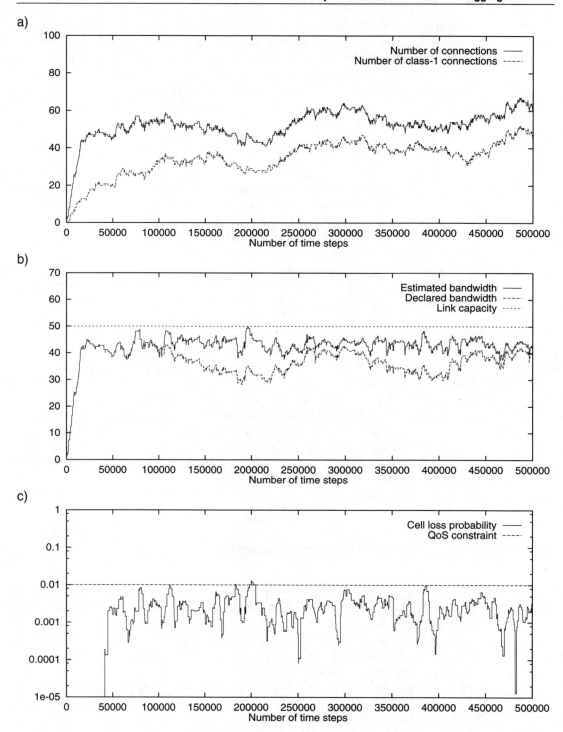

Figure 4.6. *Trajectories of the system variables vs. simulation time (Example 5).*

4.5 Trade-off Between Relaxed and Strict Source Policing

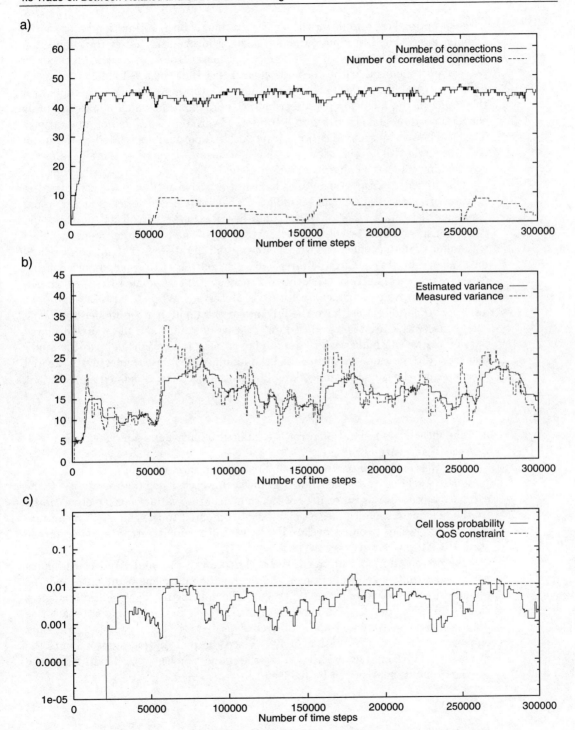

Figure 4.7. *Trajectories of the system variables vs. simulation time (Example 6).*

provides optimal conditions for the aggregate "bandwidth" estimation based on the Kalman filter. The introduction of the policing mechanism causes the connection process entering the link to be different in general from the source process. In particular, it means that the declaration error distributions are no longer Gaussian and, on top of that, they are not symmetrical (no zero mean). To cope with this issue, in the following, we describe an approximation which allows the basic connection admission algorithm to be preserved with only a few minor adaptations. This is achieved by a modification of the declared parameters related to the mean cell rate so that the declaration error has zero mean. Thereafter these parameters are used in the original algorithm which assumes Gaussian error distributions.

The influence of the policing mechanism (leaky bucket) on the mean rate declaration error distribution can be modeled as follows. First notice that the upper Gaussian distribution tail of the original distribution is cut off at the threshold value which is defined by the leak cell rate, LCR, of the policing mechanism. Then we assume that all connections with source mean cell rate higher than the threshold have mean cell rate equal to the threshold at the output of the policing mechanism. Concerning the connections with the source mean cell rate smaller than the threshold we assume that their mean cell rate is preserved. While the two assumptions are approximations, they have the advantage of simplifying the model and being safe (conservative approach). Based on these assumptions and the source declared parameters we can numerically evaluate mean $m_{e,k}^{m'}$ and variance $v_{e,k}^{m'}$ of the modified error distribution. In such a case the modified declared connection mean cell rate, used by the original connection admission procedure, is defined as

$$m'_k = m_k + m_{e,k}^{m'} \qquad (4.50)$$

and the mean rate declaration error is assumed to have Gaussian distribution with zero mean and variance $v_{e,k}^{m'}$.

Note that one could try to apply an analogous procedure to modify the declared parameters related to the cell rate variance. However, since the sensitivity of the estimation procedure with respect to the change in rate variance parameters caused by the policing mechanism is less significant compared to the change in mean rate parameters, we use source declared rate variance and its error variance in the studied CAC procedure.

The presented modifications of the declared mean rate and the declaration error variance influence evaluation of the aggregate equivalent bandwidth and the "bandwidth" reserved for estimation error. The policing mechanism also influences the "bandwidth" reserved for the declaration error, ζ. In the original approach this "bandwidth" was evaluated as the difference between the connection peak rate and equivalent bandwidth, Equation (4.49). Since the policing mechanism limits the *worst case* equivalent bandwidth used by the connection, the "bandwidth" reserved for the declaration error can be reduced to

$$\zeta'_k = d_k^{\max} - d_k^d \qquad (4.51)$$

where $d_k^{\max}$ denotes the equivalent bandwidth required by the policing mechanism's *worst case* source.

4.5 Trade-off Between Relaxed and Strict Source Policing

TABLE 4.5 Numerical results (Example 3).

Policed mean cell rate	1.0	0.51	0.41
Leak rate	1.0	0.62	0.50
Average reserved "bandwidth"	1.54	1.34	1.25
Average aggregate equivalent bandwidth	22.2	22.8	22.9
Average cell loss probability	3.5×10^{-3}	4.7×10^{-3}	5.7×10^{-3}
Average number of connections	46.2	49.0	52.4

TABLE 4.6 Numerical results (Example 4).

Policed mean cell rate	1.0	0.51	0.41
Leak rate	1.0	0.62	0.50
Average reserved "bandwidth"	1.39	1.25	0.56
Average aggregate equivalent bandwidth	22.3	22.8	23.5
Average cell loss probability	1.8×10^{-3}	3.2×10^{-3}	5.0×10^{-3}
Average number of connections	39.2	41.6	44.6

Accuracy and performance of the described approximations for different policing thresholds are illustrated in Table 4.5 and Figure 4.8 for Example 3 (Gaussian distribution of the declaration error) and in Table 4.6 and Figure 4.9 for Example 4 (deterministic and non-stationary declaration error). Three different mean cell rate thresholds are applied to each example: 1.0 (peak rate limitation only), 0.51 and 0.41 (the declared mean cell rate). The design of the policing mechanism was based on the model presented in (Liao et al., 1992) (cell rejection rate for conforming source: 10^{-3}). The corresponding leak rates are 1.0, 0.62, and 0.5 respectively. The cell loss probabilities are estimated in the windows used for the cell variance measurements. The average and the distribution of cell loss probabilities are given in Tables 4.5 and 4.6, and Figures 4.8 and 4.9, respectively. The performance of the CAC is very close to the constraints ($\mathcal{B}^c = 0.01$, $\epsilon_1 = 0.01$) in all cases. This result shows that the described approximation is accurate and that the adaptive CAC can work efficiently with the policing mechanism.

As could have been predicted, the more strict the policing is, the more connections can be accepted on average, especially in the case of a large positive error (Example 4). Nevertheless, this feature is not necessarily advantageous. First of all, the increase in the number of accepted connections is achieved at the expense of the source parameters' limitation, which can be seen as an unwanted restriction from the user's point of view. This restriction can force users to declare increased parameters, since in many cases the sources do not know well their parameters in advance. Thus this behavior can reduce resource utilization. Secondly, observe that the average aggregate equivalent bandwidth is similar in all cases, especially in the Example 3. Thus if the tariff is proportional to the actual "bandwidth" usage, the network revenue is not affected significantly. Note that introduction of a tariff, which charges more for the traffic exceeding the declaration, can motivate users to declare parameters as accurately as possible (Kelly, 1994).

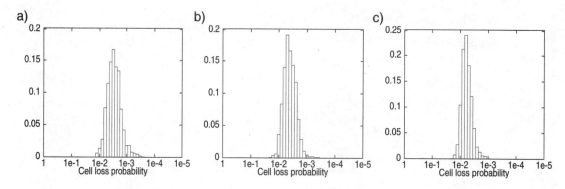

Figure 4.8. *Cell loss probability distributions (Example 3): (a) $LCR = 1.0$; (b) $LCR = 0.62$; (c) $LCR = 0.50$.*

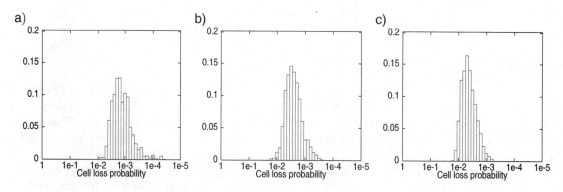

Figure 4.9. *Cell loss probability distributions (Example 4): (a) $LCR = 1.0$; (b) $LCR = 0.62$; (c) $LCR = 0.50$.*

4.6 Discussion and Bibliographic Notes

Most of the resource management and traffic control algorithms for ATM based networks presented in the literature focus on particular types of services. In this chapter we have described a unified framework which bridges the gap between the algorithms for real-time services and controllable data services. The framework is based on the rate-based schemes for controllable data services (ABR, ABT), recommended by the ATM-Forum and ITU-T, and the adaptive connection admission control (CAC_{QoS}) algorithm based on aggregate equivalent bandwidth estimation for real-time services. Both react to measurements of the cell rate process at the switch output ports and use the same database structure and signaling protocol. Besides reduction of the algorithms complexity and cost, this feature can also increase resource utilization by close coordination of both algorithms.

The aggregate equivalent bandwidth estimation is based on a linear Kalman filter. The Kalman filter approach takes into account the connection level dynamics of the system in an optimal way, and gives information for evaluation of "bandwidth" reserved for possible estimation error. This "bandwidth" ensures a statistical guarantee for quality of service. The numerical study of the CAC_{QoS} algorithm based on the aggregate equivalent bandwidth estimation, under non-stationary conditions and large declaration errors, demonstrated that the approach is very robust and copes very well with undeclared changes in traffic parameters. The study of the trade-off between relaxed and strict source policing indicated that application of relaxed policing can be advantageous from both user and network viewpoints. The inclusion of a policing mechanism also shows that the described framework can coexist with the ATM Forum and ITU recommendations.

There are several further model extensions which could be investigated. For example, the measurement process is an important area of study where declared and measured autocorrelation functions can play an important role. More complex filters can be tried to check whether it is possible to estimate some characteristics related more directly to the QoS metrics. Finally an integrated model for resource management for CTP and NCTP services based on a partly common estimation algorithm can be studied.

In the literature one can find several works which try to achieve adaptiveness of the CAC_{QoS} procedure for real-time services. Although they have some common elements with the framework presented in Section 4.1, they differ in many aspects from it. The approach from (Tedijanto and Gun, 1993) is based on declarations and measurements of the source process at UPC so the "bandwidth" allocation and policing mechanism parameters can be adjusted (no measurements in the switch output ports). The algorithm proposed in (Saito, 1991) uses measurements of the cell process parameters on the network links in order to estimate the cell process distribution (no equivalent bandwidth allocation). In (Gibbens, Kelly, and Key, 1995.) an adaptive algorithm based on a threshold for the current link load is presented. The thresholds are evaluated in the framework of the Bayesian decision theory. In (Li, Chong, and Hwang, 1995) it is shown that cell process measurements in the low frequency band can constitute a basis for adaptive connection admission control. Another approach to measurement based connection admission is presented in (Jamin et al., 1996) in the context of predictive services which allow occasional delay constraint violations.

References

Boyer, P.E., and Tranchier, D.P. 1992. A reservation principle with applications to the ATM traffic control. *Computer Networks and ISDN Systems*, 24:321–334.

Chemouil, P., and Filipiak, J. 1989. Kalman filtering of traffic fluctuations for real-time network management. *Annales des Telecommunications*, 44(11-12):633–640.

COST 224, 1992. *Performance evaluation and design of multiservice networks,* ed. Roberts J.W. COST 224 final report. Commision of the European Communities.

Dziong, Z., Montanuy, O., and Mason, L.G. 1994. Adaptive traffic admission in ATM networks–Optimal estimation framework. In *Proceedings of the 14th International Teletraffic Congress,* pp. 1065–1076. North-Holland.

Dziong, Z., Shukhman, B., and Mason, L.G. 1995. Estimation of aggregate effective bandwidth for traffic admission in ATM networks. In *Proceedings of IEEE INFOCOM'95,* pp. 810–818. IEEE Computer Society Press.

Dziong, Z., Juda, M., Jajszczyk, A., and Mason, L.G. 1995. Influence of UPC on connection admission based on measurements in ATM switch output ports. In *Proceedings of the First Workshop on ATM Traffic Management WATM'95,* pp. 77–84. Paris December.

Dziong, Z., Juda, M., and Mason, L.G. 1997. A framework for bandwidth management in ATM Networks–Aggregate equivalent bandwidth estimation approach. *IEEE/ACM Transactions on Networking,* in press (considered for February 1997 issue).

Gach, A., Mialaret, C., and Allard, P.E. 1992. An experimental evaluation of call acceptance management algorithms in ATM based networks. In *Proceedings of Canadian Conference on Electrical and Computer Engineering CCECE-92,* pp. TA1.161-4. Toronto.

Gelb, A. 1974. *Applied Optimal Estimation,* The MIT Press.

Gibbens, R.J., Kelly F.P., and Key, P.B. 1995. A decision-theoretic approach to call admission control in ATM networks. *IEEE Journal on Selected Areas in Communication* 13(6):1101–1114.

Jamin, S., Danzig, P.B., Shenker, S., and Zhang, L. 1996. A measurement-based admission control algorithm for integrated services packet networks. *IEEE/ACM Transactions on Networking,* in press.

Kelly, F. 1994. Tariffs and effective bandwidths in multiservice networks. In *Proceedings of the 14th International Teletraffic Congress,* pp. 401–410. North-Holland.

Li, S.-Q., Chong, S., and Hwang, C.L. 1995. Link capacity allocation and network control by filtered input rate in high-speed networks. *IEEE/ACM Transactions on Networking,* 3(1):10-25.

Liao, K-Q., Dziong, Z., Mason, L.G, and Tetreault, N. 1992. Effectiveness of leaky bucket policing mechanism. In *Proceedings of IEEE ICC'92,* IEEE Computer Society Press.

Pack, C.D., and Whitaker, B.A. 1982. Kalman filter models for network forecasting. *B.S.T.J.,* 61(1):1–14.

Saito, H. 1991. Dynamic call admission control in ATM networks. *IEEE Journal on Selected Areas in Communication,* 9(7):982–989.

Tedijanto, T.E., and Gun, L. 1993. Effectiveness of dynamic bandwidth management mechanisms in ATM networks. In *Proceedings of IEEE INFOCOM'93,* pp. 358–367. IEEE Computer Society Press.

Yamada, H., and Sumita, S. 1991. A traffic measurement method and its application for cell loss probability estimation in ATM networks. *IEEE Journal on Selected Areas in Communication,* 9(3):315–324.

Chapter 5

CAC & Routing Strategies

CONNECTION admission control and routing functions are key elements of resource management and traffic control in broadband networks. They can influence significantly several important network features: utilization, robustness, reliability, fairness, stability, and quality of service (QoS). Historically the *connection admission control and routing* (CAC & routing) algorithms were developed separately for packet and circuit switched networks resulting in different solutions. These differences stem from many causes. In particular, in circuit switched environment the connections are physically separated resulting in a bandwidth constrained routing optimization problem. In packet switched networks all connections share the same resources, making the routing design sensitive to some functions of the end-to-end packet delay. Also, both network categories are oriented towards different services (real-time vs. data applications) characterized by very different performance constraints.

In ATM based networks the two historically separate CAC & routing domains are combined. This fusion makes the general problem of CAC & routing design more difficult since it has to take into account both bandwidth and packet delay (QoS) constraints at the same time. Moreover, the methodology used in traditional packet switched networks cannot be applied directly due to a significantly higher bandwidth-delay product in fast packet switched networks.

The issue of CAC & routing in ATM based networks can be substantially simplified by using a logical "bandwidth" allocation to connections, as described in Chapters 3 and 4. In this case the QoS constraints of a virtual circuit are determined by the QoS class selected to establish the connection (e.g., selection of the corresponding QoS virtual network) and the number of links in the considered path. As a result the CAC & routing problem can be essentially treated as a bandwidth constrained problem. In this chapter we focus on this approach. Note that the logical "bandwidth" allocation to connections makes the operation of fast packet switched networks, at the connection layer, similar to that of multi-rate circuit switched networks. As a consequence the material presented in this chapter is applicable to both ATM and STM broadband networks.

In Section 5.1 we define in more detail the CAC & routing problem including the definition of four general categories of routing algorithms: fixed, adaptive, macro-state-dependent, and state-dependent. We also formulate a generic CAC & routing problem as maximization of the reward from carried connections (each connection class is characterized by a reward parameter). The advantage of this formulation is that by changing the reward parameters one can achieve any required operating point defined by the connection class rejection probabilities. To achieve this feature a near-optimal control is required. In Section 5.2 such a solution is presented based on decomposition of a Markov decision problem (a short introduction to Markov decision theory is given in Appendix B). In this model the CAC & routing decisions are based on state-dependent link shadow prices interpreted as predicted prices for seizing link "bandwidth" by the connection. The important feature of this approach is that by decomposing the network reward process into a set of separable link processes one can introduce economic considerations in network resource management. The basis for these considerations is setting a relation between the link cost, reward from the link, and connection reward parameters. These issues will be discussed in Chapter 9. In Section 5.2.5 we analyze the important features of the reward maximization approach showing that this approach also provides a general framework for studying, constructing, and optimizing other CAC & routing strategies. Implementation issues connected with CAC & routing algorithms are discussed in Section 5.3. The problem of CAC & routing for multi-point connections is described separately in the next chapter. Application of the CAC & routing algorithms for achieving a fair and efficient network operating point is presented in Chapter 8. Performance evaluation models for networks employing the considered CAC & routing algorithms are discussed in Chapter 7.

5.1 Problem Formulation

While the connection admission control function and the routing function are often treated as two separate issues, in fact they are parts of the same problem which is the selection of the best path in the network to carry a new connection. Nevertheless, the decomposition of this problem into two above mentioned functions can significantly simplify the algorithms. In this case the objective of the routing is to find a path which optimizes some criteria (such as network resource utilization or average cell delay), while the connection admission procedure verifies whether certain constraints are not exceeded (such as maximum link "bandwidth" available to the connection class or maximum end-to-end cell delay), on the selected path. This decomposition combined with some other decomposition techniques, to be described in the following, can be very useful in fast packet switched broadband networks where the general formulation of the path selection results in a multi-criterion optimization problem with several constraints, which is difficult to solve in real time.

To make the optimization problem easier to manage it is convenient to transform it into a single criterion optimization problem. In general the potential optimization criteria can be divided into two categories: the ones associated with the cell layer (functions of QoS characteristics such as cell losses and delay) and the ones linked to the connection layer (functions of GoS characteristics such as connection rejection probabilities). Observe that in broadband networks the optimization on the packet layer is less critical, compared to traditional packet switched networks, due to more stringent packet delay and loss constraints. Thus by applying the logical "bandwidth" allocation to connections, described in the previous chapters, one can decompose the CAC function into two parts associated with cell and connection layers, respectively.

The first part, referred to as CAC_{QoS}, deals with the connection QoS requirements and is realized by allocating "bandwidth" within the corresponding QoS virtual network. Since logical "bandwidth" allocation is usually based on the link QoS constraint, the end-to-end QoS performance depends on the number of links in the selected path. Thus, the CAC_{QoS} function can be decomposed into two sub-functions. The first one verifies whether the required end-to-end QoS constraints are met on the considered path, and is referred to as CAC^p_{QoS}. This function can be realized by limiting the number of links in the selected path since the end-to-end QoS performance can be easily bounded by a function of the link QoS constraints. The second sub-function, referred to as CAC^l_{QoS}, evaluates the "bandwidth" required by a connection in such a way that the required link QoS constraints are met.

The second part of the CAC function manages the GoS requirements (distribution of the connection class loss probabilities) and is henceforth referred to as CAC_{GoS}. The objectives of this function are strongly coupled with the objectives of the routing function. In fact the decomposition of the CAC function based on the logical "bandwidth" allocation to connections causes the routing function objectives to be separated from the cell layer. As a result the CAC_{GoS} and routing problem becomes a bandwidth constrained problem which is similar to the one in circuit switched (STM) networks. There are two main optimization criteria in this problem: network resource utilization (efficiency) and network access fairness. While high resource utilization is important from the network operator's viewpoint, equitable access to the network is important from the network users' viewpoint. Having two optimization criteria makes the optimization problem still difficult. To make the problem manageable, the optimization process can be decomposed into two parts operating in different time scales. In the short time scale of the CAC & routing decision, the optimization algorithm attempts to maximize an objective function associated with network resource utilization. This objective function includes, or is constrained by, a set of "fairness" parameters. In general the relation between the fairness parameters and the fairness objective is a complex function. Thus, in the longer time scale the fairness criteria are verified (based on measurements or an analytical model) and the fairness parameters are updated by the second part of the optimization algorithm in order to achieve an optimal operating point. In this chapter we discuss the first part of the algorithm. The fairness criteria and corresponding optimization procedures are discussed in Chapter 8.

5.1.1 CAC & routing classification

Let us describe the network as a set of nodes and a set of links connecting these nodes. Each link is allocated a certain "bandwidth" which is also referred to as the link capacity, L^s, where s denotes the link index. The network is offered many connection classes characterized by origin-destination (OD) node pair, mean holding (service) time, μ_j^{-1}, and "bandwidth" requirement, d_j, where j denotes the class index. In general the connection "bandwidth" requirement can be link dependent, d_j^s (e.g., equivalent bandwidth on ATM links with different speed). Nevertheless, for the sake of presentation simplicity we limit our considerations to fixed "bandwidth" allocation along the connection path. The extension of the presented models to the general case is straightforward (see, for example, Section 9.6.1). The above network definition can be used to model CAC & routing algorithms in multi-rate circuit-switched (STM) networks, ATM based networks, and virtual networks established within the first two categories.

The process of CAC & routing can be interpreted as a selection of the best path from a set of alternative paths. The set of alternative paths, W_j, can contain all potential paths or can be limited for technical or performance reasons. The path selection can be a function of link metrics, path metrics, and the applied signaling protocol. The link and path metrics can contain static and/or dynamic information (e.g., cost, connection state, average traffic, QoS constraints) and in most cases the path metrics are a function of the link metrics. The organization, access, and updating of the database used for the path selection, as well as other implementation issues influencing the path selection process, are discussed in Section 5.3. In the mean time we consider a system where all information is available instantly to the CAC & routing algorithm.

Let us define the routing function as a choice, from the set of alternative paths W_j, of a path recommended for carrying a new connection. Then the connection admission control function decides whether the recommended path under consideration can be used for carrying the connection. If, according to the CAC algorithm, the recommended path is inadmissible, the routing algorithm can recommend another path or the connection can be rejected. As will be shown in Section 5.3, this decomposition of the path selection function makes the algorithm implementation simpler and more flexible.

Classification of routing strategies

The routing strategies can be divided into four general categories depending on network state information based on which path recommendation is done.

Fixed. In this category the path recommendation is done without any information about the current traffic conditions in the network. The fixed *load sharing strategy* fits into this category. In this case path k from W_j is recommended with probability h_j^k. If the connection cannot be accepted on the chosen path, it is lost. The routing probabilities are evaluated at the network design stage based on the predicted offered traffic matrices. Since in this type of routing the traffic flow distribution in the network is directly determined by the routing probabilities, this strategy is attractive from the viewpoint of stability and analytical performance models. Nevertheless its performance is usually inferior to strategies which are based on some feedback from the network.

Adaptive. In this case the path recommendation algorithm utilizes traffic statistics, **a**, related to traffic flow and GoS distributions. These statistics are associated with a time scale which allows one to approximate quasi-stationary characteristics of the network traffic process. The adaptive load sharing strategy falls into this category. In this case the routing probabilities are a function of traffic statistics, $h_j^k(\mathbf{a})$. This feature increases strategy effectiveness when the offered traffic matrix is variable. The adaptation of routing probabilities can be based on learning as described in (Narendra, Wright, and Mason, 1977).

Macro-state-dependent. Here the term macro-state refers to the path under consideration. If, according to the CAC mechanism, the path can carry the new connection, its macro-state is *admissible*; otherwise the path is in the *inadmissible* macro-state. In macro-state-dependent routing the path selection process depends on the macro-states of the paths in W_j. There are two well known strategies falling into this category. The first one is *sequential routing*. In this case the paths from W_j are ordered into a sequence and the first admissible path in the sequence is selected to carry a connection. Obviously if all paths are inadmissible the connection is rejected. The path sequence can be determined by the path costs where each path cost is the sum of its link costs. The second example is called *dynamic alternative routing* (DAR), which has been developed by British Telecom for telephone networks. In this routing the connection is first offered to the path consisting of one link between origin and destination nodes (*direct-link-path*). If this path is inadmissible, a singled out alternative path is tried. Also, if this path is inadmissible the connection is lost, but for the next connection the singled out alternative path is changed to another alternative path chosen randomly from W_j (the direct-link-path is not considered in this step).

State-dependent. Here the routing decision is a function of detailed states of links constituting the path under consideration. For example, the link state can be expressed as residual capacity available for the class j connections, C_j, or by the vector $\mathbf{x} = [x_j]$, where x_j, $j \in J^s$ denotes the number of class j connections carried on the considered link. There are two important examples of state-dependent routing strategies. The first is called *least loaded path* (LLP). The basic idea of this approach is to carry the new connection on the path with the maximum residual capacity

$$C_j^{\max} = \max_{k \in W_j} \{C_j^k\} \tag{5.1}$$

where the path residual capacity, C_j^k, is defined as the minimum link residual capacity over the links constituting path k

$$C_j^k = \min_{s \in S^k} \{C_j^s\} \tag{5.2}$$

Obviously the maximum path residual capacity should be larger (or equal) than the connection "bandwidth" requirement. Otherwise the connection is rejected. The second example of the state-dependent routing strategy is based on the concept of a state-dependent link cost, $p_j(\mathbf{x})$. This cost can be interpreted as a price for using the link "bandwidth." In this case the path with a minimal cost, defined as the sum of link costs, is selected. The nice feature of this formulation is that, as it will be shown later, the same link costs can be used for CAC decision.

Some of the above four categories can be mixed together. In particular the state-dependent or macro-state-dependent categories can also be adaptive. For example, the sequential routing strategy can adapt the sequence to the changes in the traffic flow distribution in the network. Also, the strategy based on link cost can adapt these costs to the changes in the link traffic levels.

Classification of CAC mechanisms

As previously stated, the CAC objective is to verify which of the paths are admissible (i.e., can carry the new connection). In general one can consider three modes of connection admission control operation. In the first one the states of the considered paths are verified at the moment of the new connection demand, and if there are no admissible paths the connection is rejected. This mode of operation is referred to as *loss system*. If the demand for connection can be delayed (buffered) until one of the paths becomes admissible due to another connection departure, the network operates in the mode with *delayed admission*, see e.g. (Dziong, Liao, and Mason, 1990). The third mode of operation is called *advanced reservation* and can be used for connections which are planned well in advanced as conference connections; see, e.g. (Roberts and Liao, 1985; Liang et al., 1988). All of these modes can exist in one system. Although in this book we focus on the networks operating in the loss system mode (expected to be predominant) the basic ideas of CAC & routing are similar for all three operation modes.

The criteria for CAC decisions can be divided into the following categories: feasibility, utilization, and fairness. Feasibility criteria define whether the connection can be accepted on a path so that the QoS constraints are fulfilled (CAC_{QoS} function). Once the logical "bandwidth" allocation to connection is applied these criteria are easy to verify by comparing each link's total residual capacity, $C^s = L^s - \sum_{j \in J^s} x_j d_j$, with the new connection "bandwidth" demand, d_j, and by checking the end-to-end QoS constraints based on the link QoS constraints. It is important to emphasize that although the path is feasible it may be still inadmissible due to the criteria associated with network utilization and access fairness (CAC_{GoS} function). To give an intuitive explanation for this claim, which will be proved formally later on, let us consider two scenarios. Assume overload conditions in the network and a new connection demand which could be accepted on a feasible *multi-link-path* consisting of l links. If this connection is rejected, it is likely that, in the "bandwidth" space released by the rejection, l connections can be accepted on the links constituting the path under consideration which will be used as direct-link-paths. Thus, from the network utilization viewpoint it may be beneficial to reject the connection. Now consider an overloaded network which is offered two connection classes: narrow-band and wide-band. Moreover, assume that the narrow-band traffic is dominant and that the CAC is only based on a feasibility condition (CAC_{QoS}). In this case the wide-band connections will encounter significantly higher rejection rate due to the fact that any capacity available after a wide-band connection departure will be most likely quickly filled by the dominant narrow-band connections. Thus it may be beneficial to reject narrow-band connections when the link residual capacity is equal to the "bandwidth" required by the wide-band connections. In the following we discuss three main categories of CAC_{GoS} mechanisms which can be used for optimization of network resource utilization and access fairness.

Set of alternative paths. The set of alternative paths can be used as a simple tool for optimization of network utilization and access fairness. By reducing the number of alternative multi-link-paths considered for recommendation one can reduce the risk of having too many multi-link-path connections under overload conditions. Access fairness can also be improved by reducing the set W_j for the connection classes with very low rejection probability and increasing this set for the connection classes with high rejection probability. Obviously the membership of W_j can be made adaptive to changes in traffic conditions.

Link "bandwidth" reservation. Link "bandwidth" reservation is another simple yet very powerful mechanism for CAC_{GoS}. In this case different priorities for different connection classes are achieved by reserving a part of the link "bandwidth" for some connection classes. For example, the admissible residual link capacity for class j can be defined as

$$C_j^s = C^s - \mathcal{R}_j \tag{5.3}$$

where $\mathcal{R}_j$ denotes "bandwidth" reserved for connections with higher priorities. This mechanism can be used to provide higher priority for connections using direct-link-paths and for wide-band connections. Thus the network utilization and access fairness issues can be addressed. The "bandwidth" reservation level can be a function of link traffic statistics and/or link state. Note that in the case of link "bandwidth" reservation the path connection admission control ($\text{CAC}_{\text{GoS}}^p$) is decomposed into a set of link connection admission control ($\text{CAC}_{\text{GoS}}^l$) decisions.

Link cost. The link cost can be a function of the connection class, link connection state, and link traffic statistics. In general, the higher the link traffic level and/or the smaller the link residual "bandwidth," the higher link cost. The CAC_{GoS} decision is made based on the path cost defined as a sum of the link costs over the links constituting the path. The new connection is admitted if the path cost is smaller than a threshold. To achieve access fairness objectives, the threshold can be a function of connection class. The priority for direct-link-path connections is an inherent feature of this mechanism since the cost of multi-link-paths is a sum of direct-link-path costs.

The above three classes of CAC$_{GoS}$ mechanisms can be mixed together. In this case different mechanisms focus on different objectives. The motivation for such an approach can arise from the fact that optimization process of some mechanisms can be simplified if the number of objectives is limited. Also, the implementation cost can be a factor since the implementation complexity is smallest for the approach based on the set of alternative paths and highest for the one based on link costs.

5.1.2 Reward maximization as an objective function

The presented classification of CAC & routing strategies shows a large variety of objectives and possible solutions. In the remainder of this chapter we focus on a generic formulation of the CAC & routing problem which is based on the reward maximization objective. This formulation has several advantages. It is derived from Markov decision theory, which provides an optimal solution to the problem. While this solution is difficult to implement, after applying some approximations a sub-optimal implementable solution is achieved. This sub-optimal solution can be classified as CAC & routing based on state-dependent and adaptive link costs. The model also provides a framework for analyzing, synthesizing, and optimizing other CAC & routing strategies. These capabilities will be illustrated with regard to sequential and load sharing routing strategies (this chapter) and CAC$_{GoS}$ mechanisms based on link "bandwidth" reservation (Chapter 8).

Let us consider a system where each connection carried in the network gives to the network control manager a reward with the rate q_j during the connection service time. The reward rate can be defined as a function of the connection reward parameter, $r_j \in (0, \infty)$, interpreted as the average reward for carrying class j connection:

$$q_j = r_j \mu_j \tag{5.4}$$

where μ_j^{-1} is the connection mean service time.

For such a system, under stationary offered traffic conditions, we formulate the CAC & routing objective as follows: find the optimal control policy π^* which maximizes the mean value of reward from the network defined as

$$\overline{R}(\pi) = \sum_{j \in J} r_j \overline{\lambda}_j \tag{5.5}$$

where $\overline{\lambda}_j$ denotes the average rate of accepted class j connections. Assuming that the optimal policy is found, the reward maximization objective provides a very convenient tool for controlling almost independently and continuously the grade of service of different connection classes defined as the rejection probability, B_j. This is achieved by simply increasing the reward parameters of connection classes with too large rejection probabilities and vice versa. This capability fits very well into the already mentioned two level CAC & routing optimization approach where the reward parameters are optimized over a longer time scale to meet the access fairness criteria. Evaluation of reward parameters for objectives which can be defined as maximization of a certain quantity being a linear function of carried traffic is easy. In particular minimization of traffic rejection probability (equivalently traffic maximization) is achieved when all reward parameters, normalized by bandwidth and time units, $r'_j = r_j \mu_j / d_j$, are equal to each other since in this case the average reward can be expressed as

$$\overline{R}(\pi) = \sum_{j \in J} d_j \mu_j^{-1} \overline{\lambda}_j \qquad (5.6)$$

Analogously, maximization of network revenue is achieved when the reward parameters are proportional to the connection tariff.

As shown in Section 5.2.1, the optimal CAC & routing policy can be found using Markov decision processes theory. The solution is a function of the connection state in the whole network and as such is, in general, very complex. An implementable sub-optimal solution can be achieved by decomposition of the network reward process into a set of statistically independent link reward processes (see Section 5.2.2). In this case the CAC & routing decision is a function of the path net-gains, g_j, defined as the difference between the connection reward parameter, r_j, and the link state-dependent shadow price, $p_j^s(\mathbf{x})$, interpreted as the expected price for accepting a jth type connection on link s in state $\mathbf{x}$:

$$g_j = r_j - \sum_{s \in S^k} p_j^s(\mathbf{x}) \qquad (5.7)$$

where S^k denotes the set of links forming path k. The CAC & routing decision is to carry a new connection on the path from W_j with maximum positive net-gain. If the maximum net-gain is negative, the connection is rejected. This result, based on the theoretical model, illustrates well the decomposition of the path selection problem into the CAC_{GoS} function and the routing function. Namely, although the decision is formally based on one function, Equation (5.7), it can be executed in two steps. In the first one, the path with maximum gain is recommended (the routing function). In the second step the CAC_{GoS} function verifies whether the recommended path is admissible (positive net-gain condition).

The state-dependent link shadow prices can be evaluated using link traffic statistics in order to adapt to changes in network traffic conditions (see Section 5.2.2). Thus, the considered approach can be classified into a category where both CAC & routing functions are based on state-dependent and adaptive link costs.

5.2 Reward Maximization Approach

In this section[1] we show how the reward maximization problem, defined in the previous section, can be solved within the framework of continuous-time *Markov Decision Processes* (MDP) theory.

---

[1]Portions of this section are reprinted, with permission, from (Dziong and Mason, 1994). ©1994 IEEE.

5.2.1 Optimal solution*

In the considered model the jth connection class is characterized by the following: origin-destination (OD) node pair, required "bandwidth," d_j, intensity of arrival process (assumed to be Poissonian), λ_j, mean holding time (assumed to be exponentially distributed), μ_j^{-1}, set of alternative paths, W_j, and reward parameter r_j.

The state of the considered system can be described by a matrix $\mathbf{z} = [z_j^k]$, where z_j^k denotes the number of class j, $j \in J^s$, connections carried on path k, $k \in W_j$, and $\mathbf{z} \in Z$. For each state the rate of reward from the network, $q(\mathbf{z})$, is given by

$$q(\mathbf{z}) = \sum_{j \in J} \sum_{k \in W_j} r_j \, z_j^k \, \mu_j \tag{5.8}$$

In the case of new connection acceptance the state transition is described as $\mathbf{z} \to \mathbf{z} + \mathbf{\Delta}_j(\mathbf{z}, \pi)$, where $\mathbf{z} + \mathbf{\Delta}_j(\mathbf{z}, \pi)$ denotes the state after accepting the class j connection on path k recommended by policy π in state $\mathbf{z}$. If the control policy rejects the connection, the decision is defined by $\mathbf{\Delta}_j(\mathbf{z}, \pi) = [0]$. In the case of connection departure the state transition is described as $\mathbf{z} \to \mathbf{z} - \delta_j^k$, where $\mathbf{z} - \delta_j^k$ denotes the state after the departure of class j connection from path k in state $\mathbf{z}$. The rates of the transitions are λ_j and $z_j^k \cdot \mu_j$, respectively.

From the MDP theory it follows that since our system is ergodic, the optimal policy π^* is *deterministic*. In other words for a given network state and connection class the action is always the same (rejection or acceptance on a given path); for details see Appendix B. This policy can be found by applying one of the well known algorithms such as policy iteration, linear programming, or value iteration. As will become obvious later, the policy iteration algorithm is most convenient for our problem because it converges very fast to the optimal solution, and the number of iteration is practically independent of the number of states (see Appendix B). After application of this algorithm to our system, the following iteration cycle is obtained:

1. *For given policy π, solve the set of value-determination equations [for relative values $v(\mathbf{z}, \pi)$]*

$$\overline{R}(\pi) = q(\mathbf{z}) + \sum_{j \in J} \lambda_j \left[v(\mathbf{z} + \mathbf{\Delta}_j(\mathbf{z}, \pi), \pi) - v(\mathbf{z}, \pi) \right]$$
$$+ \sum_{j \in J} \sum_{k \in W_j} z_j^k \, \mu_j \left[v(\mathbf{z} - \delta_j^k, \pi) - v(\mathbf{z}, \pi) \right], \quad \mathbf{z} \in Z \tag{5.9}$$

by setting the relative value for an arbitrary reference state $\mathbf{z}_r$ to zero.

2. *For each state $\mathbf{z}$ find the alternative set of decisions, $\mathbf{\Delta}_j(\mathbf{z}, \pi')$, that maximizes the expression*

$$q(\mathbf{z}) + \sum_{j \in J} \lambda_j \left[v(\mathbf{z} + \mathbf{\Delta}_j(\mathbf{z}, \pi'), \pi) - v(\mathbf{z}, \pi) \right] + \sum_{j \in J} \sum_{k \in W_j} z_j^k \, \mu_j \left[v(\mathbf{z} - \delta_j^k, \pi) - v(\mathbf{z}, \pi) \right]$$
$$\tag{5.10}$$

using the relative values from the previous policy. This set of decisions constitute an improved policy π' to be used again in the first step.

The theory of MDP ensures that starting from an arbitrary initial policy this procedure converges to π^* in a finite number of iterations.

For further consideration it is convenient to introduce the notion of path net-gain, $g_j(\mathbf{z}, \pi)$, defined as

$$g_j(\mathbf{z}, \pi) = v(\mathbf{z} + \mathbf{\Delta}_j(\mathbf{z}, \pi), \pi) - v(\mathbf{z}, \pi) \qquad (5.11)$$

where $\mathbf{\Delta}_j(\mathbf{z}, \pi)$ determines the path on which the jth class connection is accepted in state $\mathbf{z}$. Based on the physical interpretation of the relative values (see Appendix B), the path net-gain can be also expressed as

$$g_j(\mathbf{z}, \pi) = \lim_{T \to \infty} [R(\mathbf{z} + \mathbf{\Delta}_j(\mathbf{z}, \pi), \pi, T) - R(\mathbf{z}, \pi, T)] \qquad (5.12)$$

where $R(\mathbf{z}, \pi, T)$ denotes the expected reward from the network in the interval $(t_0, t_0 + T)$, assuming state $\mathbf{z}$ in t_0. Note that the separable form of the objective function, Equation (5.10), to be maximized in the second step of the iteration procedure, assures that this step is equivalent to the separate maximization of $g_j(\mathbf{z}, \pi)$ over π, for each $\mathbf{z}, j$, pair.

5.2.2 Strategy based on link shadow price concept

For most telecommunication networks, the policy iteration procedure based on the exact network state description is intractable due to the enormous cardinality of the state and policy spaces. In this section we describe an approach where the network reward process is decomposed into a set of separable link reward processes. While this approximation reduces computational and memory requirements to manageable levels, the policy iteration algorithm is preserved.

First let us decompose the network Markov process. This can be done by assuming that the connection arrival process seen by a link is a link state-dependent Poisson process and that the link state distributions are statistically independent. These assumptions are commonly made in network performance analysis models. In particular, they imply that a connection, established on a path consisting of l links, is decomposed into l independent link connections characterized by the same mean holding time as the original connection. Then the Markov process, under policy π, can be described separately for each link in terms of the link state $\mathbf{x} = [x_j]$ and the transition rates defined by the link arrival rates $\lambda_j^s(\mathbf{x}, \pi)$ and departure rates μ_j, where x_j denotes the number of class j, $j \in J^s$, connections carried on the considered link. To simplify our notation, it is assumed that the alternative paths for class j connections have no common links (this is not a limitation of the approach). The evaluation of $\lambda_j^s(\mathbf{x}, \pi)$ can be done in two ways. One is to develop an analytical model for performance analysis of the network with the given routing policy π. The second method is to estimate $\lambda_j^s(\mathbf{x}, \pi)$ based on some simple statistics measured in the network. The second option is more attractive due to its smaller complexity and automatic adaptation to changes in the traffic patterns. The details of this approach are given in Section 5.2.3.

Although the above assumptions provide a decomposition of the Markov process, it is not sufficient to decompose the analysis of the Markov decision problem. To do that one can decompose the network reward process into a set of separable link reward processes. This can be done by dividing the multi-link connection reward parameter among the link connections, so that each link connection is characterized by the link connection reward parameter, $r_j^s(\pi)$. It is clear that the division rule should provide maximization of the mean value of reward from the network with the obvious constraint:

$$r_j = \sum_{s \in S^k} r_j^s(\pi) \tag{5.13}$$

where path k is selected by policy π for carrying class j connection. The division of the reward parameter is discussed in Section 5.2.3.

Now, for a given routing policy, each link reward process can be described independently by the set $\{r_j^s(\pi), \lambda_j^s(\mathbf{x}, \pi), \mu_j\}$, enabling definition of the link net-gain, $g_j^s(\mathbf{x}, \pi)$, as the expected increase in the reward from the link caused by accepting class j link connection:

$$g_j^s(\mathbf{x}, \pi) = \lim_{T \to \infty} [R^s(\mathbf{x} + \delta_j, \pi, T) - R^s(\mathbf{x}, \pi, T)] \tag{5.14}$$

where $R^s(\mathbf{x}, \pi, T)$ denotes the expected reward from the link in the interval $(t_0, t_0 + T)$, assuming state $\mathbf{x}$ in t_0, and δ_j is the J-vector with 1 in the j position and zeros in all other positions. Observe that under the link independence assumption we have

$$\lim_{T \to \infty} R(\mathbf{y}, \pi, T) = \sum_{s \in S} \lim_{T \to \infty} R^s(\mathbf{x}, \pi, T) \tag{5.15}$$

where $\mathbf{y} = \{\mathbf{x}^s\}$ denotes the network state in the decomposed model. Thus from Equations (5.12) and (5.14) it follows that the path net-gain for the decomposed model is given by

$$g_j^k(\mathbf{y}, \pi) = \sum_{s \in S^k} g_j^s(\mathbf{x}, \pi) \tag{5.16}$$

This separable form of the path net-gain constitutes our basis for decomposition of the Markov decision problem since it was shown in the previous section that the values of path net-gain define the policy improvement procedure.

For subsequent simplifications of our model it is convenient to define a state-dependent link shadow price $p_j^s(\mathbf{x}, \pi)$ which is related to the link net-gain and the link connection reward by the equation

$$p_j^s(\mathbf{x}, \pi) = r_j^s(\pi) - g_j^s(\mathbf{x}, \pi) \tag{5.17}$$

This value can be interpreted as the expected price for accepting class j connection on link s in state $\mathbf{x}$. Notice that from Equations (5.13), (5.16), and (5.17) it follows

5.2 Reward Maximization Approach

that the path net-gain from carrying class j connection on path k can be expressed as

$$g_j^k(\mathbf{y}, \pi) = r_j - \sum_{s \in S^k} p_j^s(\mathbf{x}, \pi) \tag{5.18}$$

The decomposition of the network problem into the link problems is illustrated in Figure 5.1. Models for evaluation of the link shadow prices (and link net-gains) are given in Section 5.2.3.

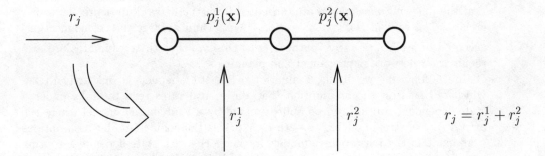

Figure 5.1. *Decomposition of the network problem.*

By using the concept of state-dependent link shadow prices, the policy iteration procedure from the previous section can be rewritten in the following steps:

1. *Collect the statistics in the network operating under a given routing policy π, then evaluate $\lambda_j^s(\mathbf{x}, \pi)$ and, using these values, compute the new values of link shadow prices $p_j^s(\mathbf{x}, \pi)$.*

2. *For each arrival of class j connection find the maximum net-gain over all feasible paths*

$$g_{\max} = \max_{k \in W_j}[r_j - \sum_{s \in S^k} p_j^s(\mathbf{x}, \pi)] \tag{5.19}$$

using the new values of link shadow prices. If $g_{\max}$ is positive, carry the connection on the path with the maximum net-gain, otherwise reject the connection.

Assuming that starting from an arbitrary policy this procedure (henceforth called MDPD) converges to a limit policy π^*, we could treat this solution as the optimal policy if all used assumptions are exact. Since there are approximations, policy π^* is, in general, sub-optimal.

The main advantage of the presented decomposition approach (compared to the exact model) is that the evaluation of the path net-gains is decomposed into the link analysis problems (state space reduction) and that the policy need not to be stored for all network states; instead the decisions can be easily computed at the instant of connection arrivals on the basis of link shadow prices (memory requirement reduction). It is also important that the evaluation of shadow prices can be based on real-time measurements, which implies that the CAC & routing policy will adapt to a time variable traffic demand.

5.2.3 Link shadow price evaluation

Link Connection Parameters

In this section we consider several variants for evaluation of link arrival rates $\lambda_j^s(\mathbf{x}, \pi)$, and link connection reward parameters, $r_j^s(\pi)$. The approach assumes that the following parameters are estimated from the network measurements: average arrival rate of class j connections, $\lambda_j(\pi)$, and average rate of connections accepted on each alternative path, $\overline{\lambda}_j^k(\pi)$. Observe that these estimates are also required for network management and planning.

In the following we discuss evaluation of the state-dependent link arrival rates which is based on the assumption that the arrival rates seen by links under a state-dependent routing can be approximated by a load sharing model where the arrival rate of stream offered to path k, $\lambda_j^k(\pi)$, is Poissonian (similar assumptions were made in the performance models for networks with state-dependent routing presented in (Mason and Girard, 1982; Girard, 1990; Krishnan, 1990)). The load sharing probabilities, h_j^k, are assumed to be proportional to the rate of accepted class j connections on path k:

$$\lambda_j^k(\pi) = \lambda_j \frac{\overline{\lambda}_j^k(\pi)}{\sum_{k \in W_j} \overline{\lambda}_j^k(\pi)} \quad (5.20)$$

Then, under the link independence assumption, the link state-dependent arrival rates in the non-blocking states can be found from the path model and are given by

$$\lambda_j^s(\mathbf{x}, \pi) = \lambda_j^k(\pi) \cdot f_j^s(\mathbf{x}, \pi) \prod_{o \in S^k \setminus \{s\}} (1 - b_j^o(\pi)) \quad (5.21)$$

where $b_j^o(\pi)$ denotes the probability that link o has not enough residual capacity to accept class j connection (blocking state) and $f_j^s(\mathbf{x}, \pi)$ denotes a filtering probability defined as

$$f_j^s(\mathbf{x}, \pi) = Pr\left\{ \sum_{o \in S^k \setminus \{s\}} p_j^o(\mathbf{x}, \pi) < r_j - p_j^s(\mathbf{x}, \pi) | \mathbf{y} \in Y_{j,nb}^k \right\} \quad (5.22)$$

where $Y_{j,nb}^k$ denotes a set of non-blocking path states for connection class j [observe that $p_j^s(\mathbf{x}, \pi)$ is constant in Equation (5.22)]. In other words $f_j^s(\mathbf{x}, \pi)$ is the probability that the path net-gain is positive (on condition that there is enough path capacity to carry the connection). This probability can be computed using the link state probability distributions. Note that for evaluation of the link shadow prices, the link arrival rates in the link blocking states are not required. Nevertheless, to simplify the presentation of link models, it is assumed that $\lambda_j^s(\mathbf{x}, \pi) = 0$ in these states.

5.2 Reward Maximization Approach

TABLE 5.1 Reward losses [%] for different reward allocation rules.

network example	nominal conditions			overload conditions		
	D1	D2	D3	D1	D2	D3
W7A	0.96 ±.15	1.03 ±.12	1.02 ±.17	4.84 ±.32	4.83 ±.30	4.73 ±.32
N7A	1.44 ±.09	1.42 ±.09	1.41 ±.09	6.12 ±.10	6.10 ±.10	6.15 ±.10
N11A	0.51 ±.07	0.51 ±.06	0.53 ±.06	5.21 ±.10	5.19 ±.11	5.23 ±.11

Concerning evaluation of the link connection reward parameters, $r_j^s(\pi)$, the rule for allocation of r_j among the path's links can have an influence on the average reward from the network. The exact solution, maximizing the reward from the network, is quite complex since the optimization procedure would require a network performance model. Instead we focus on a simple solution that, due to its economic interpretation, is very attractive for network management and dimensioning. Namely, it is assumed that the reward parameter $r_j^s(\pi)$ assigned to link s should be proportional to the connection average link shadow price, $\overline{p}_j^s(\pi)$, paid for carrying class j link connection on this link:

$$r_j^s(\pi) = r_j \frac{\overline{p}_j^s(\pi)}{\sum_{o \in S^k} \overline{p}_j^o(\pi)} \tag{5.23}$$

where $\overline{p}_j^s(\pi)$ is defined by

$$\overline{p}_j^s(\pi) = E_c[p_j^s(\mathbf{x}, \pi)] = \sum_{\mathbf{x} \in X} Q_j(\mathbf{x}) p_j^s(\mathbf{x}, \pi) \tag{5.24}$$

where $E_c[\,]$ denotes the connection expected value and $Q_j(\mathbf{x})$ denotes the probability that class j connection is accepted in state $\mathbf{x}$. Note that $\overline{p}_j^s(\pi)$ can be estimated in a real network by averaging the values of the link shadow price at the instants of class j connection arrivals. It can be easily shown that the proposed rule is optimal in the case of a fully symmetrical path and when $\overline{p}_j^s(\pi) = 0$.

In order to verify the sensitivity of network performance to the division rule, we compare performance [expressed as the network reward losses, $H = 1 - \overline{R}/\sum_{j \in J} r_j \lambda_j$] for three different division rules. The first rule, D1, is described above. The second rule, D2, assumes that all link connection reward parameters are equal to each other. The last rule, D3, is an adaptation of the reward allocation rule from the model developed in (Kelly, 1988) for the load sharing strategy:

$$r_j^s(\pi) = r_j - \sum_{o \in S^k \setminus \{s\}} \overline{p}_i^o(\pi) \tag{5.25}$$

In this case the sum of link connection reward parameters is not equal to the connection reward parameter [Equation (5.13) does not hold]. The explanation of this fact and relation of the MDPD approach to Kelly's model are described in Section 5.2.4. The simulation results for network examples, W7A, N7A and N11A (described in Table 5.3), are presented in Table 5.1. The performance of all three versions is very close to each other and falls within the confidence intervals of the other versions. This result indicates that, in the practical range of network parameters, the reward from the network is not very sensitive to the division rule. Thus, from the control performance viewpoint, the simple rule D2 is sufficient. Nevertheless, as shown in Chapter 9, the natural economic interpretation of the D1 rule can be very attractive from the management, planning, and dimensioning viewpoint. It can be also shown that in some limiting cases the network performance under the D2 rule is worse than the one under the D1 rule.

Exact Link Models*

Once the values of link arrival rates, $\lambda_j^s(\mathbf{x}, \pi)$, and link connection reward parameters, $r_j^s(\pi)$, are given, the link net-gains can be evaluated by solving the following set of equations [achieved by applying Equations (5.9) and (5.11) to the link reward process]

$$\overline{R}^s(\pi) = q(\mathbf{x}) + \sum_{j \in J^s} \lambda_j^s(\mathbf{x}, \pi)\, g_j^s(\mathbf{x}, \pi) - \sum_{j \in J^s} x_j\, \mu_j\, g_j^s(\mathbf{x} - \delta_j, \pi), \quad \mathbf{x} \in X^s \quad (5.26)$$

where $\overline{R}^s(\pi)$ denotes average reward from the link and $q(\mathbf{x}) = \sum_{j \in J^s} r_j^s(\pi) x_j \mu_j$ is the rate of link reward in state $\mathbf{x}$.

The value-iteration algorithm (see Appendix B) constitutes an attractive alternative to the solution of Equations (5.26). In general, it is a convenient method for solving large Markov problems due to the numerical simplicity. Since this algorithm is applicable directly only to discrete time Markov processes, the uniformization technique with a certain average length of the transition time, τ, must be applied (for details see Appendix B). Then the basic recurrence for our link model is stated as follows:

$$V_n^s(\mathbf{x}, \pi) = q(\mathbf{x}) \cdot \tau + \sum_{j \in J^s} \lambda_j^s(\mathbf{x}, \pi)\, \tau\, [V_{n-1}^s(\mathbf{x} + \delta_j, \pi) - V_{n-1}^s(\mathbf{x}, \pi)]$$

$$+ \sum_{j \in J^s} x_j\, \mu_j\, \tau\, [V_{n-1}^s(\mathbf{x} - \delta_j, \pi) - V_{n-1}^s(\mathbf{x}, \pi)] + V_{n-1}^s(\mathbf{x}, \pi), \quad \mathbf{x} \in X^s \quad (5.27)$$

where the value function, $V_n^s(\mathbf{x}, \pi)$, can be interpreted as the expected reward from the link within n transition periods assuming state $\mathbf{x}$ at the beginning of the considered time and terminal reward of $V_0^s(\mathbf{x}, \pi)$ at the end of this time. It can be proven (see Appendix B) that starting from an arbitrarily chosen vector $[V_0^s(\mathbf{x}, \pi), \mathbf{x} \in X]$ the difference $V_n^s(\mathbf{x}, \pi) - V_{n-1}^s(\mathbf{x}, \pi)$ will be as close as needed to the average reward from the link, $\overline{R}^s(\pi) \cdot \tau$, for sufficiently large n.

Having the value functions, the link net-gain can be expressed as

$$g_j^s(\mathbf{x}, \pi) = \lim_{n \to \infty} [V_n^s(\mathbf{x} + \delta_j, \pi) - V_n^s(\mathbf{x}, \pi)] \quad (5.28)$$

It is important that in the particular case of a single-rate system (one connection class) the set of equations (5.26) can be rewritten in the form of two recurrence

5.2 Reward Maximization Approach

TABLE 5.2 Reward losses, H, and shadow price error, $\overline{e}$, for different link models.

example	link model	λ_d	λ_m	$\overline{e}$	H [%]
N7S nominal	exact	42.0	3.6-0.1		0.28 ±.05
	aggreg.	42.0	3.6-0.1	0.036	0.30 ±.05
	SAR	42.0	3.6	0.001	0.27 ±.03
N7S overload	exact	46.2	6.6-0.0		4.37 ±.12
	aggreg.	46.2	6.6-0.0	0.063	4.43 ±.16
	SAR	46.2	6.6	0.053	4.38 ±.12

λ_d = total arrival rate of direct-link-path stream (Poissonian)
λ_m = total arrival rate of multi-link-path stream (state-dependent)

relations which provide a very efficient solution for link net-gains. Namely after solving recurrences

$$u(x) = \frac{1 + x \cdot \mu \cdot u(x-1)}{\lambda(x, \pi)}, \quad x = 1, .., N-1 \quad (5.29)$$

$$w(x) = \frac{x \cdot \mu \cdot w(x-1) - q(x)}{\lambda(x, \pi)}, \quad x = 1, .., N-1 \quad (5.30)$$

with initial values $u(0) = 1/\lambda(0, \pi)$ and $w(0) = 0$, we have

$$\overline{R}^s(\pi) = \frac{q(N) - N \cdot \mu \cdot w(N-1)}{1 + N \cdot \mu \cdot u(N-1)} \quad (5.31)$$

$$g^s(x, \pi) = \overline{R}^s(\pi) u(x) + w(x), \quad x = 0, .., N-1 \quad (5.32)$$

where N denotes maximum number of connections that can be carried on the link.

Link Model Simplifications

Despite a significant reduction of the state space in the MDPD model, one can still encounter some numerical problems or time constraints during the solution of Equations (5.26) or (5.27) if the link state space X is very large. In this section we discuss several possible approximations which can reduce the complexity of the link model. Such an approach is encouraged by the fact that the control performance is hardly affected even if the optimal values of shadow prices are changed by several percent. This feature is illustrated in Figure 5.2 where the performance of the networks W7A and N7A (described in Table 5.3) is presented versus parameter γ by which all shadow prices in the decision algorithm are multiplied. The reward loss function $H(\gamma)$ is shallow over quite a large neighborhood of the optimal points. In the following we present several simplifications that can significantly reduce the cardinality of the link state space.

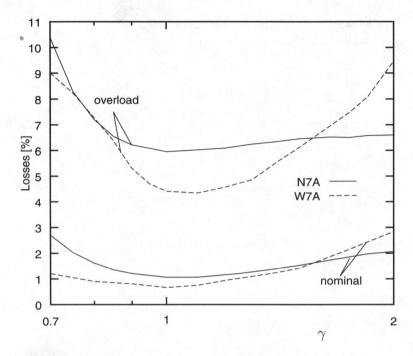

Figure 5.2. *Reward losses vs. shadow price factor γ.*

Aggregation of link connection classes. Let us construct a modified link reward process in which the link connection classes with the same "bandwidth" requirement and mean holding time are aggregated into one class i with reward parameter defined as

$$r_i^s(\pi) = \frac{\sum_{j \in J_i} r_j^s(\pi) \overline{\lambda}_j^s(\pi)}{\sum_{j \in J_i} \overline{\lambda}_j^s(\pi)} \qquad (5.33)$$

where $\overline{\lambda}_j^s(\pi)$ denotes the average rate of class j connections accepted on the link. Since most of the system parameters are unchanged, the aggregated reward process is statistically close to the original process so one can expect that

$$p_j^s(\mathbf{x}, \pi) \cong p_i^{s'}(\mathbf{x}, \pi), \quad j \in J_i \qquad (5.34)$$

where J_i denotes a set of connection classes aggregated in class i. In many cases this relation becomes an equality. In particular it can be easily shown that this is true when all link streams have steady Poissonian arrival rates. A sample of the results for state-dependent arrival rates are presented in Figure 5.3 and Table 5.2. To simplify the presentation, in the studied example (N7S, described in Table 5.3) each connection is offered to the direct-link-path first, and if the direct-link-path is inadmissible, another path is selected according to the maximum positive netgain criterion. In this case each direct-link-path connection stream has a steady Poissonian arrival rate and the multi-link-path streams are state-dependent [modeled according to Equation (5.21)]. The values of link shadow prices $p(x')$, where

5.2 Reward Maximization Approach

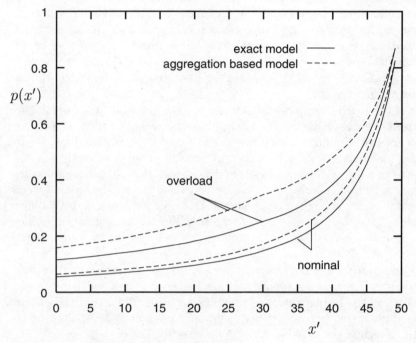

Figure 5.3. *Shadow price vs. link state.*

x' is the total number of carried connections, presented in Figure 5.3 indicate that the model based on aggregation overestimates the shadow price and the error (compared to the exact state description) is in the range of a few percent of the average link connection reward (in the considered case the shadow prices are the same for direct-link-path and multi-link-path connections). Despite this overestimation the differences in the reward losses H for the exact and aggregated link model, presented in Table 5.2, are within the confidence intervals of the simulation model. The connection average of the absolute error in the values of shadow prices, $\overline{e} = E_c[p_i^{s'}(\mathbf{x}, \pi^*) - p_j^s(\mathbf{x}, \pi^*)]$, is also given in Table 5.2.

Steady arrival rates (SAR). In the model for state-dependent link arrival rate evaluation, Equation (5.21), the most significant part of the complexity is attributed to the evaluation of the filtering probabilities. To simplify the evaluation one can try a simplified model (henceforth called SAR) where the filtering probability is neglected for all link states. The results for the network example N7S are presented in Table 5.2. The comparison between the exact and SAR link models shows that although the error in shadow price values, $\overline{e}$, is as large as a few percent of the connection reward parameter value ($r_j = 1$), its influence on the network reward losses H is negligible.

Modification of the link process structure. The complexity of the shadow price evaluation is determined, to a large extent, by the link state dimensionality. Although the presented aggregation technique for connection classes with the same "bandwidth" requirements reduces this dimensionality significantly, the number of different "bandwidth" requirements can still cause numerical and time problems in solving Equations (5.26) or (5.27). There are two basic approaches to reduce further the cardinality of the system.

In the first approach the connection classes requiring similar "bandwidth" are aggregated into one class with the "bandwidth" requirement resulting in the same traffic load offered to the link. This process, also called quantization, is especially attractive for modeling ATM based networks which can support services in the continuous "bandwidth" domain. Obviously the quantization error depends on the connection "bandwidth" requirements before and after the quantization process but in general it is not as large as one might imagine, see e.g. (Lea and Alyatama, 1995). Moreover, the shadow price values corresponding to the original connection classes can be approximated by linear interpolation between the two "quantized" values.

The second approach is based on modification of the link process structure in such a way that the link process can be described by one or several independent uni-dimensional processes. In this case, besides reducing the cardinality of the system, one can also apply an efficient recurrence solution, Equations (5.29)-(5.32). In the following, three techniques based on this concept are discussed.

Decomposition of the link process, DLP: Let us divide the link connection classes into two groups: narrow-band (NB) class with the smallest "bandwidth" requirement (say $d_n = 1$ after quantization) and wide-band (WB) classes with $d_i \gg d_n$. The main idea of the link model decomposition is based on the assumption that since the transition rates of WB connections are expected to be significantly smaller than the ones for NB connections, the NB connections reach a steady state distribution instantaneously after each change of the WB connection state. This feature, also known as *near complete decomposability*, allows to analyze the stationary properties of the system separately for the NB and WB Markov processes; for more details see (Liao, Dziong, and Mason, 1989). After applying this decomposition technique, the shadow prices for NB connections can be evaluated from the recurrence solution, Equations (5.29)-(5.32), and the evaluation of the shadow prices for WB connections is simplified by neglecting the NB connections (in the case of one WB connection class the recurrence solution can also be used). The efficiency of this approach (henceforth called DLP) is illustrated in Section 5.2.5.

Class oriented transformation, COT: The main idea of this approach is to evaluate link shadow prices for each connection class from a separate, class oriented, modified link model. The class oriented transformation is done by an approximate aggregation of all traffic streams offered to the link into one class with the "bandwidth" requirement and mean holding time the same as the connection class under consideration. The arrival rate, λ'_j, and link connection reward parameter, $r_j^{s'}$, are chosen to preserve the traffic and reward offered to the link in the modified system; more details can be found in (Dziong et al., 1991). After this aggregation the recurrence solution for evaluation of the link shadow prices in single-rate systems can be applied. The evaluated shadow prices are used in the actual system as a function of residual capacity. Linear interpolation is used for states which do not exist in the modified system. The efficiency of this approach (henceforth called COT) is illustrated in Section 5.2.5.

Aggregation into uni-dimensional link state, UDL: In this case all traffic classes are aggregated into one class with the smallest "bandwidth" requirement, $d_a = 1$. The resulting model allows us to apply a recurrence solution to evaluate the shadow prices for the considered class. The shadow prices for other connection classes are derived by assuming that a new connection consists of d_j independent connections with $d_1 = 1$. In this case the shadow prices are approximated by

$$p_j^s(x) = \sum_{x'=x}^{x'=x+d_j-1} p_a^s(x')\mu_a/\mu_j \tag{5.35}$$

where μ_a^{-1} denotes the average service time of the connections after aggregation. Note that the arrival rate in the aggregate system can be modeled as state-dependent to ensure that variance of the traffic offered in the original and modified systems are the same. Such a model, based on Pascal distribution, is derived in (Hwang, Kurose, and Towsley, 1992). Application of this model to the performance evaluation of networks with routing based on reward maximization is presented in Chapter 7.

Observe that in all three presented techniques the shadow prices are evaluated from models where the link multi-class state structure is removed. This feature has one important consequence. Namely, although these shadow prices can well serve for the routing function (path recommendation) and the CAC_{GoS} function connected with direct-link-path connection protection, they may have limited ability in providing access fairness and optimal resource utilization since interaction between different connection classes is not taken into account. This drawback can be compensated for by adding a simple $\text{CAC}_{\text{GoS}}^l$ mechanism based on the link "bandwidth" reservation (see Section 5.1). Detailed description and optimization of link "bandwidth" reservation mechanisms is discussed in Chapter 8.

Another possible approach to reduction of the complexity of the link shadow price evaluation is approximation of the link shadow price function by a simple heuristic function of the link input parameters. Such a model is presented in Section 6.3.3. It is based on the inverse of available link capacity function. The numerical study presented in Section 6.3.4 indicates that this approach is comparable with the one based on explicit link shadow price evaluation.

5.2.4 Average shadow price as a sensitivity measure*

In this section we will show that the sensitivity of average network reward with respect to connection class arrival rates can be expressed as a function of the average shadow prices. This result has several practical implications. In particular it will be shown that the average shadow price can be used to optimize the load sharing strategy. Moreover, in Section 5.2.5, the average shadow prices will be used to improve the path selection process in MDPD strategy.

Let us define the time average net-gain for connection clas j, in the exact network model, as

$$\overline{g}_j^t(\pi) = E[g_j(\mathbf{z},\pi)] = \sum_{\mathbf{z} \in Z} Q(\mathbf{z},\pi) g_j(\mathbf{z},\pi) \qquad (5.36)$$

where $Q(\mathbf{z},\pi)$ denotes the state probability and $g_j(\mathbf{z},\pi) = 0$ for $\mathbf{z}$ where connection j is rejected.

Theorem 1. *In the exact network model the derivative of average network reward with respect to arrival rate of class j is given by*

$$\frac{d\overline{R}(\pi)}{d\lambda_j} = E[g_j(\mathbf{z},\pi)] \qquad (5.37)$$

Proof: From the sensitivity analysis models presented in (Reiman and Simon, 1989; Bremaud, 1990), it follows that in the case of our system the right-hand derivative is given by

$$\frac{d^+ \overline{R}(\pi)}{d\lambda_j} = \lim_{T \to \infty} E\left[\int_{t_0-T}^{t_0+T} (q'(\mathbf{z},\pi) - q(\mathbf{z},\pi)) dt\right] \qquad (5.38)$$

where $q'(\mathbf{z},\pi)$ denotes the reward rate of the process with one additional connection added randomly at time t_0. By using Equation (5.12) it can be shown that Equation (5.38) is equivalent to

$$\frac{d^+ \overline{R}(\pi)}{d\lambda_j} = E[g_j(\mathbf{z},\pi)] \qquad (5.39)$$

Analogous proof holds for the left-hand derivative. In this case one connection is removed randomly from the system ∎

For further considerations it is convenient to define the derivative of the average network reward with respect to the average rate of connections accepted on the path:

$$\frac{d\overline{R}(\pi)}{d\overline{\lambda}_j^k} = \lim_{\Delta \lambda_j \to 0} \frac{\overline{R}(\overline{\lambda}_j^k + \Delta \overline{\lambda}_j^k, \pi) - \overline{R}(\overline{\lambda}_j^k, \pi)}{\Delta \overline{\lambda}_j^k} \qquad (5.40)$$

where $\Delta \overline{\lambda}_j^k$ corresponds to the change of the average rate of connections accepted on path k in case λ_j is changed by $\Delta \lambda_j$ but all connections from $\Delta \lambda_j$ are ignored except ones that would be carried on path k. In the same manner as proven for Theorem 1 one can show that

$$\frac{d\overline{R}(\pi)}{d\overline{\lambda}_j^k} = E_c[g_j^k(\mathbf{z},\pi)] = \sum_{\mathbf{z} \in Z} Q_j^k(\mathbf{z},\pi) g_j^k(\mathbf{z},\pi) \qquad (5.41)$$

where $E_c[\ ]$ is the connection average and $Q_j^k(\mathbf{z},\pi)$ denotes the probability that class j connection offered to path k is accepted in state $\mathbf{z}$.

In the MDPD model the time average net-gain can be expressed as

$$\overline{g}_j^t(\pi) = E[g_j(\mathbf{y},\pi)] = \sum_{k \in W_j} \frac{\overline{\lambda}_j^k}{\lambda_j}\left(r_j - \sum_{s \in S^k} \overline{p}_j^s(\pi)\right) \tag{5.42}$$

Since the Theorem 1 is also valid for the MDPD model, based on Equations (5.37), (5.41), and (5.42) we have

$$\frac{d\overline{R}(\pi)}{d\lambda_j} = E[g_j(\mathbf{y},\pi)] = \sum_{k \in W_j} \frac{\overline{\lambda}_j^k}{\lambda_j}\left(r_j - \sum_{s \in S^k} \overline{p}_j^s(\pi)\right) \tag{5.43}$$

and

$$\frac{d\overline{R}(\pi)}{d\overline{\lambda}_j^k} = E_c[g_j^k(\mathbf{y},\pi)] = r_j - \sum_{s \in S^k} \overline{p}_j^s(\pi) \tag{5.44}$$

The results from the sensitivity analysis can be applied to optimization of the load sharing policy, π^f, where class j connection is offered to path k with probability h_j^k so $\lambda_j^k = h_j^k \lambda_j$. We do not impose any policy concerning connection admission to the path. In case the connection cannot be accepted by the chosen path it is lost.

Theorem 2. *Equalization of the derivatives $\partial \overline{R}(\pi^f)/\partial \lambda_j^k$, on all paths used by connection class j, is a necessary condition to maximize $\overline{R}(\pi^f)$ over the matrix of load sharing probabilities, $[h_j^k]$.*

Proof: By changing the optimization variables from h_j^k to λ_j^k and transforming the first order Kuhn-Tucker conditions one can achieve the following optimality conditions

$$\frac{\partial \overline{R}(\pi^f)}{\partial \lambda_j^k} = -v_j, \quad k \in \{k : \lambda_j^k > 0\} \tag{5.45}$$

where v_j denotes the associated Lagrangian multiplier. This condition also implies that a path with negative $\partial \overline{R}(\pi^f)/\partial \lambda_j^k$ should not be offered any traffic of type j ∎

Note that due to Equation (5.43), in the MDPD model, the optimality conditions can be rewritten as follows:

$$\frac{\partial \overline{R}(\pi^f)}{\partial \lambda_j^k} = \frac{\overline{\lambda}_j^k}{\lambda_j^k}\left(r_j - \sum_{s \in S^k} \overline{p}_j^s(\pi)\right) = -v_j \tag{5.46}$$

If we restrict the model to single-rate networks ($d_j = 1$ and $\mu_j = 1$ for all j) with load sharing strategy, where all connections are accepted if there is sufficient residual capacity, it can be shown that the average shadow prices in Equation (5.46) are the unique solutions to the equations

$$\overline{p}^s(\pi^f) = \frac{\lambda^s}{\overline{\lambda}^s}(E(\lambda^s, N^s - 1) - E(\lambda^s, N^s)) \sum_{j \in J^s} \overline{\lambda}_j^k r_j^s(\pi^f) \tag{5.47}$$

where N^s is the link capacity expressed as the maximum number of connections which can be carried on the link, λ^s, $\overline{\lambda^s}$ denote the rates of the superposition of all connections offered to and accepted on the link s, respectively, and

$$E(\lambda, N) = \frac{\frac{\lambda^N}{N}}{\sum_{x=0}^{N} \frac{\lambda^x}{x}} \qquad (5.48)$$

is the Erlang's formula that defines the connection loss probability on link with capacity N (assuming Poissonian arrival rate). Almost identical results as Equations (5.46) and (5.47) were obtained in (Kelly, 1988) for a telephone network with the load sharing policy but based on a different model (no application of Markov decision theory). In fact the only difference is in the definition of the link reward parameter which in (Kelly, 1988) is given by Equation (5.25) in place of Equation (5.23). The explanation of this difference is that although in both cases the evaluation of the average shadow prices [shadow price in (Kelly, 1988)] is based on the decomposition of the Markov process, resulting in the Erlang fixed-point approximation (see Chapter 7), the reward process is treated differently. Namely, in the MDPD model the reward process is also decomposed providing that $r_j = \sum_{s \in S^k} r_j^s(\pi)$ and consequently $\overline{R}(\pi) = \sum_{s \in S} \overline{R}^s(\pi)$. This feature enables one to incorporate economic considerations into network dimensioning and the routing optimization problems by relating the average link reward with the link real cost (see Chapter 9).

TABLE 5.3 Description of network examples.

network example	W7A	W7S	W8A	W6A	N7A	N7S	N11A	W7A'	W8A'
fully symmetrical	no	yes	no	no	no	yes	no	no	no
number of nodes	7	7	8	6	7	7	11	7	8
required bandwidth	1, 6	1, 12	1, 12	1, 6	1	1	1	1,5,13,22	1,5,13,22
average service time	1, 10	1, 10	1, 10	1, 1	1	1	1	1,3,8,10	1,3,8,10
class i traffic [%]	50,50	51,49	65,35	67,33				9,15,46,32	25,25,25,25
total traffic [Erl.]	2466	2058	394	203	1137	861	10355	2708	1014
overload [%]	+10	+10	+20	+20	+10	+10	+10	+20	+31
link capacity	0-200	120	120	120	7-170	50	0-439	50-500	240
$r_j' = r_j \mu_j / d_j$	1	1	1	1	1	1	1	1	1

5.2.5 Important features of MDPD model

In this section we present several important features of the CAC & routing model based on state-dependent link shadow prices. The study also includes comparisons with the sequential and least loaded path routing strategies. The features are illustrated by numerical results obtained from a simulation model. The parameters of studied network examples are described in Table 5.3. To facilitate identification of the examples the following name code is used: the first letter distinguishes the narrow-band traffic examples (N), with one bandwidth allocation, from the

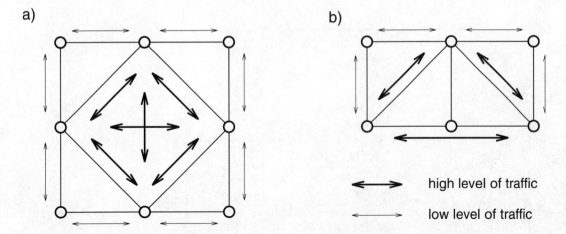

Figure 5.4. *Network connectivity and traffic structures: (a) Examples W8A and W8A'; (b) Example W6A.*

examples with narrow-band and wide-band traffic (W); the number in the middle identifies the number of nodes in the network example; the last letter indicates whether the example is fully symmetrical (S) or assymetrical (A) from traffic and connectivity viewpoints. The levels of traffic and network structures in examples W7A, N7A, N11A, and W7A' are non-symmetrical but well connected and well dimensioned. The examples W7S, N7S, and W7A' are fully connected and symmetrical. Finally the examples W8A, W6A, and W8A' (illustrated in Figure 5.4) are not well connected and not well dimensioned (e.g., specific hour in a multi-hour case). In all examples the set W_j is limited to two-link alternative paths. The policy iteration procedure is implemented with the direct-link-path routing, π^d (only the direct-link-path is in W_j) as the initial policy. The link shadow prices are evaluated by the value iteration algorithm for examples with two different connection "bandwidth" allocations and by the class oriented transformation for examples with four different connection "bandwidth" allocations. The applied link model employs the steady arrival rate (SAR) model, D2 model for connection reward parameter division, and aggregation of all link connection classes with the same "bandwidth" requirements. The 95% confidence intervals for simulation results are presented in the tables. For the sake of the presentation clarity they are omitted in the figures; nevertheless they can be approximately assessed from the corresponding examples in tables.

Convergence. The convergence of the MDPD policy iteration scheme can be influenced by the accuracy of statistics used for evaluation of the link arrival rates. To minimize the influence of this factor, in this study, the policy updating period is set to ten maximum mean holding times. The convergence is illustrated in Figure 5.5, where the average reward losses, H, are given as a function of the number of iteration cycles, i. Two iterations are sufficient to achieve a policy close to the

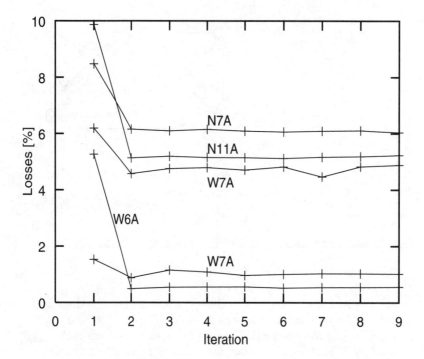

Figure 5.5. *Convergence of the MDPD policy iteration algorithm.*

limiting one. This result is very important from the implementation viewpoint. This follows from the fact that in order to provide reliable estimates of the link input parameters for the link shadow price evaluation, the time between two subsequent iterations cannot be too small. Thus the fast convergence of the policy iteration is critical for fast tracking of the network's offered traffic variations. This fact explains well the choice of policy iteration for the MDPD strategy.

The results also indicate that by neglecting the multi-link-path connection flows, the network performance can deteriorate significantly. This is supported by the fact that, as previously mentioned, the policy from the first iteration uses direct-link-path routing for shadow price evaluation and the performance for this step is in many cases very poor.

Path selection optimality. It is obvious that, especially in the case of state-dependent routing, the link occupancy distributions in the network are correlated. Here we try to assess the influence of the link's statistical independence assumption on the optimality of the path selection process. Since the exact solution is not feasible for any reasonable example, we study the issue indirectly by considering two modifications of the path selection algorithm. In the first one, the direct-link-path has priority. It means that a new connection is offered to the direct-link-path whenever the direct-link-path net-gain is positive; otherwise a path with maximum positive

5.2 Reward Maximization Approach

TABLE 5.4 Impact of different path selection schemes on reward losses [%].

network example	nominal conditions				overload conditions			
	MDPD $\alpha = 0$	priority $\alpha = 0$	MDPD' $\alpha = 0.6$	SEQ $\alpha = 1.0$	MDPD $\alpha = 0$	priority $\alpha = 0$	MDPD' $\alpha = 0.6$	SEQ $\alpha = 1.0$
W7A	1.03 ±.12	0.79 ±.14	0.83 ±.20	0.88 ±.15	4.83 ±.30	4.34 ±.24	4.57 ±.37	4.27 ±.30
W7S	1.10 ±.17	0.97 ±.23	0.90 ±.15	1.21 ±.27	5.54 ±.43	4.80 ±.40	5.13 ±.45	5.46 ±.44
W8A	0.91 ±.17	25.81 ±.37	0.70 ±.14	2.29 ±.22	5.02 ±.31	33.87 ±.25	4.92 ±.29	10.50 ±.26
N7A	1.42 ±.09	1.02 ±.09	1.07 ±.09	1.18 ±.10	6.10 ±.10	6.08 ±.11	6.00 ±.11	6.18 ±.10
N7S	0.76 ±.07	0.27 ±.03	0.26 ±.04	0.48 ±.05	4.60 ±.12	4.38 ±.12	4.42 ±.11	4.90 ±.10
N11A	0.51 ±.06	0.22 ±.03	0.20 ±.05	0.26 ±.05	5.19 ±.11	5.34 ±.12	5.11 ±.13	5.28 ±.10

net-gain is chosen. In fact the priority is commonly given to the direct-link-paths in most existing and proposed routing schemes for circuit-switched networks. The second modification utilizes the result from sensitivity analysis presented in Section 5.2.4. In this case the probability of choosing a path with high value of $d\overline{R}/d\overline{\lambda}_j^k$, is increased by modifying the path net-gain as follows:

$$g_j^{k'}(\mathbf{y}, \pi) = (1-\alpha)g_j^k(\mathbf{y}, \pi) + \alpha \frac{d\overline{R}}{d\overline{\lambda}_j^k} = r_j - \sum_{s \in S^k}[(1-\alpha)p_j^s(\mathbf{x}^s) + \alpha \overline{p}_j^s] \quad (5.49)$$

where $\alpha = [0, 1]$ is a weighting factor and $d\overline{R}/d\overline{\lambda}_j^k$ is substituted using (5.44). The modified net-gain is used for path recommendation purpose only while the CAC function utilizes the original values. The rationale of this modification can be explained as follows. Firstly, by increasing the connection flow on the paths with higher values of $d\overline{R}/d\overline{\lambda}_j^k$, the algorithm tries to equalize, whenever possible, the average net-gain of the used paths (the optimality condition for load sharing strategy). Secondly, the model for evaluation of the average shadow prices is not directly influenced by the link independence assumption thus the derivative $d\overline{R}/d\overline{\lambda}_j^k$ is less biased by this assumption compared to the state dependent path net-gain.

The results presented in Table 5.4 show that priority for the direct-link-paths provides marginal improvement in the performance of well connected and well dimensioned network examples, especially under nominal conditions. Nevertheless, the modification significantly deteriorates the performance of the W8A example where some of the direct-link-paths are not the best choice for connection routing. These results indicate that in the MDPD model the flow distribution can be slightly biased by the link independence assumption. On the other hand, it is clear that giving priority to the direct-link-paths should not be used as a general solution.

Performance of the strategy with the modified path net-gains is illustrated in Figure 5.6 as a function of α. In the first phase of increasing α the control performance is approaching the performance of the case with priority for direct-link-path (if this priority provides performance improvement). Thus, in this range, the incorporated objective of adaptive load sharing strategy is correcting the flow

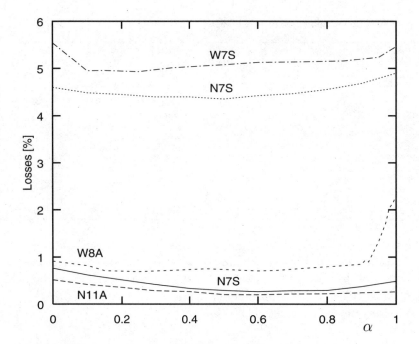

Figure 5.6. *Reward losses vs. weighting factor α.*

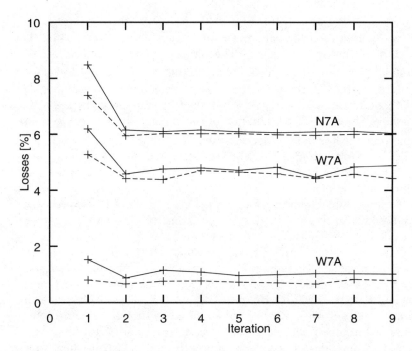

Figure 5.7. *Convergence of the algorithm with $\alpha = 0.0$ (continuous) and $\alpha = 0.6$ (dashed).*

distribution. In the second phase, when α approaches unity, the network reward losses the increase in most cases. This effect can be explained by eliminating the state-dependent feature of the routing function. From a practical point of view it is important that the optimal value of α falls in the interval $[0.3, 0.7]$ and that the function $H(\alpha)$ is shallow in this interval. This indicates that one value of α can provide close to optimal performance for all networks. Based on this premise, $\alpha = 0.6$ is used in all subsequent considerations (henceforth this case is referred to as MDPD'). Besides improving performance of the final policy ($i \geq 2$), the net-gain modification can also significantly improve the performance of the first iteration policy. This feature is illustrated in Figure 5.7. On can expect that the discussed connection flow distribution error is negligible under overload conditions since in this case the multi-link-paths are much less likely to be chosen. This phenomenon is confirmed in Figure 5.8.

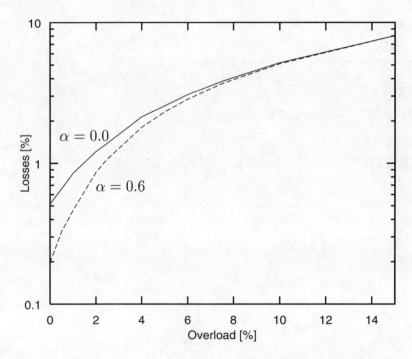

Figure 5.8. *Reward losses vs. overload factor (example N11A).*

The performance for $\alpha = 0.6$ (MDPD') and $\alpha = 1.0$ is also presented in Table 5.4. The case of $\alpha = 1.0$ is of special interest since it illustrates how the MDPD model can be used to analyze and optimize other routing strategies. Namely, in this case all paths for given OD are ordered according to the value of $d\overline{R}/d\overline{\lambda}_j^k$. This corresponds to sequential routing (SEQ) that is optimized from reward maximization viewpoint.

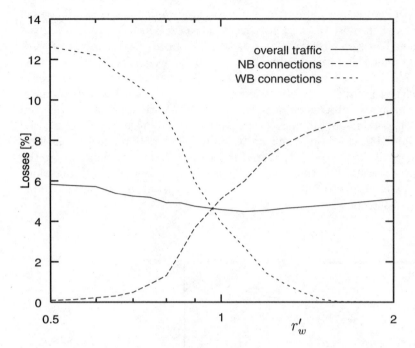

Figure 5.9. *Traffic losses vs. reward parameter r'_w (example W7A).*

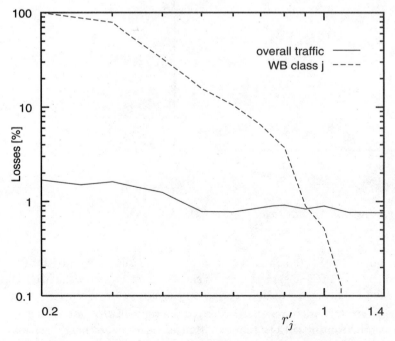

Figure 5.10. *Traffic losses vs. reward parameter r'_j (example W7S).*

5.2 Reward Maximization Approach

GoS distribution control. The powerful influence of connection reward parameters on the GoS distribution among services is illustrated in Figures 5.9 and 5.10. Figure 5.9 presents the overall traffic loss probability of NB connections, B_n, the overall traffic loss probability of WB connections, B_w, and the overall network traffic loss probability, $\overline{B}$, versus normalized reward parameter of WB connections, $r'_w = r_w \mu_w / d_w$. The reward parameters provide a tool for controlling the ratio of WB to NB traffic losses over a very wide range including their equalization. The control of a particular connection class loss probability is illustrated in Figure 5.10. In this case the connection loss probability, B_j, of WB connections offered to the jth origin-destination node pair can be controlled over a wide range by its normalized reward parameter, $r'_j = r_j \mu_j / d_j$, while the overall network traffic losses, $\overline{B}$, are little influenced. Optimization of the reward parameter values is discussed in Chapter 8 from the access fairness viewpoint and in Chapter 9 from the network dimensioning with economic constraints angle.

Efficiency comparison of different strategies. In this paragraph we compare efficiency of several MDPD model options and two other strategies (least loaded path, LLP, and sequential, SEQ). We begin with more precise definition of the considered strategies.

- MDPD′ — This is the MDPD strategy with modified path net-gains for path recommendation function ($\alpha = 0.6$). The link model for shadow price evaluation utilizes the aggregation of connection classes with the same "bandwidth" requirements. Then, the exact solution is applied.

- MDPD′ with DLP — This is the MDPD′ strategy where shadow prices are evaluated from the DLP (decomposition of the link process) model. To compensate for the lost structure of the link state, the DBRA dynamic "bandwidth" reservation mechanism (see Section 8.2) is added to the CAC$_{\text{GoS}}$ decision.

- MDPD′ with COT — This is the MDPD′ strategy where shadow prices are evaluated from the COT (class oriented transformation) model. To compensate for the lost structure of the link state, the DBRH dynamic "bandwidth" reservation mechanism (see Section 8.2) is added to the CAC$_{\text{GoS}}$ decision.

- LLP — This is an extension of the least loaded path approach proposed in (Cameron, Galloy, and Graham, 1980). In the extended model a connection from class i is offered to the direct-link-path first and if it is blocked a recommended path is tried. The recommended path is chosen randomly with the probability proportional to the path's available capacity defined as

$$\overline{x}_i^k = \min_{s \in S^k} \left[L^s - \sum_{j \in J^s} x_j^s d_j - t_i^s - h_i^s(\mathbf{x}) \right] \quad (5.50)$$

where t_i^s denotes link s "bandwidth" reserved to protect direct-link-path connections against multi-link-path connections and $h_i^s(\mathbf{x})$ denotes the "bandwidth" reserved to provide fair access for services with different "bandwidth"

TABLE 5.5 Traffic losses [%] for different routing strategies.

network example	nominal conditions				overload conditions			
	MDPD'	MDPD' DLP	LLP	SEQ	MDPD'	MDPD' DLP	LLP	SEQ
W7A	0.83 ±.20	0.96 ±.16	1.20 ±.08	0.88 ±.15	4.57 ±.37	5.29 ±.28	5.71 ±.18	4.27 ±.30
W7S	0.90 ±.15	0.89 ±.12	1.17 ±.08	1.21 ±.27	5.13 ±.45	5.31 ±.33	5.97 ±.29	5.46 ±.44
W8A	0.70 ±.14	0.89 ±.30	9.93 ±.20	2.29 ±.22	4.92 ±.29	5.58 ±.41	13.15 ±.23	10.50 ±.26
N7A	1.07 ±.09		1.26 ±.09	1.18 ±.10	6.00 ±.11		6.42 ±.07	6.18 ±.10
N7S	0.26 ±.04		0.37 ±.02	0.48 ±.05	4.42 ±.11		4.53 ±.07	4.90 ±.10
N11A	0.20 ±.05		0.34 ±.07	0.26 ±.05	5.11 ±.13		6.93 ±.03	5.28 ±.10

requirements. The threshold t_i^s is equal to the class i traffic overflowing from the direct-link-path with the constraint $t_i^s \geq d_i$. The threshold $h_i^s(\mathbf{x})$ is evaluated from the DBRA (two aggregated connection classes) or DBRH (more than two aggregated connection classes) "bandwidth" reservation mechanisms (see Section 8.2).

- SEQ — Two implementations of the sequential routing are considered. The first is achieved by setting $\alpha = 1$ in the MDPD strategy with modified path net-gain. In the second option (SEQ'), as in the case of LLP, a new connection is offered to the direct-link-path first. If the direct-link-path is blocked, the order of the paths considered for recommendation is established according to the average path net-gain values (the higher the value, the higher the priority) which are evaluated at the beginning of simulation. In SEQ' the CAC_{GoS} functions are executed by the link "bandwidth" reservation mechanisms (the same as in the LLP strategy).

Since the exact solution is not computable for any reasonable example, in the following we analyze the relation between the listed strategies by means of simulation. A sample of the results is presented in Table 5.5 for MDPD', MDPD' with DLP, and LLP strategies. Here the average traffic losses are given together with confidence interval for examples with one or two aggregated connection classes. The examples with four aggregated connection classes are presented in Table 5.6 where MDPD' with COT, MDPD' with COT+LE, LLP and SEQ strategies are compared. Here LE stands for the loss equalization objective, and in this case the reward parameters are adapted to achieve an operating point where the connection loss probabilities for each aggregated connection class, B_i, are similar. The B_i values are also presented in Table 5.6 for all strategies. Except from the LE option, all compared CAC & routing strategies (including DBRA and DBRH mechanisms) are optimized to achieve the traffic maximization objective.

The results show that the average traffic losses are similar for all the MDPD options. In particular the performance of MDPD' with DLP is very close to that of MDPD', which confirms that the simplified link models for shadow price evaluation

5.2 Reward Maximization Approach

TABLE 5.6 Traffic losses [%] for different routing strategies.

network example		nominal conditions				overload conditions			
		MDPD' COT	MDPD' COT+LE	LLP	SEQ'	MDPD' COT	MDPD' COT+LE	LLP	SEQ'
W7A'	$\overline{B}$	0.39 ±.48	0.38 ±.59	0.77 ±.29	0.77 ±.35	4.90 ±.23	5.02 ±.17	5.11 ±.23	6.28 ±.10
	B_1	0.35	0.43	0.30	0.37	3.07	5.04	3.31	4.87
	B_2	0.04	0.34	0.0	0.0	1.49	4.47	0.05	0.05
	B_3	0.17	0.40	0.26	0.30	3.31	5.08	2.42	3.45
	B_4	0.87	0.37	1.97	1.90	9.34	5.19	11.8	13.6
W8A'	$\overline{B}$	0.64 ±.57	0.40 ±.63	9.43 ±.12	8.36 ±.13	4.77 ±.28	3.93 ±.27	18.6 ±.03	17.9 ±.04
	B_1	0.79	0.50	5.9	5.65	7.05	3.39	22.2	21.7
	B_2	0.50	0.29	0.87	0.64	1.04	4.07	7.01	6.83
	B_3	0.42	0.31	7.89	6.7	3.13	3.97	15.5	15.9
	B_4	0.95	0.51	23.0	20.5	7.73	4.28	29.8	27.5

can provide good performance. The close results of MDPD' with COT and MDPD' with COT+LE also confirm that in the reward maximization approach fair access can be achieved without sacrificing resource utilization.

The comparison of MDPD' and LLP strategies indicates that in all tested examples the performance of the MDPD' model is better than that of the LLP model although the differences are relatively small in most cases with the exception of examples W8A and W8A' which will be discussed later. Obviously the advantage of the MDPD' model can be explained by the fact that it is derived from the optimal model. Nevertheless this statement can also be supported by two simple arguments. Firstly, although both the MDPD' and the LLP schemes use all link states to make a decision, in the case of LLP only the state of the path's link with smallest residual capacity counts. Thus the probability of choosing the path is not influenced by a whole range of states on the other links (providing that their residual capacity is larger than or equal to the bottleneck). In the case of MDPD', the state of each link influences the path net-gain; thus more information about the network state is used to arrive at a decision. Secondly, a decision in the MDPD' model takes into account the current traffic flow distribution in the network by using the traffic measurements in the evaluation of the shadow prices. In the case of LLP the network state information is not influenced by the traffic flow distribution. Concerning examples W8A and W8A' the unsatisfactory performance of the LLP strategy is caused by the fact that the direct-link-path is not the best choice for many OD pairs, as it is easy to find from Figure 5.4. This result confirms once again that the direct-link-path priority approach should not be used as a general solution.

The performance analysis of the sequential strategies indicates that the macro-state routing strategies can be quite efficient if the path sequence is well chosen. This follows from the fact that the SEQ' (priority given to direct-link-path) performance is comparable with that of LLP while the performance of SEQ (no priority for direct-link-path) is in most cases better, including the example W8A. Note that both the LLP and SEQ strategies give a large variance of GoS for classes with different "bandwidth" requirements.

In summary the presented study showes that the main advantage of the MDPD strategies is the ability to optimize the connection flow distribution and to control the GoS distribution among the connection classes. These features can provide high resource utilization and access fairness. While the other strategies can also provide very good performance, they require additional mechanisms for adaptation to traffic changes and for efficient control of access fairness.

5.3 Implementation Issues

The study of CAC & routing algorithms presented in the previous section assumed instant access to all information at the moment of CAC & routing decision making. Implementation of these algorithms in a real network has to take into account geographical distribution of required information and time needed to synthesize and access this information. In the following we will discuss these issues for the CAC & routing category based on state-dependent and adaptive link costs. Observe that this category is most complex from the implementation viewpoint; therefore, the discussion will also cover implementation elements of other categories.

One of the basic implementation issues in link cost based routing algorithms is the evaluation of minimum cost paths, often referred to as shortest paths. Dijkstra's algorithm (Dijkstra, 1959) is commonly used for this purpose when all required information is available at one location. For distributed implementation one can apply Ford's algorithm (Ford and Fulkerson, 1962). In this case each node only requires information from neighboring nodes.

In general an adaptive state-dependent CAC & routing process can be decomposed into three functions: estimation of parameters used to achieve adaptiveness, path recommendation, and connection admission control. Each of these functions is associated with different time scales and operates on information distributed over the whole network. This geographical distribution of the database gives rise to many implementation scenarios. In the following we present three generic strategies: centralized, distributed and hierarchical.

5.3.1 Centralized implementation

An example of centralized implementation is illustrated in Figure 5.11 for the first two functions of the CAC & routing process. While the general schemes from Figure 5.11 are applicable to any adaptive and state-dependent strategies, in brackets we have indicated the functions related to the reward maximization approach. In the estimation cycle, performance and traffic statistics are sent periodically from the network nodes to the centralized CAC & routing manager. Based on these statistics the required parameter estimates are evaluated. The set of alternative paths for each OD pair and connection class can also be determined in this cycle.

5.3 Implementation Issues

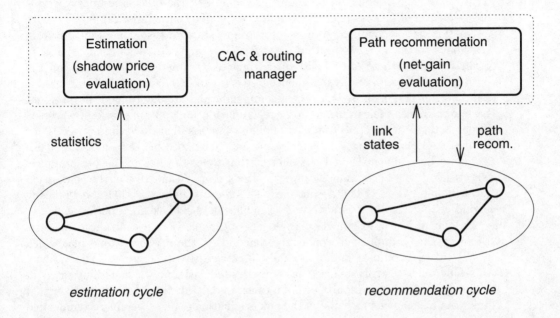

Figure 5.11. *Centralized implementation.*

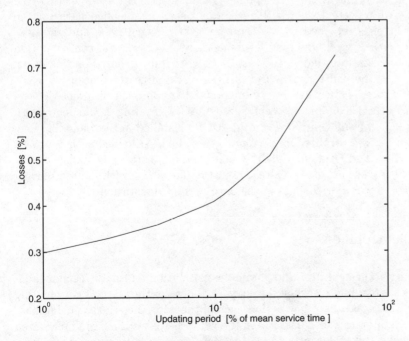

Figure 5.12. *Influence of the recommendation updating period.*

The updating period should be commensurable with the quasi-stationary periods of the measured process (e.g., few minutes for traffic characteristics).

In the path recommendation cycle the link states are sent periodically from the nodes to the CAC & routing manager. Routing path recommendations are evaluated using these states and the estimated parameters. Depending on the particular algorithm, one or a sequence of paths can be recommended for each OD pair and connection class. Once the recommendations are determined they are sent back to the nodes. The influence of the recommendation updating period (expressed as percentage of the connection mean holding time) is illustrated in Figure 5.12 for Example N11A; the detailed updating model is described in (Dziong et al., 1988). As could be easily predicted the shorter the cycle, the better the performance. Nevertheless, the performance gain for periods shorter than 10% of the connection mean holding time is not significant. This feature indicates that the updating period of the order of several seconds is sufficient to achieve good performance.

In general the CAC function could also be realized in a centralized manner. However a large number of connection demands in the network would make such a system very complex and potentially very sensitive to failures. That is why this function is usually realized in a distributed fashion, which is illustrated in Figure 5.13. Here the resources can be reserved only at the physical link or virtual link resource managers located in the link originating nodes. When a recommended path is tried, the CAC protocol attempts to reserve the required "bandwidth" in each of the transit node resource managers. The resource managers, based on the link state and the link CAC algorithm, decide whether the connection can be accepted. If at least one of the link managers cannot accept the connection, the connection can be rejected (e.g., if there is only one recommended path) or the connection control is returned to the connection originating node and the next recommended path is tried. In circuit-switched networks the latter control option is called crank-back. Note that in the case of CAC & routing based on link costs, the path cost can be controlled in the process of resource reservation which is illustrated in Figure 5.13 for the reward maximization approach (in brackets).

There are several examples of centralized CAC & routing implementations. For example, *dynamically controlled routing* (DCR) installed in the Canadian long distance network is a centralized implementation of the least-loaded path approach (Cameron, Galloy, and Graham, 1980). One of the important advantages of this solution is that it can be added to already existing networks, with limited routing capabilities, without major changes in the existing implementations.

5.3.2 Distributed implementation

In a distributed implementation the parameter estimation and path recommendation functions are realized in each of the network nodes, for the OD pairs originating in this node. This approach requires that each of the nodes has the same database concerning the statistics and link states in the whole network. This can be realized by periodically sending relevant information from each node to every other node, as illustrated in Figure 5.14 for parameter estimation and path recommendation

5.3 Implementation Issues

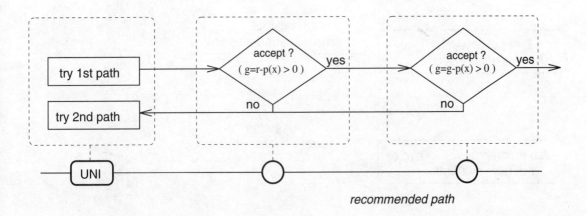

Figure 5.13. *CAC protocol.*

functions. The resource reservation function is realized also in distributed fashion, illustrated in Figure 5.13. The main advantage of the distributed CAC & routing implementation is potentially faster updating of routing recommendations since some of the information is locally available and the rest is going through only one communication phase in the updating cycle. Moreover, a greater node autonomy provides higher reliability in view of the possible central CAC & routing manager failures. A critical element of the distributed implementation is the protocol used for information exchange. One possibility is to apply a flooding algorithm, which is a reliable advertising mechanism (Bertsekas and Gallager, 1987). This type of mechanism is applied in the PNNI protocol recommended by the ATM Forum for routing in and between private ATM networks (ATM Forum, 1994).

5.3.3 Hierarchical implementation

In the case of large networks or networks composed of several sub-networks belonging to different operators, updating one database covering the whole network may be not convenient or even not possible due to the complexity increasing with the network size or different policies of different operators. To overcome these limitations one can implement a hierarchical structure for CAC & routing. The main idea of this concept is illustrated in Figure 5.15. In this case the physical network is composed of several network domains. Each of the domains supports a detailed database (centralized or distributed) for CAC & routing within the domain. For communication between the domains, logical hierarchical levels are created where each logical node represents a domain (or a logical domain) from the lower level and each logical link indicates that there is a physical connection between domains represented by the logical nodes connected by this link. The database of a logical domain is updated by information exchange with other logical nodes belonging to

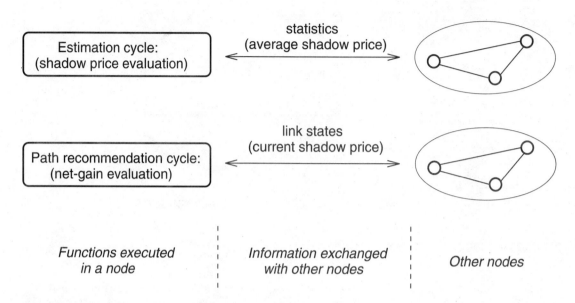

Figure 5.14. *Distributed implementation.*

this logical domain. A node in a physical domain has access to databases of logical domains which cover this physical domain. To establish a connection between two nodes belonging to different domains, the originating node selects first the recommended path in the lowest logical domain which covers the destination node. Then the recommended paths in all lower domains, visible from the originating node, are selected. Obviously, each of these paths has to be consistent with the recommended path in the parent domain. Once the connection path is established in the physical domain of the originating node, the connection control is transfered to the first node in the neighboring domain on the path in the parent domain. This node selects recommended paths in all visible domains which are on the path to the destination domain but were not visible from the originating domain. Then the procedure is repeated until the destination node is reached. If any of the connection links on the recommended paths is blocked, the crank-back procedure can be invoked to try an alternative path. This procedure is illustrated in Figure 5.15 where the recommended paths and the communication between different levels are indicated by the thick lines (only domains used in the connection are shown). There are two important advantages of the hierarchical implementation: scalability and security. Scalability allows one to consider a network of any size by adding additional logical levels without a significant increase of the database size. Security provides that each domain can have different internal CAC & routing policies which are not visible from other domains. A particular, distributed implementation of the hierarchical concept is recommended by the ATM Forum for CAC & routing in private ATM networks - PNNI protocol (ATM Forum, 1994).

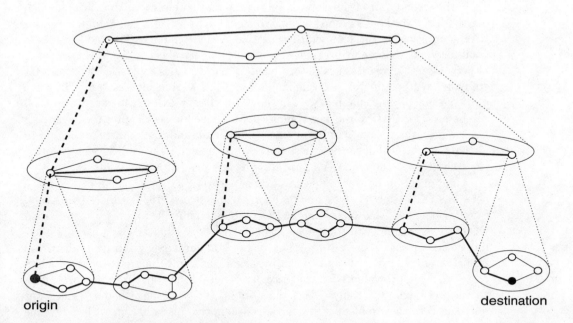

Figure 5.15. *Hierarchical implementation.*

5.4 Discussion and Bibliographic Notes

In the first part of this chapter we formulated the CAC & routing issue as a bandwidth constrained optimization problem, by applying logical "bandwidth" allocation to connections. This approach makes this issue similar to the one in circuit switched loss networks. In Section 5.1.1 we classified the CAC & routing strategies and discussed their main features. There is an extensive literature on algorithms for CAC & routing in circuit switched networks. In view of the anticipated high variability of service demands in broadband networks and bandwidth constrained control optimization, much valuable information can be found in works describing adaptive and state-dependent CAC & routing algorithms implemented in circuit switched networks. Most of them are based either on the least-loaded path concept introduced in (Grandjean, 1967), e.g. (Cameron, Galloy, and Graham, 1980; Ash et al., 1991), or on the dynamic alternative routing variations, e.g. (Gibbens, Kelly, and Key, 1988; Inoue et al., 1991).

There is also extensive literature on routing algorithms in datagram networks. Due to the connection oriented nature of ATM networks, these works apply less to the ATM layer. Nevertheless, they can be useful in an overlay packet switched network implemented on top of an ATM network by means of connectionless servers (see Section 1.3). For classification and a survey of packet routing algorithms in high-speed wide-area networks interested readers can refer to (Maxemchuk and El Zarki, 1990).

In the second part of the chapter we focused on the reward maximization model, derived from Markov decision theory, which can be seen as a generic approach for CAC & routing in the bandwidth constrained environment. The model ensures good network resource utilization and provides a very effective tool for fair access to network resources. Besides these advantages, it provides a general framework for studying, synthesizing, and optimizing other CAC & routing strategies. In some parts of the following chapters we will show how the reward maximization approach can be used for CAC & routing of multi-point connections (Chapter 6), for finding a fair-efficient network operating point (Chapter 8), and for resource allocation to virtual and physical networks where the key instrument is the ability to incorporate the resource cost into the CAC & routing algorithms (Chapters 9 and 10).

An application of Markov decision theory to the routing problem in telephone networks (exact model) was first introduced in (Benes, 1966). In (Lazarev and Starobinets, 1977) and (Krishnan and Ott, 1986) a simplified model was presented where the separability of state-dependent link costs is achieved by applying direct-link-path routing as the initial policy in the standard policy iteration algorithm from Markov decision theory. Thus by neglecting multi-link-path flows the link costs can be easily evaluated at the price of restricting the policy iteration procedure to one iteration. In (Kelly, 1988) the notion of average link shadow prices was introduced to the problem of decentralized adaptive load sharing with a revenue maximization objective. In (Dziong et al., 1988) it was shown how this concept can be generalized to state-dependent link shadow prices. Decomposition of the network reward process into a set of separable link processes, enabling execution of a full policy iteration procedure from Markov decision theory and extension to multi-service environment, was presented in (Dziong and Mason, 1992; Dziong and Mason, 1994). Simplified models for link shadow price evaluations, enabling application of this approach in ATM networks, were proposed in (Dziong et al., 1991) and (Hwang, Kurose, and Towsley, 1992).

The implementation issues presented in this chapter focused on functional and geographical decomposition of the CAC & routing algorithms. There are still several outstanding issues which are critical from a practical viewpoint. In particular, most CAC & routing models assume a limited number of connection classes with regard to the required "bandwidth." On the other hand, in an ATM environment, this number can be very large (if not infinite). This issue can be resolved by "bandwidth" quantization which reduces the number of connection classes and link states. While this procedure introduces the certain loss of resource utilization, this loss is not significant as shown in (Lea and Alyatama, 1995). Another important issue is the addressing structure. In Chapter 9 we discuss an application of virtual path addresses for the purpose of routing. A complete addressing structure for hierarchical implementation is given in ATM Forum PNNI recommendations (ATM Forum, 1994).

References

ATM Forum. 1994. PNNI Draft Specification. 94-0471R9.

References

Ash, G.R., Cardwell, R.H., and Murray, R.P. 1981. Design and optimization of networks with dynamic routing. *Bell System Technical Journal* 60:1787–1820.

Ash, G.R. 1985. Use of a trunk status map for real-time DNHR. In *Proceedings of the 11th International Teletraffic Congress*, Kyoto, Japan.

Ash, G.R., Blake, B.M., and Schwartz, S.D. 1988. Integrated network routing and design. In *Proceedings of the 12th International Teletraffic Congress*, Torino, Italy.

Ash, G.R., Chen, J.-S., Frey, A.E., and Huang, B.D. 1991. Real-time network routing in a dynamic class-of-service network. In *Proceedings of the 13th International Teletraffic Congress*, Copenhagen, Denmark.

Benes, V.E. 1966. Programming and control problems arising from optimal routing in telephone networks. *B.S.T.J.*, (9):1373–1438.

Bertsekas, D., and Gallager, R. 1987. *Data Networks*. Prentice-Hall, Inc.

Bremaud, P. 1990. On computing derivatives with respect to the rate of a Poisson process trajectorywise: the phantom method. Submitted to QUESTA.

Cameron, W.H., Galloy, P., and Graham, W.J. 1980. Report on the Toronto advanced routing concept trial. In *Proceedings of Telecommunication Networks Planning Conference*, Paris, France.

Dijkstra, E.W. 1959. A note on two problems in connection with graphs. *Numer. Math.*, 1:269–271.

Dziong, Z., Pioro, M., Körner, U., and Wickberg, T. 1988. On adaptive call routing strategies in circuit switched networks– maximum revenue approach. In *Proceedings of the 11th International Teletraffic Congress*, Torino, Italy.

Dziong, Z., and Liao, K.-Q. 1989. Reward maximization as a common basis for routing, management and planning in ISDN. In *Proceedings of Networks'89*. Spain.

Dziong, Z., and Mason, L.G. 1989. Control of multi-service loss networks. In *Proceedings of the 28th IEEE Conference on Decision and Control*, pp. 1099–1104. IEEE Computer Society Press.

Dziong, Z., Liao, K-Q., and Mason, L.G. 1990. Flow control models for multi-service networks with delayed call set up. In *Proceedings of IEEE INFOCOM'90*, pp. 39–46. IEEE Computer Society Press.

Dziong, Z., Liao, K-Q., Mason, L., and Tetreault, N. 1991. Bandwidth management in ATM networks. In *Proceedings of the 13th International Teletraffic Congress*, Copenhagen, Denmark.

Dziong, Z., and Mason, L.G. 1992. An analysis of near optimal call admission and routing model for multi-service loss networks. In *Proceedings of IEEE INFOCOM'92*, pp. 141–152. IEEE Computer Society Press.

Dziong, Z., and Mason, L.G. 1994. Call admission and routing in multi-service loss networks. *IEEE Transactions on Communications* 42(2/3/4):2011–2022.

Ford, Jr., L.R., and Fulkerson, D.R. 1962. *Flows in Networks*. Princeton University Press.

Gersht, A., and Lee, K.J. 1988. Virtual-circuit load control in fast packet-switched broadband networks. In *Proceedings of IEEE GLOBECOM'88*, Hollywood, Florida.

Gibbens, R.J., Kelly, F.P., and Key, P.B. 1988. Dynamic alternative routing-modeling and behaviour. In *Proceedings of the 12th International Teletraffic Congress*, pp. 3.4A.3.1–7, Torino, Italy.

Girard, A. 1990. *Routing and Dimensioning in Circuit-Switched Networks*. Addison-Wesley.

Gopal, I.S., and Stern, T.E. 1983. Optimal call blocking policies in an integrated services environment. In *Proceedings of Conf. Inform. Sci. Syst.*, pp. 383–388. The Johns Hopkins Univ.

Grandjean, C. 1967. Call routing strategies in telecommunication networks. In *Proceedings of the 5th International Teletraffic Congress*, New York, USA.

Howard, R. A. 1960. *Dynamic Programming and Markov Process*. The M.I.T. Press, Cambridge, Massachusetts.

Hwang, R.-H., Kurose, J.F., and Towsley, D. 1992. State dependent routing for multirate loss networks. In *Proceedings of IEEE GLOBECOM'92*, pp. 565–570. IEEE Computer Society Press.

Inoue, I., Yamamoto, H., Ito, H., and Mase, K. 1991. Advanced call-level routing schemes for hybrid controlled dynamic routing. *IEICE Transactions*, E74(12).

Kelly, F.P. 1988. Routing in circuit switched networks: optimization, shadow prices and decentralization. *Adv. Appl. Prob.* 20:112–144.

Krishnan, K.R., and Ott, T.J. 1986. State dependent routing for telephone traffic: Theory and results. In *Proceedings of the 25th IEEE Conference on Decision and Control*, Athens, Greece.

Krishnan, K.R., and Ott, T.J. 1988. Forward-looking routing: A new state-dependent routing scheme. In *Proceedings of the 12th International Teletraffic Congress*, Torino, Italy,

Krishnan, K.R. 1990. Performance evaluation of networks under state-dependent routing. In *Proceedings of Seminar on Design and Control of a Worldwide Intelligent Network*, Paris, France.

Lazarev, W.G., and Starobinets, S.M. 1977. The use of dynamic programming for optimization of control in networks of commutations of channels. *Engineering Cybernetics (Academy of Sciences, USSR)*, (3).

Lea, C.-T., and Alyatama, A. 1995. Bandwidth quantization and states reduction in the broadband ISDN. *IEEE/ACM Transactions on Networking*, 3(3):352–360.

Liao, K.-Q., Dziong, Z., and Mason, L.G. 1989. Dynamic link bandwidth allocation in an integrated services network. In *Proceedings of IEEE ICC'89*, Boston.

Liang, Y., Liao, K.-Q., Roberts, J., and Simonian, A. 1988. Queuing models for reserved set up telecommunications services. In *Proceedings of the 12th International Teletraffic Congress*, pp. 4.4B.1.1-7, Torino, Italy.

Lindberger, K. 1987. Blocking for multislot heterogeneous traffic streams offered to a trunk group with multislot traffic streams. In *Proceedings of the 5th ITC Seminar*, Lake Como, Italy.

Mason, L.G. 1985. Equilibrium flows, routing patterns and algorithms for store-and-forward networks. *Large Scale Systems*, 8:187–209.

Mason, L.G., and Girard, A. 1982. Control techniques and performance models for circuit-switched networks. In *Proceedings of the 21st IEEE Conference on Decision and Control*, pp. 1374–1383, Orlando, Florida, December.

Maxemchuk, N.F., and El Zarki, M. 1990. Routing and flow control in high-speed wide-area networks. *Proceedings of the IEEE*, 78(1):204–221.

Miyake, K. 1988. Optimal trunk reservation control for multi-slot connection circuits. *Transactions of IEICE*, (11).

Narendra, K.S., Wright, E.A., and Mason, L.G. 1977. Application of learning automata to telephone traffic routing and control. *IEEE Trans. Systems, Man and Cybernetics*, SMC-7(11).

Oda, T., and Watanabe, Y. 1990. Optimal trunk reservation for a group with multislot traffic stream. *IEEE Transactions on Communications* 38(7):1078–1084.

Reiman, M.I., and Simon, B. 1989. Open queueing systems in light traffic. *Math. of Oper. Res.*, 14(1):26–59.

Roberts, J., and Liao, K.-Q. 1985. Traffic models for telecommunications services with advance capacity reservation. *Computer Networks and ISDN Systems*, 10:221–229.

Ross, S. M. 1983. *Introduction to Stochastic Dynamic Programming*. Academic Press.

Ross, K.W., and Tsang, D.H.K. 1989. Optimal circuit access policies in an ISDN environment: A Markov decision approach. *IEEE Transactions on Communications*, 37:934–939.

Szybicki, E., and Lavigne, M.E. 1979. The introduction of an advanced routing system into local digital networks and its impact on networks' economy, reliability and grade of service. In *Proceedings of ISS'79*, Paris, France.

Szybicki, E. 1985. Adaptive, tariff dependent traffic routing and network management in multi-service networks. In *Proceedings of the 11th International Teletraffic Congress*, Kyoto, Japan.

Tijms, H.C. 1986. *Stochastic Modelling and Analysis: A Computational Approach*. New York: Wiley.

Chapter 6

Multi-point Connections

MULTI-POINT connections are likely to make up a significant share of traffic in broadband networks due to a large number of potential multi-media group applications such as conferences, broadcasting, teaching, distributed processing, etc. Conventional algorithms for creation and management of multi-point connections, developed for narrow-band services, are not suitable for the new environment. This is caused by differences in the networks' capabilities and connection traffic characteristics. In particular the optimal connection tree for wide-band connections depends not only on the network configuration and fixed costs assigned to the links, but also on the current state of connections. Another important factor is that under overload conditions the wide-band multi-point connections will be more prone to rejection unless some kind of priority provides a fair access to network resources for all services. Additionally, in many cases a multi-point connection will be dynamic. Thus expansion of an existing connection should have priority over new connections.

In the following we discuss the main issues which are critical to CAC & routing of multi-point connections in broadband networks. As in the case of point-to-point connections we treat CAC & routing as a bandwidth constrained problem with connections requiring "bandwidth" reservation on each link. We start with a discussion of the main issues which have to be addressed by CAC & routing algorithms for multi-point connections in broadband networks (Section 6.1). Then we describe several heuristics for the design of the minimum cost tree and compare their features by means of numerical study (Section 6.2). In Section 6.3 we describe three CAC & routing strategies based on different link cost metrics. The first strategy is an extension of the reward maximization approach that results in link shadow prices being used as a cost metric. The second is an application of the least loaded path approach (LLP). While the LLP approach is relatively simple it does not fit the multi-point routing problem very well. In order to provide a simple yet efficient solution we propose a third strategy where the link shadow prices are approximated by a simple heuristic function of the inverse of the available link capacity. Important features of these strategies are presented and compared by means of simulation.

6.1 Classification and Main Issues

In general the notion of multi-point connection refers to connections that involve a group of users with more than two members. One can recognize three basic categories of multi-point connections illustrated in Figure 6.1. In the first category, referred to as point-to-many connections, there is one source and several destinations. Broadcasting of the same information to several users is a typical application in this category. In the opposite case, many-to-point connections, the information from several sources is collected by one destination. Data collection and distributed measurement are possible applications within this category. Many-to-many connections constitute the third category where each of the group members can send and receive information to or from all other members. Conference connections are basic applications in this category. Obviously a multi-point connection can consist of a mixture of the three connection categories. It is also important to note that a many-to-many connection can be realized by a set of point-to-many or many-to-point connections.

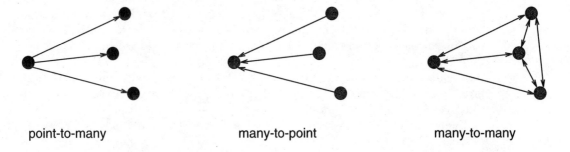

point-to-many   many-to-point   many-to-many

Figure 6.1. *Multi-point connection categories.*

One of the important distinctions between the point-to-point and multi-point connections is the possible connection dynamism. It is quite likely that in some applications the group membership can vary during the connection's life time. This variability adds additional complexity to the already difficult problem of CAC & routing of multi-point connections. In the following we discuss the main issues which are specific to multi-point connections. As in the case of point-to-point connections we concentrate first on the system where all information is available instantly to the CAC & routing algorithm.

Let us first consider the routing issue. The routing of multi-point connections depends strongly on multi-cast capabilities available in the network. The basic options are illustrated in Figure 6.2 for a point-to-many connection. When the network switches have no multi-cast capabilities, the multi-point connections can be established by a set of point-to-point connections (Figure 6.2a) originating at the source node. An alternative is to use a central server, associated with one node, that can multi-cast the received multi-point connection signals to other group members (Figure 6.2b). The server location should be optimized to minimize the cost of the routing tree. In general this approach can be extended to a network of multi-cast servers organized in a hierarchical or non-hierarchical structure. This approach, used in circuit switched environment, is also based on point-to-point connections but can provide superior resource utilization. Moreover, the connection control algorithm can also be located in the server which simplifies the connection management protocol. The obvious price for the improvements is the cost of the servers. This drawback can be avoided if the switches have internal multi-cast capability (Figure 6.2c). In this case the minimum cost tree can be implemented. In the following we concentrate on the last option since the other two can be treated as restricted cases of the latter.

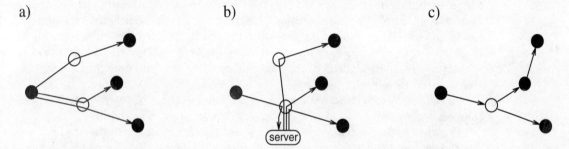

Figure 6.2. *Multi-cast alternatives.*

Let us focus on point-to-many connections. In this case the objective of the routing function is to recommend an optimal multi-cast tree on which a *multi-cast virtual channel* (MVC) can be established. As in the case of point-to-point connections, the main optimization criterion is maximization of the network resource utilization function which may include fairness parameters or constraints. Nevertheless there is an important difference between the routing procedures for point-to-point and multi-point connections. Namely, due to a very large number of possible multi-cast groups and trees it is not practical to consider a set of pre-established alternative multi-cast trees for a particular user group. Instead the tree has to be constructed for each connection. This feature causes the classification of routing strategies for multi-point connections to be modified into the following two categories:

- *Link cost based strategies:* Here a multi-cast tree is designed based on link costs. The link cost can be fixed (e.g., distance), adaptive (e.g., average shadow price or average delay) or state-dependent (e.g., inverse of free capacity or shadow price).

- *Point-to-point path based strategies:* In this category the multi-cast tree design is based on routing information for point-to-point connections. The main idea of this approach is to use, as much as possible, information which already exists in the system in order to speed up the process of the tree design. Note that a strategy from this category can give the same result as a strategy from the previous category if both the point-to-point and point-to-many routing strategies are based on the same link cost metrics.

Consider routing strategies based on link costs. The objective of the multi-cast tree design is minimization of the tree cost. When considering this issue it is convenient to distinguish two basic network types: undirected cost network and directed cost network. In the undirected case the cost of the link between two nodes is the same for both transmission directions. In the directed network case the link cost can be different for each transmission direction. Since in most real systems each transmission direction uses separate resources, the undirected case basically corresponds to fixed link costs (distance, real bandwidth cost). The adaptive or state-dependent link costs can be treated as undirected only if the traffic distribution is symmetrical on each link direction, or if each link connection is symmetrical (reserves the same "bandwidth" in both directions). Observe that the multi-cast connections can introduce an asymmetry in link states for each transmission direction. Thus, in general, the broadband networks should be treated as directed cost networks if adaptive or state-dependent routing is considered. As a consequence the designed tree is also directed in the sense that the origin node has to be distinguished in the design process. In Section 6.2.1 we consider heuristics for directed cost networks (obviously they can be applied for undirected cost network as well).

The algorithms for undirected cost networks do not distinguish the origin node, and the designed tree is independent of the identification of the origin node amongst the connection group nodes. Optimization of such a tree corresponds to the Steiner tree problem in graph theory. While this case is less practical in some applications, it has the advantage of being more disposed to mathematical analysis, which gives the possibility of exact bound evaluation and worst case analysis. Such an analysis provides better insight into the main features of the algorithms. Moreover, in most cases, the algorithms derived for the undirected case can be easily extended to the directed case; see, e.g. (Barnebei, Coppi, and Winkler, 1996). For these reasons in Section 6.2.2 we discuss a class of heuristics for undirected cost networks.

Dynamic connections introduce additional complexity to the already difficult problem of the multi-point tree optimization. This follows from the fact that an addition or removal of a group member can result not only in adding or removing of a tree branch but also in changing the optimal (or sub-optimal) tree structure. Since re-routing of the existing connection without interrupting the connection continuity is quite complex, it is desirable that the addition or removal of a new member not affect the part of the tree which is used by other members. In particular, if a new member is located in a node which is not part of the tree, a new branch can be added that connects the new node with one of the tree nodes. Concerning the choice of the new branch, two options can be considered. The first is to chose the least cost path between the node and the existing tree. While in the short run this solution is optimal, there is a risk that, if the group membership is variable, after many additions and removals of the group members, the tree structure can become very far from optimal. In such cases it may be more efficient to connect a new member to a node that is likely to be used by an optimal tree in any future instant of the connection.

A many-to-many connection can be represented as a set of point-to-many connections. This feature gives certain flexibility in routing of many-to-many connections. Let us consider first an approach where all multi-cast virtual channels are established on the same minimum cost tree. While this approach is used successfully in narrow-band networks, it may be problematic when multi-point connections involve wide-band services due to a potential "bandwidth" requirement outburst. Namely, if all MVC's require "bandwidth" reservation for a wide-band service (d_j) and the group size ($|\tilde{V}'|$) is large, the "bandwidth" required on links constituting the minimum cost tree is $|\tilde{V}'| \cdot d_j$. This "bandwidth" requirement outburst can increase significantly the loss probability of such connections, and may create bottlenecks on some links. To overcome the problem one can realize MVC's on different trees. In particular, if state-dependent link costs (e.g., shadow price) are used, the MVC's can be optimized sequentially with the link state updated after each MVC creation.

While the "bandwidth" requirement outburst should be avoided, in some cases application of a common minimum cost tree for all MVC's in many-to-many connections can be beneficial. This is the case when there is a significant gain from multiplexing of the connection signals transmitted on the same link. An ideal situation is when there is a strong negative correlation between the signals transmitted in different MVC's. For example, consider a conference connection where only one member is active at a time. In this case it is sufficient to reserve only "bandwidth" d_j on all links constituting the minimum cost tree. When the negative correlation is not so strong the gain may be smaller but still worth considering.

Many-to-point connections can be established on a minimum cost tree even when the network switches do not have multi-cast capabilities. This is the case in ATM networks where a standard switch can transfer cells from many incoming VC's onto one outgoing VC. Nevertheless this option is worth considering only when there is a negative correlation between the signals generated by the connection sources. Otherwise, the best resource utilization is achieved when the connection is established as a set of point-to-point connections due to more uniform traffic distribution.

While the routing problem for multi-point connections includes several new challenges, the CAC function and CAC algorithms are similar to the ones for point-to-point connections. In particular the decision criteria of the CAC function: (connection feasibility, network utilization, and access fairness) are the same. Concerning the CAC_{GoS} mechanism categories, the one based on limiting the set of alternative solutions cannot be applied since this concept is not used for routing of multi-point connections. The remaining two CAC_{GoS} mechanism categories which can be applied for multi-point connections are the same as for the point-to-point connections: the algorithms based on link costs and the algorithms based on link bandwidth reservation.

Concerning implementation of CAC & routing algorithms for multi-point connections, the general options are similar to the ones described for point-to-point connections: centralized, distributed and hierarchical (see Section 5.3). The important difference lies in connection addressing, which is much more complex in the case of multi-point connections. In particular, in ATM networks, the recommended path for a point-to-point connection can be defined by a pre-established virtual path. Thus, for example, the resources on the path can be reserved by simply sending one resource management cell along the path (see Section 4.1). In the case of point-to-many connections the group address has to be included in the signaling cell information field. Moreover, such a cell has to be analyzed in each of the nodes where the path splits in order to establish a multi-cast connection in the switch and to generate subsequent signaling cells addressed to the next nodes where the recommended path splits or ends.

6.2 Heuristics for Minimum Cost Tree Design

In the following we discuss several heuristics for design of a minimum cost tree; a more detailed study is given in (Dziong et al., 1995). We concentrate on the algorithms that use routing information evaluated for point-to-point connections. This approach integrates the CAC & routing algorithms for all connection types and can reduce complexity of the multi-point connection routing. In particular it is assumed that for any node pair the least cost path structure (transit nodes and link costs) is available. To simplify presentation the least cost path will be also referred to as the shortest path.

6.2.1 Heuristics for directed cost networks

The simplest approach is to create a multi-cast tree from a set of shortest paths between the origin and destination nodes. If the paths are used independently, the tree cost is equal to the sum of the path costs. If any link is used by more than one path, the tree cost can be reduced by carrying only one signal on this link (switch multi-cast capability required) and is equal to the sum of link costs over the links composing the tree. Henceforth the two cases are referred to as MPC1 and MPC2, respectively (in the literature the MPC2 algorithm is also referred to as "naive" algorithm). The basic complexity order of both algorithms is $O(|V'|)$, where $|V'|$ denotes the multi-cast group size.

An alternative approach is to create the connection tree in a step-by-step manner by connecting each destination node to a partially constructed tree one at a time, using the shortest path. This path is defined as the minimum cost path over the shortest paths between the destination node and any of the nodes included in the partial tree (origin, destination, and transit nodes). A very efficient algorithm based on this concept was presented in (Takahashi and Matsuyama, 1980):

1. Define the initial partial tree, PT, as a single node of the multi-point group (can be the origin node).

2. Find another origin or destination node, v_i, not included in PT (PT can include transit nodes), with the lowest cost path between v_i and PT.

3. Add to PT the shortest path between v_i and PT.

4. If there are group nodes not included in PT, go to Step 2.

The complexity order of this algorithm (henceforth referred to as TAK) is $O(|V||V'|^3)$. It was proven (Takahashi and Matsuyama, 1980) that the the ratio of the worst case solutions to the optimal ones is bounded by the factor $2(1 - 1/|V'|)$, which is the same as for the algorithms based on minimum spanning tree described in Section 6.2.2.

In the following we describe another heuristic (henceforth called SPT algorithm) based on the partial tree concept which can be seen as a modification of the TAK algorithm:

1. Define the initial partial tree, PT, as the highest cost path over the shortest paths between the origin and destination nodes (PT can include transit nodes).

2. Find a destination node, v_i, not included in PT, with the lowest (SPT_{min}) or the highest (SPT_{max}) cost of the shortest OD path.

3. Add to PT the shortest path between v_i and PT.

4. If there are destination nodes not included in PT go to Step 2.

The complexity order of the SPT algorithm is $O(|V||V'|)$, which is smaller than that of the TAK heuristic. The algorithm simplification is caused by the fact that the choice of the next node to be included in the partial tree, in step 2, can be determined by the origin node based on local information. It could be considered paradoxical to choose the highest cost path over the shortest OD paths in step 1 and, for the option SPT_{max}, in step 2, since choosing the lowest cost path instead, as proposed in (Takahashi and Matsuyama, 1980), would give a less expensive tree in some cases. The idea is to introduce a hidden secondary objective aiming at reducing the maximum and average length of connection paths over the OD pair paths in the created tree. Effectiveness of this approach will be illustrated latter by means of numerical examples.

Observe that the heuristics based on the partial tree concept have several attractive features from implementation viewpoint. They can be easily implemented as distributed algorithms. They can be applied directly to dynamic connections since an addition of a new destination node does not modify existing connections. Finally they are applicable to directed cost networks without any modifications.

Numerical examples

To compare the important features of the TAK, SPT_{min}, SPT_{max}, and MPC2 heuristics we designed multi-cast trees for several hundred examples. Each example corresponds to a randomly generated link cost matrix and multi-cast group, similarly to (Waxman, 1988). We did simulations for the network size $|V| = 40$

and group size $|V'| = 10$. The complexity and average values of chosen parameters, over more than 300 examples, are given in Table 6.1.

TABLE 6.1 Complexity order and average values of the parameters for different heuristics.

	TAK	SPT_{min}	SPT_{max}	MPC2
complexity order	$O(|V||V'|^3)$	$O(|V||V'|)$	$O(|V||V'|)$	$O(|V'|)$
tree cost	7.1	+1.8%	+14%	+28%
maximum # transit nodes	+90%	+76%	+63%	3.7
average # transit nodes	+84%	+67%	+60%	2.1

As could be expected, the smallest cost is provided by the TAK algorithm although the SPT_{min} heuristic gives comparable results. The simplest MCT2 algorithm gives the most expensive trees while the SPT_{max} option is situated somewhere between the TAK and MPC2 heuristics. To give more insight into the features of the considered heuristics, Table 6.1 also contains the maximum and average length of the connection paths, expressed as the number of transit nodes in the designed multi-cast tree. This length can be critical from the perspective of quality of service (QoS) of some broadband services sensitive to transmission delay and its variation. In terms of the number of transit nodes the order is reversed with MPC2 providing the smallest number of transit nodes and the TAK heuristic giving almost double the number. This large difference is caused by the fact that usually the minimization of the tree cost gives a tree in a form of chained destination nodes. Large number of transit nodes for some OD pairs can also be seen as detrimental to robustness since a node or a link failure in the tree can influence many destinations. The SPT heuristics provide a reduced length of connection paths, compared to the TAK heuristic, with SPT_{max} generating less transit nodes by selecting the shortest paths with the highest cost in step 2. A subjective comparison of the three characteristics suggests that the SPT_{min} algorithm provides the best trade-off between the complexity and the performance since its cost performance is comparable with the TAK algorithm while the complexity and the number of transit nodes are substantially smaller.

6.2.2 Heuristics based on minimum spanning tree

Minimization of the multi-cast tree cost in undirected networks corresponds to the Steiner tree problem in graph theory. Unfortunately the exact solution constitutes a NP-complete problem which is not solvable during the connection setup in a realistic environment. To circumvent the problem, numerous heuristics have been proposed which can significantly reduce the complexity at the expense of providing a sub-optimal solution. Many of them are based on a *minimum spanning tree* (MST) algorithm which for a given undirected graph, defined by a set of the nodes and edges with assigned cost, can be executed as follows:

1. Choose the least cost edge from the set of edges not yet assigned to the MST.

2. If the chosen edge joins two nodes which are not joined by the existing MST, then add this edge to the MST.

3. If all nodes are joined by the MST then stop; otherwise go to Step 1.

One of the best known heuristics based on the MST algorithm is the KMB algorithm (Kou, Markowsky, and Berman, 1981). In our application this algorithm can be described as follows:

1. Construct the complete (fully connected) undirected graph composed of all origin and destination nodes (OD nodes) where the costs of the edges are defined by the least cost paths currently recommended for point-to-point connections between the considered nodes.

2. Find a minimum spanning tree, MST_1 for the graph from Step 1.

3. Construct an extended graph by replacing each edge of the MST_1 by the corresponding minimum cost path (new nodes can be added to the graph).

4. Find a minimum spanning tree, MST_2, for the graph from Step 3.

5. Remove the MST_2 edges whose leaf nodes are not a member of the multi-cast group.

The complexity order of the KMB algorithm is $O(|V|^2)$ (if the shortest paths are known). As proved in (Kou, Markowsky, and Berman, 1981) the ratio of the worst case solution to the optimal one is bounded by the factor of $2(1 - 1/|V'|)$. To improve the performance of the KMB algorithm, Rayward-Smith proposed the RS-algorithm (Rayward-Smith, 1983; Rayward-Smith and Clare, 1986), which is based on a heuristic function used to find potential Steiner Vertices (transit nodes). The difference between the KMB and the RS heuristics is that the MST algorithm does not take into consideration nodes that are not part of V', while the RS heuristic does. Statistically, the RS heuristic performs better than KMB; however, the theoretical worst case performance of RS is still the same (Tanaka, 1993).

The KMB and RS algorithms represent two typical approaches to heuristic solutions: In RS, likely Steiner nodes are selected and then a minimum spanning tree of these nodes, together with the nodes in V', is constructed; in KMB, a minimum spanning tree of the whole graph is pruned to remove superfluous nodes. In the following we describe a heuristic algorithm which is based on the minimum spanning tree of the whole graph, but in the realization of the tree the algorithm also tries to select likely Steiner nodes to reduce the overall cost of the final tree:

1. Construct the complete (fully connected) undirected graph composed of all origin and destination nodes (OD nodes) where the costs of the edges are defined by the least cost paths currently recommended for point-to-point connections between the considered nodes.

2. Find a minimum spanning tree, MST_1 for the graph from Step 1.

3. Define the initial tree MST_2, as the path corresponding to the maximum cost edge in MST_1 (can include transit nodes).

4. Find an OD node, v_i, not included in MST_2, such that its minimum cost edge, e_i, in MST_1 joining an OD node included in MST_2 has lowest value.

5. Find a transit node (not OD node) in MST_2 that has the minimum cost path joining v_i. If the cost of this path is smaller than the cost of e_i, add this path to MST_2. Otherwise, add the path corresponding to e_i to MST_2.

6. If there are OD nodes not included in MST_2, go to Step 4.

Observe that the first two steps of this algorithm (henceforth called DJM) are the same as the ones for the KMB heuristic. Notice also the similarity with the step-by-step technique based on the partial tree concept applied in the SPT algorithm. The complexity of the DJM algorithm is $O(|V||V'|)$, which is smaller than the one of the KMB algorithm under the assumption that the shortest paths are given (if the shortest paths are not known the complexity of the DJM algorithm is of the same order as the KMB algorithm). Since both KMB and DJM algorithms are based on the minimum spanning tree concept, the ratio of their worst case solutions to the optimal ones is bounded by the same factor of $2(1 - 1/|V'|)$. Nevertheless this bound does not give a complete picture of the cost performance. In the next section we present several numerical results which allow us to compare the two heuristics more thoroughly.

Numerical results

The comparison of tree cost averages for the KMB and DJM heuristics is given in Table 6.2 for the same examples as used in Table 6.1. The results for heuristics presented in Section 6.2.1 are also included to give direct comparison of the two families of algorithms. Note that although the complexity order of the DJM and SPT algorithms is the same, the SPT heuristic is simpler and the order of the complexity difference is $O(|V'|^2)$. The average cost performance of the DJM algorithm

TABLE 6.2 Complexity order and averages of tree cost for different heuristics.

	KMB	DJM	TAK	SPT_{min}	SPT_{max}	MPC2
complexity	$O(|V|^2)$	$O(|V||V'|)$	$O(|V||V'|^3)$	$O(|V||V'|)$	$O(|V||V'|)$	$O(|V'|)$
tree cost	+1.3%	7.1	7.1	+1.8%	+14%	+28%

is slightly better than that of the KMB algorithm and identical to that of the TAK heuristic. However, the complexity of DJM is smaller than that of KMB and TAK (under assumption that the shortest paths are given). Still the SPT_{min} algorithm, which provides comparable average cost performance with lower complexity, seems to be attractive although its performance bound is unknown.

While the average difference in tree cost is important it does not give the full picture of the cost performance. To provide more insight into this performance

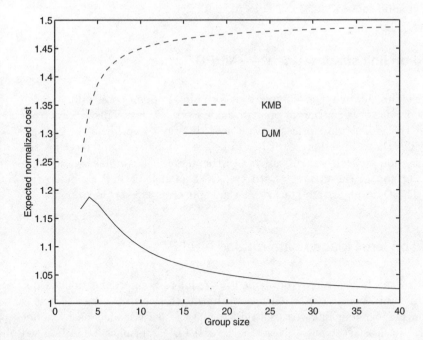

Figure 6.3. *Efficiency of the KBM and DJM algorithms.*

we compared the KMB and DJM algorithm by means of a statistical "worst case" analysis. The "worst case" corresponds to networks where an algorithm based on MST can produce a tree with cost equal to the bound $2(1 - 1/|V'|)$ (Kou, Markowsky, and Berman, 1981); for details see (Dziong, Jia, and Mason, 1997). The result of this analysis is shown in Figure 6.3, where the expected tree costs (normalized to the optimal tree cost), evaluated by the DJM and KMB algorithms, are shown as a function of $|V'|$. The figure shows that on average the cost of KMB tree approaches 150% of the optimal tree cost as the size of the multi-cast group increases. On the other hand, the DJM algorithm approaches optimal tree cost under the same conditions. This important feature indicates that besides possible application for multi-cast tree cost minimization, the DJM algorithm can serve as a reference for other heuristics when the optimal tree cost evaluation is difficult (i.e., large network examples).

6.3 CAC & Routing Strategies for Multi-point Connections

We start by defining three CAC & routing strategies for point-to-many connections. They are based on the link shadow prices, the inverse of free capacity function, and the least loaded path approach, respectively. Then the important features of these different options are illustrated by a numerical study, which also covers many-to-many and dynamic connections.

6.3.1 Strategy based on link shadow prices — MDPD

The reward maximization approach described for point-to-point connections is also very suitable for CAC & routing of point-to-many connections. This follows from the fact that the link shadow prices can be used directly to optimize the connection tree, while the connection reward parameters can be used for meeting the network utilization and fairness objectives. In the numerical examples we will use the MDPD' option of the strategy with the COT model for link shadow price evaluation. The DJM algorithm is used for the connection tree optimization.

6.3.2 Strategy based on least loaded path routing — LLP

The concept of the LLP strategy developed for point-to-point connections is not well suited for multi-point connection tree optimization due to the fact that it is not based on independent-link cost metrics. Thus the least loaded path cannot be treated as the least cost path. Nevertheless the LLP information can be used for multi-cast routing if the tree is created from the set of shortest paths between the origin and destination nodes (MPC1 and MPC2 heuristics). In this case the CAC_{GoS} function is realized only for point-to-point paths and the point-to-many connection is rejected if there is no admissible path for at least one of its OD pairs. Although the LLP approach has some limitations in the multi-point application, its attractiveness lies in a simple function determining the least loaded path.

6.3.3 Strategy based on the inverse of residual link capacity — IRCF

In the following we present a heuristic approach that tries to combine the best features of the MDPD and LLP strategies by integrating the functionality of the MDPD strategy and calculation simplicity of the LLP strategy. Let us define residual link capacity normalized by connection class i bandwidth requirement as

$$\mathcal{C}_i^s = \frac{L^s - \sum_j x_j^s d_j}{d_i} \tag{6.1}$$

Then the link cost, c_i^s (henceforth called IRCF link cost), is defined as

$$c_i^s = (\mathcal{C}_i^s)^{-\kappa} \cdot r_i^{s'} \cdot \left[1 - \left(1 - \frac{\overline{a}^s}{L^s}\right)\gamma\right], \quad \text{for } \mathcal{C}_i^s \geq d_i \tag{6.2}$$

where $\kappa > 0$ and $\gamma > 0$ are constants, $\overline{a}^s$ denotes average link traffic expressed in bandwidth units, and $r_i^{s'}$ is link connection reward parameter in the equivalent

6.3 CAC & Routing Strategies for Multi-point Connections

TABLE 6.3 Network examples.

	Example 1	Example 2	Example 3	Example 4	Example 5
d_i	1,1,6,6	1,1,6,6	1/1/10/10	1/1/1/10/10/10	1/1/1/10/10/10
categories	2 x pp/pm	2 x pp/pm	2 x pp/pm	2 x pp/pm/mm	2 x pp/pm/de
traffic[%]	40/10/40/10	25/25/25/25	40/10/40/10	40/5/5/40/5/5	40/5/5/40/5/5
L^s	320	220-260	100	100	100

system where all traffic is transformed into class i connections (COT model from Section 5.2.3):

$$r_i^{s'} = \overline{R}^s \frac{d_i}{\mu_i \sum_j x_j^s d_j} \tag{6.3}$$

It can be easily shown that the features of the IRCF link cost function are similar to the ones of the link shadow price. In particular the first factor (inverse of the available capacity) ensures that the IRCF link cost increases with decreasing link available bandwidth. The second factor, $r_i^{s'}$, assures that when the link is used by connections giving high reward to the link, the link cost is also high. The last factor reduces the link cost when the link carried traffic decreases. These features cause the IRCF link costs to be used as a substitute for link shadow prices in the reward maximization strategy for multi-point connections (MDPD). Obviously the IRCF costs can also be applied to CAC & routing of point-to-point connection.

In the following numerical study the values of κ and γ are set to 0.4 and 1.25, respectively (based on a numerical analysis of eight network examples with different traffic patterns and configurations).

6.3.4 Numerical study

The important features of the above strategies are illustrated by selected examples that are based on an 8-node network configuration shown in Figure 6.4. The network is serving narrow-band (NB) and wide-band (WB) connections which can belong to one of the following categories: point-to-point (pp), point-to-many (pm), many-to-many (mm), and dynamic extensions (de). The parameters of the considered examples are described in Table 6.3. The MDPD and IRCF strategies have normalized connection reward parameters equal to each other, unless indicated otherwise. Each connection class has the same mean holding time. For the multi-point connections, the destination node group is chosen randomly with uniform distribution. In this case the normalized reward parameter is also normalized by the number of destination nodes.

In the study the DJM algorithm was used for optimization of multi-cast trees when the MDPD and IRCF strategies were applied while the LLP strategy was tested with a tree created from independent OD paths (MPC1 algorithm). The performances of the strategies are shown in Table 6.4 for two examples under nominal and overload traffic conditions. Additionally, the performance for the IRCF strategy with the loss equalization objective is also shown (indicated by

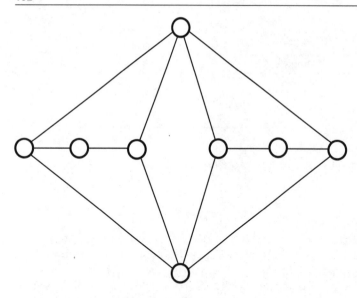

Figure 6.4. *Network configuration.*

+LE). In this case the reward parameters are adjusted in a few iterations in order to make the class rejection rates close to each other. The results indicate that the MDPD and IRCF strategies provide comparable performance while the LLP strategy gives significantly higher traffic loss probabilities.

An important advantage of the MDPD and IRCF strategies is a potential for controlling the GoS distribution among the connection classes. This is illustrated in Table 6.4 for IRCF strategy (+LE case) and in Figures 6.5 and 6.6, where the traffic loss probabilities of the WB-pm and NB-pp connection classes are controlled over a very wide range by means of the respective reward parameters. A large dispersion of the GoS for the MDPD, IRCF, and LLP strategies emphasizes the need for an efficient tool for GoS allocation control in the multi-cast environment.

One might be surprised that in the case of the MDPD and IRCF strategies the rejection rate for multi-point connections is in some cases lower than that for point-to-point connections although the multi-point connections require significantly more resources. This is explained by the fact that in these strategies the multi-cast tree is constructed not only from OD paths but also from destination-to-destination, DD, and destination-to-transit, DT, node paths. For example, assume that a direct-link-path between a particular OD pair is blocked and the best alternative path is inadmissible for point-to-point connections due to the negative net-gain. The same alternative path may be admissible for a point-to-many connection in two cases. First, if the path's transit nodes constitute at the same time destination nodes of the point-to-many connection, the reward offered to the path will be higher than in the case of a point-to-point connection. When the path transit nodes are not the destination nodes, the connection can still be accepted if the cost of other parts of the tree is small (gain from pooling the reward from all connection destinations). These features imply in general that the multi-cast

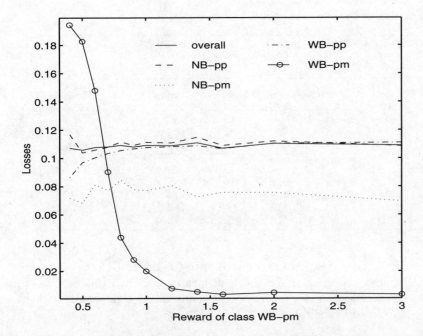

Figure 6.5. *Losses as a function of the WB-pm class reward parameter (Example 2).*

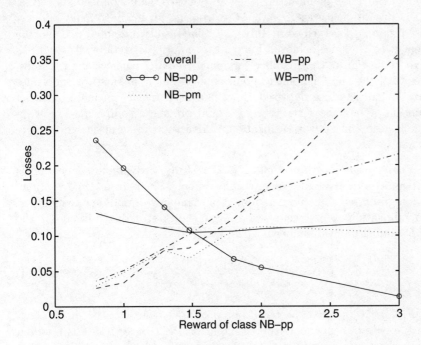

Figure 6.6. *Losses as a function of the NB-pp class reward parameter (Example 2).*

TABLE 6.4 Losses [%] under nominal and overload conditions.

Example	conn. class	nominal				overload			
		MDPD	IRCF	IRCF +LE	LLP	MDPD	IRCF	IRCF +LE	LLP
1	overall	1.27 ±0.28	0.40 ±0.21	0.50 ±0.25	6.3 ±0.26	11.6 ±0.45	9.9 ±0.46	11.7 ±0.54	16.4 ±0.49
	NB-pp	0.02	0.01	0.53	1.6	2.7	0.2	12.5	5.0
	NB-pm	0.14	0.09	0.34	4.2	0.5	0.7	13.4	12.4
	WB-pp	2.53	0.81	0.47	10.2	20.4	19.2	10.8	26.7
	WB-pm	2.30	0.63	0.57	24.8	2.3	13.9	16.9	60.5
3	overall	0.39 ±0.06	0.33 ±0.10	0.41 ±0.10	2.70 ±0.22	9.2 ±0.78	7.3 ±0.33	8.7 ±0.89	14.4 ±0.70
	NB-pp	0.00	0.01	0.49	0.23	0.0	0.1	10.4	1.7
	NB-pm	0.02	0.03	0.50	0.43	0.3	0.6	11.9	4.4
	WB-pp	0.76	0.65	0.33	5.06	18.4	14.5	6.9	26.8
	WB-pm	1.45	1.25	0.67	14.43	25.4	20.9	11.5	56.0

connection can have priority when the connection tree is not constructed from independent OD pairs. The results for the LLP strategy confirm this conclusion since in this case the rejection rate for multi-cast connection is significantly higher, especially under overload conditions.

Many-to-many connections: Here we consider many-to-many connections which are realized by a set of point-to-many connections considered in a sequence. A new many-to-many connection is admitted when all point-to-many connections can be accepted. To illustrate features of the CAC & routing for many-to-many connections, we use Example 4 described in Table 6.3. The results presented in Table 6.5 are consistent with the ones achieved for point-to-many connections although, as could be expected, the loss probability for many-to-many connections is significantly larger than the one for point-to-point connections. This feature underlines the fact that in multi-service networks there is a need for an efficient control of GoS distribution among the services which could provide a compromise between the loss equalization and network utilization. This issue will be discussed in more details in Chapter 8.

Dynamic point-to-many connections: In our study, a dynamic connection is created first as a normal point-to-many connection. Then, from time to time, the connection group membership is modified by removing or adding one member. The group modification events occur according to the Poisson distribution and the probability of adding a new member is given by

$$P_c(|\tilde{V}'|) = \frac{\alpha(|\tilde{V}| - |\tilde{V}'|)}{\alpha(|\tilde{V}| - |\tilde{V}'|) + (1-\alpha)|\tilde{V}'|} \qquad (6.4)$$

where $\alpha = E[|\tilde{V}'|/|\tilde{V}|]$ is the expected fraction of nodes participating in the connection. In the case of a new connection extension the tree structure is updated by adding the shortest path between the new member node and the existing tree (MDPD and IRCF) or by adding the shortest OD path (LLP). When a member is

TABLE 6.5 Losses under nominal and overload conditions.

Example	conn. class	nominal MDPD	nominal IRCF	nominal LLP	overload MDPD	overload IRCF	overload LLP
4	overall	0.99 ±0.31	1.94 ±0.07	5.81 ±0.16	11.74 ±0.24	11.22 ±0.39	18.74 ±0.29
	NB-pp	0.01	0.00	0.78	0.16	0.03	3.28
	NB-pm	0.19	0.54	1.94	0.72	0.55	7.40
	NB-mm	0.29	0.75	8.91	2.18	1.46	35.66
	WB-pp	1.02	3.38	8.59	22.47	22.25	30.59
	WB-pm	2.99	2.15	19.61	17.88	7.6	56.89
	WB-mm	47.80	25.59	81.33	80.71	76.38	98.06
5	overall	1.42 ±0.22	0.90 ±0.12	6.2 ±0.52	10.4 ±0.33	9.6 ±0.48	17.2 ±1.23
	NB-pp	0.003	0.01	0.8	0.0	0.1	2.7
	NB-pm	0.19	0.22	1.9	0.5	0.6	6.6
	NB-de	0.005	0.01	0.2	0.0	0.0	0.5
	WB-pp	2.75	1.79	11.2	21.5	20.0	30.7
	WB-pm	5.36	2.61	26.4	18.4	17.0	61.3
	WB-de	1.39	0.54	1.2	5.1	9.9	3.2

leaving the connection, the part of the tree which was used only by this member is pruned. It is obvious that one would like to have the loss probability for the connection extensions as small as possible. That is why the connection extensions are treated as direct-link-path connections in the LLP strategy. Example 5, chosen for illustration of the dynamic connection performance, is described in Table 6.3. The results presented in Table 6.5 indicate that the loss probability for the connection extensions is significantly smaller than the ones for the multi-cast connections and the point-to-point connections. This wanted feature is explained by the fact that a multi-cast connection extension has more alternatives than a point-to-point connection.

6.4 Discussion and Bibliographic Notes

We focused on two main issues concerning CAC & routing of multi-point connections. First, heuristics for minimum cost tree design were analyzed for given link costs. Then several CAC & routing strategies were described and studied from the network resource utilization and service access fairness viewpoint. There is a substantial literature addressing the first issue. For undirected network graphs the problem is known as the Steiner problem in networks. A solid survey of exact algorithms and heuristics for solving the Steiner problem, from a mathematical viewpoint, is given in (Winter, 1987). In broadband network implementations, the problem is more complex due to dynamic connections and "bandwidth" limitations. The issue of dynamic tree design for multi-point connection routing has been addressed in (Waxman, 1988; Waxman, 1993). The important problem of "bandwidth" limitation in case of many-to-many connections has been indicated and analyzed in (Jiang, 1992). Some of the works focus on algorithms based on distributed protocols where each node has only local information which simplifies algorithm implementation, e.g. (Ghatare and Jalote, 1993).

The second issue covered in this chapter (Section 6.3) concerns network resource utilization and access fairness in the multi-service environment. Recently this subject attracted more attention of researchers, although up to now it has not been treated extensively in the literature. Our study showed that any CAC & routing strategy in the multi-service environment with multi-point connections should be equipped with an efficient tool for controlling GoS distribution among different services and connection classes. This feature is present in the reward maximization approach which can also serve as a benchmark and a design reference for other simpler solutions. The latter quality was illustrated in Section 6.3.3 where a simple model was designed to approximate the link shadow price functions (IRCF). The numerical results show that the approximation gives comparable performance to the original model.

In this chapter we treated the CAC & routing problem as a bandwidth constrained problem. Obviously some additional constraints can be taken into account in the process of the tree design. For example, in (Kompella, Pasquale, and Polyzos, 1992) a delay bound for each OD node pair is added. In (Tode et al., 1992; Tode et al., 1993) the number of copy operations in the switches is restricted. The last feature reminds us that while in the case of point-to-point connections, the node switches can be treated as transparent on the connection level, in the case of multi-point connections the switches may have some multi-cast limitations. More information on this subject can be found in, e.g. (Turner, 1988; Lee, 1988; Shimamoto et al., 1992).

References

Ammar, M.H., Cheung, S.Y., and Scoglio, C.M. 1993. Routing multipoint connections using virtual paths in an ATM network. In *Proceedings of IEEE INFOCOM '93*, pp. 98–105. IEEE Computer Society Press.

Barnebei, F., Coppi, A., and Winkler, R. 1996. Two heuristics for multicasting in ATM networks. In *Proceedings of IFIP-IEEE Broadband Communications'96*, pp. 455–466. Chapman & Hall.

Dabbous, W., and Diot, C. 1995. A tutorial on point to multipoint communication. *HPN'95*. Spain.

Dziong, Z., Pióro, M., Körner, U., and Wickberg, T. 1988. On adaptive call routing strategies in circuit switched networks – maximum revenue approach. In *Proceedings of the 12th International Teletraffic Congress*, pp. 3.1A.5.1-8. North-Holland.

Dziong, Z., Liao K-Q., Mason, L.G., and Tetreault, N. 1991. Bandwidth management in ATM networks. In *Proceedings of the 13th International Teletraffic Congress*, pp. 821–827. North-Holland.

Dziong, Z., Wielosz, A., Ming J., and Mason, L.G. 1995. Call admission, routing and bandwidth management for multipoint connections in ATM based networks. Technical Report for General DataCom Inc., October.

References

Dziong, Z., Wielosz, A., Ming, J., and Mason, L.G. 1997. Call admission and routing for multipoint connections in ATM based networks. Submitted to the IEEE ATM'97 Workshop. Lisbon, Portugal.

Dziong, Z., Jia M., and Mason, L.G. 1997. Analysis of multi-cast routing algorithms for broadband networks. Submitted for publication.

Ferguson, M.J., and Mason, L.G. 1984. Network design for a large class of teleconferencing services. *IEEE Transactions on Communications,* 32(7):789-796.

Ghatare, S.P., and Jalote, P. 1993. Efficient multicasting in point-to-point networks. *IFIP Transcations C,* C-13:93-101.

Hwang, R.-H. 1995. Adaptive multicast routing in single rate loss networks. In *Proceedings IEEE INFOCOM'95,* pp. 571–578. IEEE Computer Society Press.

Jiang, X. 1992. Routing broadband multicast streams. *Computer Communications,* 15(1):45–51.

Kou, L., Markowsky, G., and Berman, L. 1981. A fast algorithm for Steiner trees. *Acta Informatica,* 15:141–145.

Kompella, V.P., Pasquale, J.C., and Polyzos, G.C. 1992. Multicasting for multimedia applications. In *Proceedings of IEEE INFOCOM'92,* pp. 2078–2085. IEEE Computer Society Press.

Lee, T.T. 1988. Nonblocking copy networks for multicast packet switching. *IEEE Journal on Selected Areas in Communication,* 6(9):1455-1467.

Mason, L.G., Deserres, Y., and Meubus, C. 1985. Circuit-switched multipoint service performance models. In *Proceedings of the 11th International Teletraffic Congress,* pp. 2.1A.5.1-7. North-Holland.

Noronha, Jr., C.A., and Tobagi, F.A. 1994. Optimum routing of multicast streams. In *Proceedings IEEE INFOCOM'94,* pp. 865–873. IEEE Computer Society Press.

Rayward-Smith, V.J. 1983. The computation of nearly minimal Steiner trees in graphs. *International Journal of Mathematical Education in Science and Technology,* 14(1):15–23.

Rayward-Smith, V.J., and Clare, A. 1986. On finding Steiner vertices. *Networks,* 16:283–294.

Ravindran, K. 1995. A flexible network architecture for data multicasting in multiservice networks. *IEEE Journal on Selected Areas in Communication,* 13(8):1426–1444.

Sacham, N., and Meditch, J. 1994. An algorithm for optimal multicast of media streams. In *Proceedings of IEEE INFOCOM'94,* pp. 856–864. IEEE Computer Society Press.

Shimamoto, S., Zhong, W.D., Onozato, Y., and Kaniyil, J. 1992. Recursive copy networks for large multicast ATM switches. *IEICE Trans. Commun.,* E75-B(11):1208–1219.

Takahashi, H., and Matsuyama, A. 1980. An approximate solution for the Steiner problem in graphs. *Math. Japonica,* 24(6):573–577.

Tanaka, Y. 1993. Multiple destination routing algorithms. *IEICE Trans. Commun.,* E76-B(5):544–552.

Tode, H., Sakai, Y., Yamamoto, M., Okada, H., and Tezuka, Y. 1992. Multicast routing algorithm for nodal load balancing. In *Proceedings of IEEE INFOCOM'92,* pp. 2086–2095. IEEE Computer Society Press.

Tode, H., Sakai, Y., Yamamoto, M., and Okada, H. 1993. A study on the support of multicast traffic in ATM networks. June

Turner, J.S. 1988. Design of a broadcast packet switching network. *IEEE Transactions on Communications,* 36(6):734–743.

Waxman, B.M. 1988. Routing of multipoint connections. *IEEE Journal on Selected Areas in Communication,* 6(9):1617–1622.

Waxman, B.M. 1993. Performance evaluation of multipoint routing algorithm. In *Proceedings of IEEE INFOCOM'93,* pp. 981–986. IEEE Computer Society Press.

Winter, P. 1987. Steiner problem in networks: A survey. *NETWORKS* 17:129–167.

Verma, D.C., and Gopal, P.M. 1993. Routing reserved bandwidth multi-point connections. *Proceedings of SIGCOMM'93-Ithaca,* pp. 96–105. New York.

Chapter 7

Network Performance Models

NETWORK performance models are required in many phases of network development and operation. They can serve for comparison of different design options, optimization of routing algorithms, network resource allocation, network survivability tests, identification of unexpected problems, etc. In this chapter we discuss performance evaluation models for the connection layer where the connection class rejection rates are of main interest. Note that this topic is associated with the resource management issues only indirectly. In other words, while it can influence the design and analysis of the resource management algorithms it is not directly used by them. For this reason readers interested only in application of resource management algorithms can omit this chapter without loosing continuity.

Exact performance models are not practical for any network, except the smallest ones, due to the enormous cardinality of the problem defined by the number of possible exact network states. To make the problem manageable, one has to resort to decomposition techniques which can reduce significantly the problem complexity at the expense of accuracy. In Section 7.1 we describe the basic concepts of two decomposition techniques and their important features. Decomposition into link problems assumes statistical independence of link state distributions. It brakes down the network problem into a set of quasi-independent link problems. Decomposition into a set of quasi-independent path problems may also be applied to solve the network performance problem. In this case the correlation between the links constituting the path under consideration is taken into account. In both cases the solution is achieved in an iterative procedure based on the repeated substitution principle.

In general the decomposition techniques can be applied to networks with different CAC & routing strategies, assuming that they are given in advance. In the second part of this chapter (Section 7.2) we present a performance model for CAC & routing based on the reward maximization principle and state-dependent link shadow prices (Chapter 5 — MDPD model). Although the performance model is based on generic decomposition into link problems, there is an important distinction to this model. Namely, in this case the CAC & routing policy is not given in advance since the link shadow prices are a function of the network performance which is unknown. To cope with the problem, the policy iteration procedure, from the MDPD model, is merged with the performance model. In this approach the optimal policy and the network performance are evaluated at the same time.

7.1 Decomposition Methods

In this section we consider a generic network model. Here the network is defined as a set of nodes and a set of links connecting the nodes. The sth link is described by its capacity L^s. The network serves several connection classes. Connections from class j are characterized by their origin and destination (OD) node pair, exogenous arrival rate λ_j (assumed to form independent Poisson processes), mean holding time μ_j^{-1} (in some cases assumed to have exponential distribution), and "bandwidth" d_j required by the connection. It is assumed that the CAC & routing policy, π, is given, and may belong to any of the categories defined in Chapter 5. Moreover, we assume that the connection setup time is negligible; that the rejected connections do not return to the system; that the nodes are transparent (no connection losses in the switches); and that the network is in statistical equilibrium.

Note that in general the "bandwidth" d_j can be link dependent (e.g., equivalent bandwidth is a function of the link speed; see Chapter 3). Nevertheless, in the following we assume that d_j is constant in order to simplify presentation. Extension to a link-dependent d_j is straightforward.

The objective of the performance models presented in this chapter is to find the grade of service (GoS) represented by the connection class loss probabilities

$$B_j = 1 - \frac{\overline{\lambda}_j}{\lambda_j} \tag{7.1}$$

where $\overline{\lambda}_j$ denotes the average rate of the accepted connections. In some cases the objective can be limited to some average traffic loss probabilities defined as

$$B_{J_b} = \frac{\sum_{j \in J_b} \lambda_j \mu_j^{-1} d_j B_j}{\sum_{j \in J_b} \lambda_j \mu_j^{-1} d_j} \tag{7.2}$$

where J_b denotes a set of connection classes (e.g., all connection classes or connection classes offered to a particular OD pair).

In general the exact solution of the performance problem can be found by solving a set of equilibrium equations derived for steady state probabilities where the number of equations is equal to the number of states. In our case the exact state is described by a matrix $\mathbf{z} = [z_j^k]$, where z_j^k denotes the number of class j connections, $j \in J$, carried on path k from the set of admissible paths W_j. Since the number of connection classes and maximum number of connections from a given class can be considerable, in most cases the number of states is so large that the solution of

7.1.1 Decomposition into link problems

Let us consider the system behavior from a particular network link viewpoint. This link serves several connection classes, each of them described by a set of parameters and functions, A_j^s, including link arrival process distributions. Note that if the link arrival process distributions are known, the link performance can be evaluated from a link model (henceforth referred to as *link performance function*)

$$\Pi^s = f_p(\mathbf{A}^s) \qquad (7.3)$$

where Π^s is the set of required link performance characteristics (e.g.,. state probabilities, or macro-state probabilities, or blocking state probabilities) and $\mathbf{A}^s = [A_j^s, j \in J^s]$. Then the network performance can be easily evaluated based on the link performance characteristics (e.g., carried traffic). Note that the link performance model is much simpler than the network model since the number of link states is significantly smaller than the number of network states. Nevertheless at the moment this gain is deceptive due to the fact that the exact link arrival process distribution is a function of the exact network state distribution.

To take advantage of the simpler link performance models one has to assume statistical independence of the link occupancy distributions. This key assumption allows one to evaluate the link arrival process distributions based on the performance characteristics of all links, $\Pi = [\Pi^s, s \in S]$ and the applied CAC & routing policy. From our selected link viewpoint this relation can be expressed as the following *link loading function*:

$$A_j^s = f_l(\Pi, \pi) \qquad (7.4)$$

The link loading function reflects the fact that although we assumed statistical link independence, there is still a dependence among the link distributions. To avoid confusion we refer to this effect as a functional link dependence.

Observe that the link performance function relates each link performance characteristics Π^s to the link input $\mathbf{A}^s$; conversely, the link loading function defines each link input $\mathbf{A}^s$ as a function of the performance characteristics of all links Π and a given routing policy π. Thus these two functions define a set of implicit non-linear equations:

$$\mathbf{A}^s = f_l(\Pi, \pi), \quad s \in S \qquad (7.5)$$
$$\Pi^s = f_p(\mathbf{A}^s), \quad s \in S \qquad (7.6)$$

In the literature this set of equations is often referred to as the fixed-point equations. In general finding a solution of the fixed-point equations is a difficult task. One could apply standard optimization methods such as Newton's method or one of the non-linear minimization techniques; for more details see (Girard, 1990). Unfortunatelly, in most cases these methods are not practical due to the required evaluation of gradients and low numerical efficiency. The best alternative is an iterative approach known as *repeated substitution*. In this case we start from an arbitrary initial link loading (e.g., based on direct-link-path routing policy). Then the link performance functions and link loading functions are evaluated iteratively until the variables converge. The method is simple and efficient although the convergence cannot be guaranteed due to a possibility of variable oscillations or divergence. Nevertheless, in most practical cases, the solution is achieved if the pace of convergence is regulated by a "damping factor." In this case a variable y_i evaluated in iteration i is modified as follows:

$$y'_i = y_{i-1} + \beta(y_i - y_{i-1}) \qquad (7.7)$$

where $\beta = (0,1)$ denotes the damping factor. The damping factor value and the variables affected by this factor can be chosen experimentally.

The issues of the solution's uniqueness, independence assumption errors, and influence of the CAC & routing algorithm on the model complexity are discussed in Section 7.1.3.

7.1.2 Decomposition into path problems

Let us consider one of the paths used by the routing algorithm for carrying connections offered to an OD pair. Figure 7.1a illustrates a state of the path expressed in terms of "bandwidth" x^k seized by the OD connections carried on the considered path, and "bandwidths" x^s seized by *background* connections on each of the links constituting path k. Now consider the path state as "seen" by a new connection offered to this path. One can assume that the arrival "sees" the path as if it was one link of dimension $L^e = \min_{s \in S^k}\{L^s\}$ with "bandwidth" x^k seized by the OD traffic and "bandwidth" x^e seized by the *background* traffic where $x^e = L^e - \min_{s \in S^k}\{L^s - x^s\}$. This "view" through the path is shown in Figure 7.1b.

The system depicted in Figure 7.1b suggests that to evaluate performance of the OD stream one may analyze an equivalent link shown in Figure 7.2 instead of the entire path. Here the equivalent link of dimension L^e serves the OD stream offered to path k and an equivalent background stream. To analyze this system we need a description of the equivalent background stream. To achieve exact equivalence with the original system, the equivalent background stream parameters should ensure that the state probability distribution of the OD connections is the same in the equivalent link and the original system. While it is difficult to find the exact solution to this problem, one can use an approximation based on the definition of the offered traffic distribution. This distribution is defined as the state distribution on a link with infinite capacity which is offered the considered stream. According to this definition, the state distribution of the equivalent background offered traffic on the infinite equivalent link can be defined as

$$Pr\{x^{e'} = m\} = Pr\{\max_{s \in S^k}\{x^{s'} - \Delta L^s\} = m\} \qquad (7.8)$$

7.1 Decomposition Methods

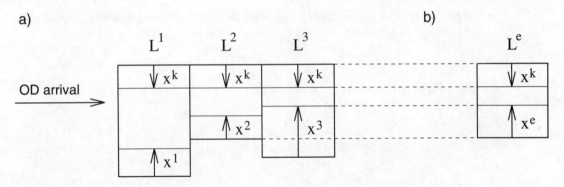

Figure 7.1. *"View" through the path.*

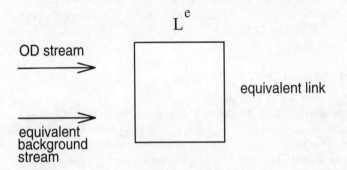

Figure 7.2. *Equivalent path model.*

where $x^{e'}$ denotes the state expressed as the overall "bandwidth" seized by the equivalent background connections, $\Delta L^s = L^s - \min_{s \in S^k}\{L^s\}$, and $x^{s'}$ is the state of background connections on link s when $L^s = \infty$.

Using Equation (7.8), the state distribution of the equivalent background offered traffic can be evaluated based on the product form solution, assuming that the arrival rates of the background streams are described by the Bernoulli-Poisson-Pascal distribution; for details see (Dziong, 1985). Having parameters of the equivalent background offered traffic, the performance of the OD pair stream can be evaluated from the equivalent link model by using a general link performance model. The above procedure defines the *path performance function*

$$\Pi^k = f_e(\mathbf{A}^k) \tag{7.9}$$

where $\mathbf{A}^k$ denotes the path input parameters including the OD stream and all background streams. Observe that the path performance function takes into account the correlation between the path's link states induced by the common OD stream and other streams using more than one link in the path.

The path performance function can be used to evaluate the network performance. Namely by assuming statistical independence of the path occupancy distributions, we can decompose the network performance problems into a set of path performance functions which are coupled by the *path loading functions*

$$\mathbf{A}^k = f_l(\mathbf{\Pi}_e, \pi) \qquad (7.10)$$

where $\mathbf{\Pi}_e = [\Pi^k, k \in \mathbf{W}]$ and $\mathbf{W} = \{W_j : j \in J\}$. This approach results in the following set of implicit non-linear equations:

$$\begin{align} \mathbf{A}^k &= f_l(\mathbf{\Pi}_e, \pi), \quad k \in \mathbf{W} & (7.11) \\ \Pi^k &= f_e(\mathbf{A}^k), \quad k \in \mathbf{W} & (7.12) \end{align}$$

As in the case of the decomposition into link problems, these equations can be solved by applying a repeated substitution algorithm.

While decomposition into path problems takes into account some of the correlation existing on the paths, this advantage is offset by increased complexity since the number of paths is usually higher than the number of links and the equivalent path model is more complex than the link model.

7.1.3 General features of the decomposition techniques

The properties of solutions based on the decomposition techniques depend on many factors, the most important being the type of loading function, and CAC & routing policy used in the network. There are three main categories of the loading functions which differ in the description of the connection streams offered to the link (or path).

- *One moment description*: In this case the connection process distributions are characterized by the first moment. Poissonian distribution is the most commonly used model in this category.

- *Two moment description*: Here the connection process distributions are characterized by the first two moments. Equivalent random theory (ERT) is a well known model falling into this category (Wilkinson, 1956).

- *Descriptions being a function of the link state*: There are three basic options in this category.

  - *Two macro-state description*: To take into account a connection admission policy algorithm based on the "bandwidth" reservation for higher priority connection classes (see Section 8.2.1) the connection stream is assumed to have no arrivals in the states where the free "bandwidth" is reserved for other connection classes (blocking macro-state). For the admissible states (admissible macro-state) a one or two moment description can be used.

  - *Linear-state dependent description*: Here the connection arrival rate is assumed to depend linearly on the number of connections carried on the link, $\lambda(x) = \alpha + x\beta$ (Bernoulli-Poisson-Pascal distribution). This approach usually overlaps with the two moment description since in most applications the Bernoulli-Poisson-Pascal distribution parameters are evaluated from the two first moments of the offered traffic distribution.

- *State-dependent description*: The state-dependent arrival rates are evaluated independently for each link state. A link loading function based on the state-dependent description is presented in Section 7.2. To reduce complexity a set of macro-states can be considered.

The choice of the connection stream description category depends on the applied CAC & routing algorithm, and the chosen compromise between the model's complexity and accuracy. For example, in general, the two moment description gives better results than the one moment description. On the other hand, if a state-dependent CAC & routing algorithm is used, the best accuracy will be achieved when a corresponding state-dependent description of the link arrival rates is applied, although one or two moment models can also be used. In the following we discuss important aspects of the solutions based on decomposition techniques.

Solution uniqueness. In general there can be more than one solution to the set of fixed-point equations if alternate routing is applied in the network. This can be seen as a surprising feature in view of the fact that in most cases the exact network model can be described as a Markov process which has a unique solution. A rough explanation is that the multiple solutions result from the link independence assumption and approximations used in the loading and performance functions. Nevertheless there is an underlying physical phenomenon which sheds more light on the issue of solution uniqueness.

Consider a network with alternate routing and a simple CAC algorithm which accepts connections whenever there is sufficient free capacity. Figure 7.3 depicts a possible relation between the overall traffic losses in the network and the level of the traffic offered to the network in a simulation experiment where the offered traffic is first gradually increased and then gradually decreased; for an example of the exact relation see (Girard, 1990). The observed hysteresis effect is explained by the fact that for a given traffic input, in the hysteresis region, the network can operate in one of the two meta-stable network modes. In the first one, most of the new connections are carried on the corresponding direct-link-paths which results in a low connection loss probability. In the other mode most of the new connections are carried on multi-link paths. Thus the connections consume more resources resulting in a high connection loss probability. Note that once the system is in the high loss mode, the situation is perpetuated by the fact that in this case the direct-link-paths are more likely to be blocked. This effect causes the system to switch from the high loss mode to the low loss mode, and vice versa, relatively rarely under stationary traffic conditions.

The existence of the hysteresis effect in some network examples provides a plausible explanation for multiple solutions of the fixed-point equations. First note that the exact analytical model based on the exact state description would give an overall loss probability somewhere between the values corresponding to the

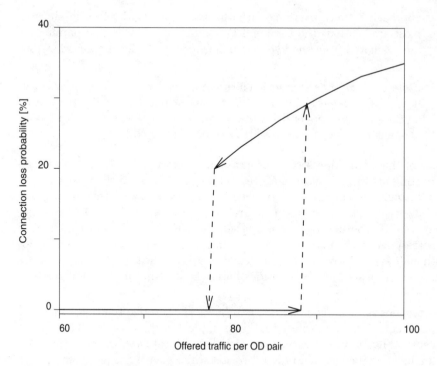

Figure 7.3. *Hysteresis effect.*

high and low loss modes. Thus one can argue that due to a certain relaxation of the performance problem constraints in the fixed-point equations (e.g., the link independence assumption) the solution can converge to one of the three values corresponding to high losses, low losses, and average losses, respectively. Obviously, the shape and location of a hysteresis evaluated from an analytical model can be different from the corresponding values in a real network due to the applied approximations.

While in the literature there is no direct proof of the above conjecture, it is supported by two phenomenona. First, as shown in (Mason and Gu, 1986; Girard, 1990), the fixed-point equations can give three solutions. Second, in the networks with the CAC_{GoS} algorithm providing sufficient priority for direct-link-path connections, both the multiple solutions and hysteresis effect cease to exist (Girard, 1990). The last feature indicates that the CAC_{GoS} algorithm can have important influence on the network performance.

Independence assumption analysis. The influence of the independence assumptions on the performance model accuracy depends on many factors, including the network configuration, offered traffic matrix, link (or path) loading and performance functions, and applied CAC & routing scheme. We start by discussing some generic features which give a qualitative understanding of the problem. Then, simple numerical examples are used for a quantitative illustration.

Let us describe the link state in terms of the available link "bandwidth." The correlation between the link states is positive in most cases. It is induced by connections carried on paths composed of many links, and by routing algorithms which distribute connections among the set of available paths until all paths are blocked. The degree of correlation depends on the share of traffic using the multilink paths and the level of network utilization. In general the smaller the share of multi-link traffic on network links, the smaller the correlation. The network utilization influences the correlation between different paths and this influence may not be monotonic. In particular, when the network utilization is very small, the routing algorithm uses direct-link-paths or best paths which are composed of a small number of links. Thus the correlation level is small. When the traffic level is increased the routing algorithm offers connections to all available paths, which increase the correlation among the paths and correlation between the links constituting the paths since more multi-link paths are used. However, when the traffic level is very high a good CAC & routing algorithm should avoid multi-link connections in order to reduce the overall traffic loss probability. In this case the correlation between paths will be reduced since most of the traffic will be carried on the direct-link-paths or the shortest paths.

The relation between the level of correlation and the accuracy of the connection class loss probabilities, B_j, evaluated from the fixed-point equations, is not straightforward except for very specific network examples. To discuss this issue let us consider a set of alternative paths, W_j, which can be used to carry a single-rate connection class offered to a particular OD pair. Let b^s and b^k denote the probabilities of link s and path k being in the blocking states, respectively. The influence of the correlation on the accuracy of the decomposition methods can be illustrated by considering two extreme cases. The first corresponds to the link independence assumption. The second assumes full synchronization between all link blocking states (maximum positive correlation). The relation between the path blocking state probabilities for these two cases, assuming the same link blocking state probabilities, is given by

$$b^k = 1 - \prod_{s \in S^k}(1 - b^s) \quad > \quad b^{k'} = \max_{s \in S^k} b^s \qquad (7.13)$$

where the prime indicates blocking for the synchronized case. The relation between the connection class blocking probabilities for these two cases, assuming the same path blocking state probabilities, is defined by

$$b_j = \prod_{k \in W_j} b^k \quad < \quad b'_j = \min_{k \in W_j} b^k \qquad (7.14)$$

Obviously, under the Poissonian assumption of the connection arrival process, the connection class loss probability equals the connection class blocking probability ($B_j = b_j$). Relations (7.13) and (7.14) indicate that the connection class blocking probabilities under the link independence assumption are influenced by two opposite effects. The assumption of link independence on the multilink path increases the path blocking probabilities. This error is becoming larger with the increase of

TABLE 7.1 OD pair loss probabilities [%].

model	stream description	nominal	overload
simulation		.25 - .69	7.3 - 10.8
decomposition into path problems	two moment	.30	9.3
decomposition into link problems	two moment	.22	13.5
decomposition into link problems	one moment	.01	12.3

the number of links on the path, the link blocking probabilities, and the level of correlation among the links. On the other hand, the assumption of path independence reduces the connection class blocking probabilities. This error grows with the increase of the number of paths, the level of correlation between the paths, and with the decrease of the link blocking probabilities. Thus, in general, one can expect a partial compensation of errors caused by the link independence assumption. The net effect depends on the number of paths and links in the paths, the levels of correlation between the links, and the levels of the link blocking probabilities.

Accuracy analysis. Obviously relations (7.13) and (7.14) cannot serve for a quantitative assessment of the errors since in realistic network examples the link blocking states are far from being fully synchronized. Moreover, the approximations used in the link (or path) loading and performance functions also introduce errors which makes the accuracy analysis difficult. Nevertheless to give a taste of such an analysis, we compare several decomposition models with a simulation in Table 7.1. The model based on decomposition into path problems (*Equivalent path model* — EPM) was presented in (Dziong, 1985) and is based on a two moment description of the connection streams. The models based on decomposition into link problems were derived in (Pióro, 1983) (two moment description) and in (Lin, Leon, and Stewart, 1978) (one moment description). The network example is fully symmetrical with five nodes and single-rate connections. Each OD pair connection can be served by its direct-link-path or one of the three two-link alternative paths, chosen according to the sequential routing scheme. There is no CAC_{GoS} mechanism to give priority for connections using direct-link-paths. The results for simulation model are given as 95% confidence interval.

It is interesting that although the decomposition into path problems gives results within the confidence interval, under both the nominal and overload conditions, the models based on decomposition into link problems give an underestimation and overestimation of the loss probabilities under the nominal and overload conditions, respectively. A possible explanation is that, under the nominal conditions with low blocking probabilities, most connections are carried on the direct-link-paths so the correlation error is small and is caused mainly by the path independence assumption. In this case the differences between the models are caused by approximations used in the loading and performance functions. The larger underestimation in the one moment model is consistent with the fact that this model underestimates the variance of the traffic offered to the alternative paths. The best performance of the

EPM model can be explained by the fact that from the particular OD pair viewpoint there are fewer approximations used in this case (four path models compared to seven link models). On the other hand, under the overload conditions, there are more multilink connections which cause the link independence error, resulting from the correlation between the links constituting a path, to be more dominant. Since the EPM model takes into account this correlation, its performance falls in the confidence interval. The models based on decomposition into link problems give an overestimation of the loss probabilities, which is consistent with relation (7.13).

7.2 Performance Model for Reward Maximization Routing

In this section[1] we describe an approach for performance evaluation of networks with state-dependent routing based on the reward maximization principle (the MDPD model described in Chapter 5). The approach is based on decomposition into link problems. Since the routing policy is defined by the matrix of state-dependent link shadow prices, $\pi^* = [p_j^s(\mathbf{x}, \pi^*), j \in J^s, s \in S]$, a consistent link model should be defined by the state-dependent Poissonian arrival rates, $\lambda_j^s(\mathbf{x})$, and by the link state probabilities, $Q^s(\mathbf{x})$. In this case the link loading and performance functions, respectively, can be described as

$$\lambda_j^s(\mathbf{x}) = f_l(\mathbf{Q}, \pi^*), \quad j \in J, \ s \in S \quad (7.15)$$
$$Q^s(\mathbf{x}) = f_p(\Lambda^s), \quad s \in S \quad (7.16)$$

where $\mathbf{Q} = [Q^s(\mathbf{x}), s \in S]$ and $\Lambda^s = [\lambda_j^s(\mathbf{x}) : j \in J^s]$. The main difference compared with the set of fixed-point equations defined by Equations (7.5) and (7.6) is that the CAC & routing policy (determined by the values of shadow prices) is not given in advance. Thus Equations (7.15) and (7.16) cannot be solved in their present form.

The natural solution would be to find the shadow prices from the policy iteration procedure used in the MDPD model. The main step in this procedure is the evaluation of the improved shadow prices. They are evaluated for each link based on the link input defined by the arrival rates, $\lambda_j^{s'}(\mathbf{x})$, and the link connection reward parameters, r_j^s, where prime indicates that the arrival rates in the shadow price model can be different from the link performance model (e.g., non-state-dependent description). In the MDPD algorithm these values are calculated based on measured statistics and a simplified analytical model. Since the measured statistics are functions of the current control policy, π, and the network state distributions, in the decomposed model we have

$$\{\lambda_j^{s'}(\mathbf{x}), r_j^s\} = f_l^\pi(\mathbf{Q}, \pi), \quad j \in J, \ s \in S \quad (7.17)$$

---

[1]Portions of this section are reprinted, with permission, from (Dziong, Mignault, and Rosenberg, 1993).

Then the improved policy, π', is defined by the new values of shadow prices evaluated from the link shadow price model

$$p_j^s(\mathbf{x}, \pi') = f_p^\pi(\Lambda^{s'}, \mathbf{r}^s), \quad j \in J, \ s \in S \tag{7.18}$$

where $\mathbf{r}^s = [r_j^s, j \in J^s]$. Observe the similarity of Equations (7.17) and (7.18) with the fixed-point equations. Namely Equation (7.17) can be interpreted as the link loading function and Equation (7.18) can be interpreted as the link performance function for shadow price evaluation. Another similarity is that the final policy, π^*, cannot be evaluated from these equations, since the model for $Q(x, \pi)$ evaluation is not included.

Note the complementary structure of the fixed-point equations [(7.5), (7.6)] and the policy iteration equations [(7.17), (7.18)] with respect to unknown variables. This feature suggests solution of both equation sets in one iterative procedure applied to the following extended set of fixed-point equations:

$$\lambda_j^s(\mathbf{x}) = f_l(\mathbf{Q}, \pi), \quad j \in J, \ s \in S \tag{7.19}$$

$$Q^s(\mathbf{x}) = f_p(\Lambda^s), \quad s \in S \tag{7.20}$$

$$\{\lambda_j^{s'}(\mathbf{x}), r_j^s\} = f_l^\pi(\mathbf{Q}, \pi), \quad j \in J, \ s \in S \tag{7.21}$$

$$p_j^s(\mathbf{x}) = f_p^\pi(\Lambda^{s'}, \mathbf{r}^s), \quad j \in J, \ s \in S \tag{7.22}$$

Assuming that, by applying repeated substitution, a solution is achieved, both the optimal policy and its performance will be evaluated.

In the following we describe a particular realization of the link functions. To reduce the model's complexity, two simplified link state descriptions are used. First, the link state is defined as the overall "bandwidth" seized by the connections carried on the link. In the second model, the state description is further simplified by introducing macro-states. The accuracy of both models is illustrated by comparison with simulations.

7.2.1 Link loading and performance functions*

Link loading function (performance model). The objective of the link loading function is to evaluate the state-dependent arrival rates, $\lambda_j^s(x)$, where x is the "bandwidth" seized by the connections carried on the link. We assume that due to the "bandwidth" quantization (Section 5.4) x and L^s are integer numbers where $x = 1$ corresponds to the "bandwidth" seized by the connection with the smallest "bandwidth" requirement. Let us define the path shadow price as

$$p_j^k(\mathbf{y}) = \sum_{s \in C^k} p_j^s(x) \tag{7.23}$$

where $\mathbf{y}$ denotes the path state. Since the new connections are offered to the path with maximum positive net-gain, the state-dependent arrival rates depend on the path shadow price distributions on all paths belonging to the set W_j. To simplify the presentation, in the following, we assume that the paths in W_j do not have common links (this is not a restriction of the approach). Let $P_j^s(x)$ denote the probability of routing class j connection on a path k which contains link s in state x. Once these routing probabilities are known, the arrival rates are given by

$$\lambda_j^s(x) = P_j^s(x) \cdot \lambda_j \qquad (7.24)$$

Let us define

$$G_j^l(v) = Pr\{p_j^l(\mathbf{y}) > v\} \qquad (7.25)$$

as the complementary shadow price cumulative distribution function for path l, under the definition that for the blocking states (no sufficient "bandwidth" to carry the connection) we have $p_j^l(\mathbf{y}) = \infty$. These distribution functions are easily obtained assuming that the link shadow price values and state probability distributions are given. Then the routing probabilities $P_j^s(x)$ can be expressed as

$$P_j^s(x) = \sum_{\mathbf{y} \in Y_s^k(x)} \delta(r_j - p_j^k(\mathbf{y})) \prod_{l \in W_j \setminus \{k\}} G_j^l(p_j^k(\mathbf{y})) \cdot \prod_{c \in k \setminus \{s\}} Q^c(x) \qquad (7.26)$$

where $\delta(a) = 0$ for $a \leq 0$ and $\delta(a) = 1$ for $a > 0$, $Y_s^k(x)$ denotes the set of all possible path k states for given state x on link s, l is the index of alternative paths except path k, and c denotes the index of links constituting path k except link s.

Link performance function (performance model). In general, an exact link performance function requires an exact multidimensional link state description, defined by the number of connections from each class, and involves solution of a set of linear equations where the number of equations is defined by the number of states. In order to reduce the problem complexity one can apply an approximation based on the one-dimensional recursion introduced in (Kaufman, 1981) and (Roberts, 1981) for steady Poissonian arrival rates:

$$x \cdot Q^s(x) = \sum_j \frac{d_j}{\mu_j} \lambda_j^s Q^s(x - d_j) \qquad (7.27)$$

This relation together with the normalizing condition

$$\sum_{x=0}^{L^s} Q^s(x) = 1 \qquad (7.28)$$

gives exact stationary probabilities for steady Poissonian arrival rates. An approximation for the state-dependent case is achieved simply by substitution of the steady arrival rates with the state-dependent Poissonian arrival rates:

$$x \cdot Q^s(x) = \sum_j \frac{d_j}{\mu_j} \lambda_j^s(x - d_j) Q^s(x - d_j) \qquad (7.29)$$

This approximation was analyzed in (Gersht and Lee, 1990; Chung, Kaspher, and Ross, 1993). In (Gersht and Lee, 1990) it was shown that the approximation gives good results over a wide range of parameters for the two macro-state description of the streams offered to the link, while in (Chung, Kaspher, and Ross, 1993) it was proven that for a particular relation between the state-dependent arrival rates, the model gives exact results.

Link loading function (shadow price model). In the considered approach we use the steady arrival rates (SAR) model from Section 5.2.3, which simplifies the formulation for the arrival rates given by (5.22) to become

$$\lambda_j^s = \lambda_j^k \prod_{c \in S^k \setminus \{s\}} (1 - B_j^c) \tag{7.30}$$

We apply rule D2 (Section 5.2.3) to evaluate the link connection reward parameters:

$$r_j^s = \frac{r_j}{|S^k|} \tag{7.31}$$

where $|S^k|$ is the number of links in path k containing link s and used by connection class j.

Link performance function (shadow price model). Several options for link shadow price evaluation were described in Section 5.2.3. In the following numerical study we use a model based on the uni-dimensional link (UDL) state description described in Section 5.2.3. In this case all traffic classes are aggregated into one class with the smallest "bandwidth" requirement, $d_a = 1$. To provide the same variance of the overall traffic offered to the link, the arrival rate of the aggregated traffic is defined by the state-dependent Pascal distribution

$$\lambda^s(x) = \frac{\xi^2}{\sigma^2} + x \cdot \left(1 - \frac{\xi}{\sigma^2}\right) \tag{7.32}$$

where $\xi = \sum_j d_j \lambda_j^s \mu_j^{-1}$ is the average aggregate offered traffic and $\sigma^2 = \sum_j d_j^2 \lambda_j^s \mu_j^{-1}$ defines the variance of the aggregate offered traffic. Having a uni-dimensional state description, one could try to apply the recurrence, Equations (5.29)–(5.32), in order to find the shadow prices. Unfortunately this approach would result in negative shadow prices. This effect is caused by the fact that due to the system transformation, the state with larger number of connections induces larger arrival intensity, which is not the case in the original system. Thus, since the recurrence, Equations (5.29)–(5.32), is based on analysis of the reward from carried connections, increasing the number of connections in the transformed system can increase the reward from future connections due to increased arrival rates. To cope with the problem the transformed system should be used only to define the state distribution, while the shadow prices can be found from the analysis of the lost reward whose rate is given by

$$q(x) = \sum_{j \in \{j : d_j < L - x\}} r_j^s \lambda_j^s \tag{7.33}$$

7.2 Performance Model for Reward Maximization Routing

Based on this concept the following equations for the shadow prices can be derived (Hwang, Kurose, and Towsley, 1992):

$$p_a^s(L-1) = \frac{\sum_{j \in J^s} r_j^s \lambda_j^s}{\lambda^s(L-1)} \frac{E(\lambda^s, L)}{E(\lambda^s, L-1)} \qquad (7.34)$$

$$p_a^s(x) = \frac{w}{\lambda^s(x) E(\lambda^s, x)}, \qquad 0 \le x \le L-2 \qquad (7.35)$$

where

$$E(\lambda^s, x) = \frac{\frac{1}{x!} \prod_{m=0}^{x-1} \lambda^s(m)}{1 + \sum_{n=1}^{x} \frac{1}{n!} \prod_{m=0}^{n-1} \lambda^s(m)} \qquad (7.36)$$

and

$$w = \sum_j r_j^s \lambda_j^s - L p_a^s(L-1) \qquad (7.37)$$

Then the shadow prices for the original connection classes can be approximated by

$$p_j^s(x) = \sum_{m=x}^{x+d_j-1} p_a^s(m) \mu_a / \mu_j \qquad (7.38)$$

To compensate for the fact that the Pascal distribution does not take into account connection rejections in states $x \in [L - d_{\max} + 1, L - 1]$, a heuristic factor in Equations (7.34) and (7.35) for these states can be added; for details see (Hwang, Kurose, and Towsley, 1992).

7.2.2 Macro-state approximation*

The complexity of the performance evaluation model for the MDPD routing policy is defined in large part by the complexity of the link loading function (performance model). In particular evaluation of the complementary cumulative distribution functions for path shadow prices, Equation (7.25), involves convolution of the link shadow price distributions, which results in $\prod_{s \in S^k} x^s$ operations for each path. Moreover, evaluation of each routing probability $P_j^s(x)$ requires addition of $\prod_{o \in S^k \setminus \{s\}} x^o$ elements, Equation (7.26). These numbers can be very large since in general the links can carry hundreds of connections with $d_j = 1$.

The problem can be significantly simplified by introduction of macro-states in the shadow price functions. In this case the nth macro-state is defined by aggregation of link states with shadow price falling into the interval $[(n-1)\Delta_p, n\Delta_p)$ where Δ_p is the interval width. The macro-state probability is given by

$$Q_j^s(n) = \sum_{x \,:\, p_j^s(x) \in [(n-1)\Delta_p, n\Delta_p)} Q^s(x) \qquad (7.39)$$

In terms of the shadow price value in macro-state n, we consider two options:

$$p_j^s(n) = \frac{\sum_{x\,:\,p_j^s(x)\in[(n-1)\Delta_p, n\Delta_p]} p_j^s(x) Q^s(x)}{Q_j^s(n)} \qquad (7.40)$$

and

$$p_j^{s'}(n) = (n - 0.5)\Delta_p \qquad (7.41)$$

While the second option is an approximation, it further reduces complexity of the complementary cumulative distribution function, $G_j^k(v)$, evaluation. Namely if the expected values are used, Equation (7.40), the convolution of the link shadow price distribution can give $n^k = \prod_{s\in S^k} n^s$ different discrete values of the path shadow price. The second option, Equation (7.41), results in $n^k = \sum_{s\in S^k} n^s - (m_k - 1)$ values which are separated by the same interval. In the following the second option is used for $G_j^k(v)$ evaluation.

The discrete convolution of link shadow price distributions could be used directly to evaluate $G_j^k(v)$, resulting in a step-wise function. To achieve a continuous approximation one can transform the discrete convolution into a probability density function. Two approximations are illustrated in Figures 7.4b and 7.4c, where convolutions for a two link path example are plotted (five link macro-states are used). Additionally, the convolution based on the original state description is presented in Figure 7.4a in the form of the continuous line joining the discrete values. The first approximation (Figure 7.4b) is achieved by a simple linear interpolation of the discrete values. The second approximation (Figure 7.4c) takes into account the fact that each discrete value corresponds to aggregation of the original path shadow prices in the two neighbouring intervals. By assuming that the original path shadow prices are distributed continuously and uniformly in these intervals, the step-wise density function is achieved. The quality of both approximations is illustrated in Figure 7.4d, where the complementary cumulative distribution functions are shown for the three cases illustrated in Figures 7.4a, 7.4b, and 7.4c. Although the first approximation gives a slightly better match, the second approximation is attractive from the complexity and memory requirement viewpoints and is used in the subsequent numerical examples.

The macro-state concept can also be used to reduce evaluation complexity of the routing probabilities. In this case we arrive at

$$P_j^s(n) = \sum_{\mathbf{n}\in N_s^k} \delta(r_j - p_j^k(\mathbf{n})) \prod_{l\in W_j\setminus\{k\}} G_j^l\left(p_j^k(\mathbf{n})\right) \cdot \prod_{c\in S^k\setminus\{s\}} Q^c(n) \qquad (7.42)$$

where $\mathbf{n} = [n^s, s \in S^k]$ denotes the macro-state of path k and N_s^k denotes the set of all possible macro-states on path k for given macro-state n on link s. The values of macro-state path shadow prices used in Equation (7.42) can be evaluated based on Equation (7.40) since in this case the approximation (7.41) has no influence on the complexity order. Note that the macro-states used for simplification of the routing probability evaluation can be based on interval Δ_p', which is different from the one used to simplify evaluation of $G_j^l(v)$.

7.2 Performance Model for Reward Maximization Routing

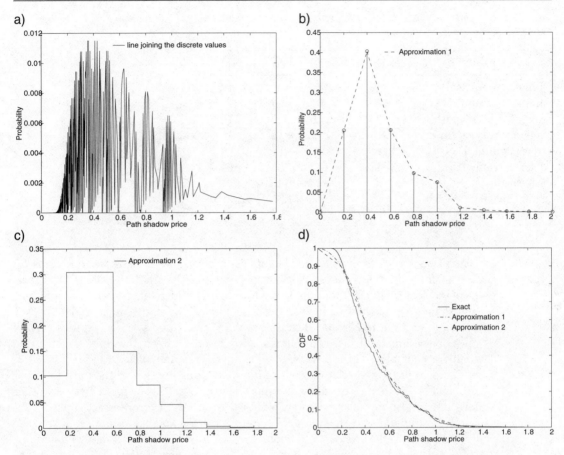

Figure 7.4. *Convolution functions (a,b,c) and corresponding complementary distribution functions, CDF, (d).*

7.2.3 Numerical study

In this section important features of the performance model for networks with reward maximization routing are illustrated by comparison with simulation. The description of the selected multi-rate and single-rate network examples is given in Table 7.2. The structure of Example W6A' is illustrated in Figure 8.9 including levels of traffic offered to OD pairs (traffic offered to the 0-3 OD pair is a modification considered in Chapter 8 only). It should be mentioned that the shadow prices in the simulation model are evaluated from the class oriented transformation (COT) model (see Section 5.2.3), which is different from the one used in the performance model (UDL model–Section 5.2.3). This may cause some additional errors in the analysis of the multi-rate examples (both models give the same result for single rate examples).

TABLE 7.2 Description of network examples.

network example	W7S′	W6A′	N7A	N7S
fully symmetrical	yes	no	no	yes
number of nodes	7	6	7	7
required bandwidth	1,5,13,22	1, 6	1	1
average service time	1,3,8,10	1, 1	1	1
class i traffic [%]	25,25,25,25	60,40		
overall traffic [Erl.]	3704	140	1137	861
overload [%]	+15	+15	+10	+10
link capacity	240	0-50	7-170	50
$r'_j = r_j \mu_j / d_j$	1	1	1	1

Overall losses. The overall traffic losses vs. overload level are shown in Figures 7.5 and 7.6 for examples N7S and W7S′, respectively. Although both network examples have the same configuration, the losses are slightly underestimated in example N7S and slightly overestimated in example W7S′, especially in the range under 10%. A plausible explanation is that in example W7S′ the wide-band connections carried on a multi-link path introduce larger correlation between the links constituting the path compared to the narrow-band connections. As indicated in Section 7.1.3, an increase of this correlation causes a larger overestimation of the path blocking probabilities in models based on decomposition into link problems. Obviously other factors can also cause the difference between the model errors in examples W7S′ and N7S. In particular the link performance model is an approximation in the case of multi-rate traffic. Also, the difference in the link shadow price model for multi-rate case in the analytical and simulation models may have an influence on the performance results.

Connection class losses. Accuracy of the network performance model with respect to the connection class loss probabilities is illustrated in Figures 7.7 and 7.8 for single rate traffic examples. Figure 7.7 shows individual loss probability values for each OD pair in the N7A network (overload conditions) while Figure 7.8 illustrates loss probability for a particular OD pair connection class as a function of its reward parameter r_j. In both cases the analytical model gives results close to the simulation although it can be observed that the larger loss probabilities are overestimated, while the smaller loss probabilities tend to be underestimated. This feature is consistent with the link independence error analysis presented in Section 7.1.3. The average loss probabilities for connection classes with different "bandwidth" requirements are given in Table 7.3 for example W7S′.

Macro-state approach. It is obvious that accuracy of the macro-state approximation depends on the number of link macro-states. The numerical results indicate that if there are ten or more macro-states in each link model the results are close to the original state model. This feature is illustrated in Figure 7.9 where narrow-

7.2 Performance Model for Reward Maximization Routing

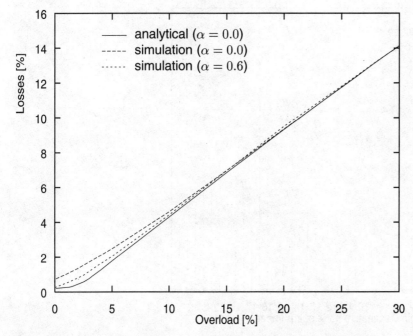

Figure 7.5. *Loss probability vs. overload level (Example N7S)*.

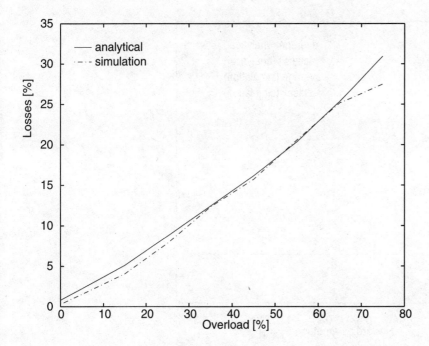

Figure 7.6. *Loss probability vs. overload level (Example W7S')*.

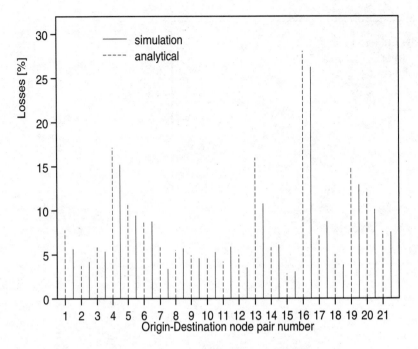

Figure 7.7. *Losses for each o-d pair connection class (Example N7A – overload).*

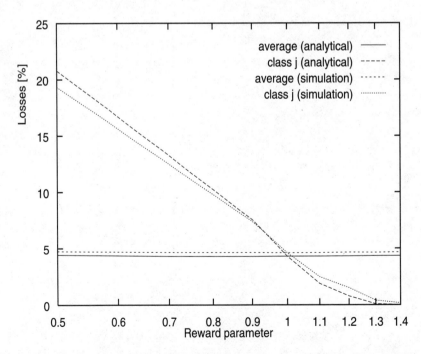

Figure 7.8. *Loss probability as a function of the reward parameter (Example N7S).*

TABLE 7.3 Loss probabilities [%] (Example W7S').

	nominal		overload	
	analytic	simulation	analytic	simulation
B_1	0.09	0.00	0.05	0.00
B_2	0.10	0.00	1.06	0.01
B_3	0.47	0.03	5.22	1.79
B_4	2.68	1.22	14.53	14.47
$\overline{B}$	0.81	0.31	5.09	4.04

band traffic losses are presented as a function of the number of macro-states in example W6A' (the values are given with step 1 until five macro-states and with step 5 afterwards).

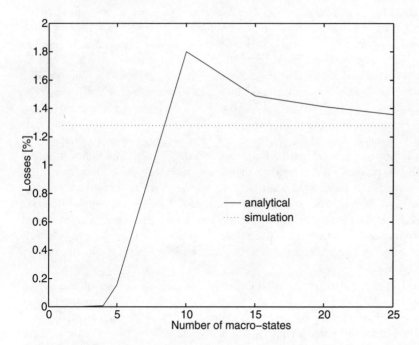

Figure 7.9. *NB traffic losses vs. number of macro-states (Example W6A').*

7.3 Discussion and Bibliographic Notes

The subject of performance evaluation on the connection layer has been studied for a long time in the context of single-rate circuit switched loss networks with sequential routing (e.g., telephone networks). Most of these works focus on decomposition of the network problem into a set of link problems and apply one or two moment traffic stream description, e.g. (Katz, 1967; Lin, Leon, and Stewart, 1978; Manfield and Downs, 1979; Pióro, 1983; Akinpelu, 1984). Other decomposition techniques (e.g., decomposition into path problems) usually give better accuracy but the increased complexity makes them less attractive in practical applications for large networks. A detailed description and analysis of different decomposition techniques resulting in the fixed-point equations is given in (Girard and Bell, 1989).

The issue of multiple solutions of the fixed-point equations, although intriguing, is not critical in practical cases due to the fact that their presence is confined to specific cases without appropriate CAC & routing mechanisms. As described in this chapter, the multiple solution phenomenon can be rationalized by existence of two meta-stable modes of network operation with very small transition rates between the modes. Under this assumption the fixed-point model was enhanced in (Marbukh, 1993) by taking into account the transition between the modes.

Implementation of state-dependent routing algorithms, based on the least loaded path (LLP) concept, in public networks created a need for performance models being able to cope with the state dependence phenomenon. The basic idea of these models is also based on the link statistical independence assumption and repeated substitution. The main difference lies in the link loading functions which can be divided into two categories. In the more detailed approach, state-dependent link arrival rates are evaluated based on the link state distributions constituting the alternative paths (Wong and Yum, 1990). To reduce complexity one can introduce macro-states as shown in (Mitra, Gibbens, and Huang, 1993). In the second category of link loading functions for state-dependent routing, the state dependence phenomenon is not modeled directly. For example, in (Girard and Bell, 1989) the average routing probabilities are used. In this case the state-dependent routing is approximated by a corresponding load sharing strategy which makes the results less accurate. In (Régnier, Blondeau, and Cameron, 1983) the cluster approach was applied for OD connection class loss evaluation, which gives better accuracy at the expense of increased complexity. More details on the cluster approach, which is basically limited to two link alternative paths, can be found in (Legall, Bernussou, and Garcia, 1985).

In general the link performance models based directly on the set of state equilibrium equations do not provide explicit functions which could be easily used to analyze and optimize network parameters. Approximations of such explicit functions can be achieved by considering the system behavior in asymptotic regimes where certain parameters become simultaneously large. In (Marbukh, 1991; Mitra, Gibbens, and Huang, 1993) asymptotic models for symmetric networks with LLP type state-dependent routing are presented. In (Kelly, 1994) a procedure for asymptotic bounding the performance of dynamic routing schemes is given. The bound is particularly appropriate for large highly connected networks. Although the asymptotic models are rather limited to large and well connected networks, they are, in general, less time consuming and can give interesting insight into some functional relations between the performance and the system parameters.

Multirate connection environments (B-ISDN, ATM) increase significantly the cardinality of the exact link or path performance models. This complexity can be significantly reduced by application of recurrence solutions proposed in (Roberts, 1981; Kaufman, 1981) for link model and (Dziong and Roberts, 1987) for path model. These models can also be used to approximate link performance in the case of state-dependent CAC & routing strategies (Gersht and Lee, 1990; Chung, Kaspher, and Ross, 1993). In (Gazdzicki, Lambadaris, and Mazumdar, 1993; Mitra and Morrison, 1994) efficient approximations based on asymptotic models are given for the case of steady Poissonian arrival rates. Although most of the network performance models for single-rate and multi-rate cases are based on the same decomposition technique (decomposition into link problems), in general the accuracy is better in the case of single-rate connections. This is caused by increased correlation between links carrying the common wide-band connections, use of simplified multi-rate link performance models, and increased variance of link aggregated multi-rate connection streams (large errors in one moment methods).

In the case of reward maximization routing based on state-dependent link shadow prices, it is important to have a good assessment of individual connection class loss probabilities in order to optimize the connection reward parameters according to the CAC & routing objectives (fairness, efficiency). For this reason we analyzed in this chapter the performance model based on the state-dependent link arrival rates evaluated from the path shadow price distributions. Evaluation of these distributions contributes significantly to the model complexity. Nevertheless this contribution can be reduced considerably if the link macro-state model is used. Applications of this model to find an optimal network operating point and to dimension virtual networks is presented in Chapters 8 and 9, respectively.

References

Akinpelu, J. M. 1984. The overload performance of engineered networks with non-hierarchical and hierarchical routing. *AT&T Bell Laboratories Technical Journal* 63(7):1261-1281.

Butto, M., Colombo, G., and Tonietti, A. 1976. On point to point losses in communication networks. In *Proceedings of the 8th International Teletraffic Congress*. North-Holland.

Chung, S., Kaspher, A., and Ross, K. 1993. Computing approximate blocking probabilities for large loss networks with state-dependent routing. *IEEE/ACM Transactions on Networking* 1(1):105–115.

Dziong, Z. 1985. Equivalent path approach for circuit switched networks analysis. In *Proceedings of the 11th International Teletraffic Congress*, pp. 4.2B.5.1–7. North-Holland.

Dziong, Z., and Roberts, J. 1987. Congestion probabilities in a circuit switched integrated services network. *Performance Evaluation* 7(4):267–284.

Dziong, Z., Mignault, J., and Rosenberg, C. 1993. Blocking evaluation for networks with reward maximization routing. In *Proceedings of IEEE INFOCOM'93*, pp. 593–601. IEEE Computer Society Press.

Gazdzicki, P., Lambadaris, I., and Mazumdar, R. 1993. Blocking probabilities for large multi-class Erlang loss systems. *Advances in Applied Probability* 25(4):997–1009.

Gersht, A., and Lee, K.J. 1990. A bandwidth management strategy in ATM networks. GET Laboratories Technical Report.

Girard, A., and Bell, M.-A. 1989. Blocking evaluation for networks with residual capacity adaptive routing. *IEEE Transactions on Communications* 37(12):1372–1380.

Girard, A. 1990. *Routing and Dimensioning in Circuit-Switched Networks*. Addison-Wesley.

Hwang, R.-H., Kurose, J., and Towsley, D. 1992. State dependent routing for multirate loss networks. In *Proceedings of IEEE GLOBECOM'92*, pp. 565–570. IEEE Computer Society Press.

Katz, S. 1967. Statistical performance analysis of a switched communication network. In *Proceedings of the 5th International Teletraffic Congress*. North-Holland.

Kaufman, J.S. 1981. Blocking in a shared resources environment. *IEEE Transactions on Communications* 29(10):1474–1481.

Kelly, F.P. 1986. Blocking probabilities in large circuit-switched networks. *Advances in Applied Probability* 18:473–505.

Kelly, F.P. 1990. Routing and capacity allocation in networks with trunk reservation. *Mathematics of Operations Research*, 15(4):771–793.

Kelly, F.P. 1994. Bounds on the performance of dynamic routing schemes for highly connected networks. *Mathematics of Operations Research* 19(1):1–20.

Krishnan, K. R. 1990. Performance evaluation of networks under state-dependent routing. In *Proceedings of the Seminar on Design and Control of a Worldwide Intelligent Network*. 26–29 June, CNET, Issy-les-Moulineaux, France.

Legall, F., Bernussou, J., and Garcia, J.M. 1985. A state dependent one-moment model for grade of service and traffic evaluation in circuit swutched networks. In *Proceedings of the 11th International Teletraffic Congress*, pp. 5.2b.2:1–6. North-Holland.

Lin, P.M., Leon, B.J., and Stewart, C.R. 1978. Analysis of circuit-switched networks employing originating-office control with spill-forward. *IEEE Transactions on Communications*, 26(6):754–765.

Manfield, D., and Downs, T. 1979. On the one-moment analysis of telephone traffic networks. *IEEE Transactions on Communication* 27:1169–1174.

Marbukh, V. 1991. Asymptotic investigation of a complete communications network with a large number of points and bypass routes. *Probl. Inf. Transmission* 16:212–216.

Marbukh, V. 1993. Loss circuit switched communication network — performance analysis and dynamic routing. *Queueing Systems* 13:111–141.

Mason, L.G., and Girard, A. 1982. Control techniques and performance models for circuit-switched networks. In *Proceedings of the 21st IEEE Conference on Decision and Control*, pp. 1374–1383, Orlando, Florida, December.

Mason, L.G. 1986. On the stability of circuit-switched networks with non-hierarchical routing. In *Proceedings of the 24th IEEE Conference on Decision and Control*, December, Athens, Greece.

Mignault, J. 1991. Analyse des performances de réseaux à commutation de circuits avec acheminement à maximisation du revenu. Master's thesis, École Polytechnique de Montréal.

Mitra, D., Gibbens, R.J., and Huang, B.D. 1991. Analysis and optimal design of aggregated-least-busy-alternative routing on symmetric loss network with trunk reservations. In *Proceedings of the 13th International Teletraffic Congress*, pp. 477–482. North-Holland.

Mitra, D., and Gibbens, R. J. 1992. State-dependent routing on symmetric loss networks with trunk reservations, II: asymptotics, optimal design. *Annals of Operations Research* 35:3–30.

Mitra, D., Gibbens, R.J., and Huang, B.D. 1993. State-dependent routing on symmetric loss networks with trunk reservations, I. *IEEE Transactions on Communication* 41(2):400–411.

Mitra, D., and Morrison, J.A. 1994. Erlang capacity and uniform approximations for shared unbuffered resources. *IEEE/ACM Transactions on Networking* 2(6):558–570.

Pioro, M.P. 1983. A uniform approach to the analysis and optimization of circuit switched communication networks. In *Proceedings of the 10th International Teletraffic Congress*. pp. 4.3A:1-7. North-Holland.

Ramos, J.M. 1996. Évaluation de la performance de réseaux commutés multi-débits avec acheminement à maximisation de la récompense. Master's thesis, INRS-Telecommunications.

Régnier, J., Blondeau, P., and Cameron, W. H. 1983. Grade of service of a dynamic call routing system. In *Proceedings of the 10th International Teletraffic Congress*, pp. 3.2.6.1–9. North-Holland.

Roberts, J.W. 1981. A service system with heterogeneous user requirements. In *Performance of Data Communications Systems and Their Applications*, ed. G. Pujolle, pp. 423–431. North-Holland.

Simonian, A., Roberts, J.W., Théberge, F., and Mazumdar, R. 1997. Asymptotic estimates for blocking probabilities in a large multi-rate loss network. *Advances in Applied Probabilities*, in press (considered for September issue).

Wilkinson, R.I. 1956. Theories for toll traffic engineering in the USA. *Bell System Technical Journal*, 35:421–514.

Wong, E.W.M., and Yum, T.-S. 1990. Maximum free circuit routing in circuit-switched networks. In *Proceedings of IEEE INFOCOM'90*, pp. 934–937. IEEE Computer Society Press.

Chapter 8

Optimal Operating Point — Fairness vs. Efficiency

THE ISSUE of fairness has gained importance in multi-service environments where diverse service characteristics can cause a very unfair resource allocation unless the issue is considered explicitly at the design stage. This chapter addresses the problem of fair-efficient resource allocation on the connection level. A framework based on cooperative game theory is presented. It can serve to synthesize and analyze connection admission control (CAC_{GoS}) policies which provide efficient use of the available resources and fair allocation of these resources to the contending heterogeneous users.

In order to consider the fairness issue explicitly, the network has to be equipped with CAC & routing mechanisms that allow one to control the GoS allocation among the services. In particular, the reward maximization approach, described in Chapter 5, provides such a mechanism for achieving an arbitrary operating point. Nevertheless there remains a question: Which operating point is both fair and efficient? The natural notion of fairness is GoS equalization but, as it will be shown later on, this approach might lead to low efficiency. On the other hand, the traditional single criterion formulations, involving overall throughput maximization or minimum average delay, yield efficient solutions but do not provide any guarantees of fairness. Indeed, as we will show in this chapter, maximizing throughput can lead to solutions where some users and/or services receive no access at all! Thus there is a need for a framework which can cope with these dual aspects.

Cooperative game theory provides an excellent framework for this problem, in that it explicitly considers efficiency through the notion of Pareto optimality and fairness by means of the axiomatic properties which a solution must possess. Another advantage is that this framework provides a precise mathematical formulation of the problem. These features give a new perspective on the connection admission problem. In particular by choosing different sets of fairness axioms one can select connection admission procedure characteristics which fit specific requirements. Moreover, different solutions (including heuristic ones) can be compared in a precise mathematical framework. In Section 8.1 we describe basic concepts from cooperative game theory focusing on arbitration solutions which provide a useful framework for analysis and synthesis of fair-efficient connection admission controls in broadband networks. Also, models for evaluation of arbitration solutions are discussed. More information about basic game theory concepts can be found in Appendix C.

As discussed in Chapter 5, in many cases, the network CAC_{GoS} mechanism is decomposed into a set of link CAC^l_{GoS} algorithms. Also the CAC_{GoS} algorithm in ATM networks based on the virtual path design (see Section 9.3) is equivalent to the link CAC^l_{GoS} algorithm. In Section 8.2 the main characteristics of fairness concepts from game theory are analyzed for a single-link system where several connection classes are competing for resources. As a reference we use two control objectives, throughput maximization and connection loss equalization, which are commonly associated with efficiency and fairness, respectively. It is shown that the arbitration solutions from cooperative game theory have several attractive features, especially in overload conditions. Also, a dynamic interpretation is presented where in case of an overload the increase in throughput is fairly distributed among the network users, using nominal conditions as a reference point. In Section 8.2.1 simplified connection admission policies are studied in the context of the game theoretic framework. These policies include complete sharing, coordinate convex policies, "bandwidth" reservation, and dynamic "bandwidth" reservation. In general, they are sub-optimal but simple to implement. The numerical results indicate that the "bandwidth" reservation and dynamic "bandwidth" reservation policies are, in many cases, near optimal in terms of efficiency and fairness and can offer a practical solution to the problem. Finding an optimal operating point in multi-link networks with reward maximization routing is discussed in Section 8.3.

8.1 Game Theoretic Framework

8.1.1 Fairness concepts

In this section[1] we discuss the basic notions of cooperative game theory. To simplify the presentation let us consider a game with two players. An outcome of the game is defined by the values of players' utilities, $\mathbf{u} = [u_1, u_2]$. It is assumed that the higher the value of the utility, the higher the satisfaction of the player with that outcome. In general each utility can be expressed in different units. The set of all possible outcomes is termed the bargaining domain, $\mathbf{U}$. In the following we consider bargaining domains that are convex, closed, and bounded sets of $R^n : R \geq 0$ where $n = |J|$ denotes the number of players. An example of such a domain is shown in Figure 8.1 for $n = 2$.

---

[1]Portions of Sections 8.1 and 8.2 are reprinted, with permission, from (Dziong and Mason, 1996). ©1996 IEEE.

8.1 Game Theoretic Framework

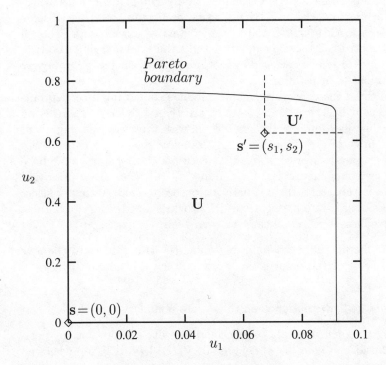

Figure 8.1. *Example of bargaining domain.*

It is obvious that the outcome of the game depends strongly on cooperation among players. That is why game theory distinguishes between cooperative and non-cooperative games. Since in our network problem we are interested mainly in a solution fair to all players (network users) as well as efficient, the cooperative setting is more appropriate. Note that in this setting the solution must provide that no player can increase its utility without adversely affecting the other's. In game theory this feature is called Pareto optimality, and henceforth the set of Pareto optimal solutions is called the Pareto boundary.

The outcome of the game also depends on the starting (conflict, disagreement) point $\mathbf{s} = [s_1, s_2]$: $\mathbf{s} \in \mathbf{U}$. This point can be interpreted as a pre-game assumption that the solution point $\mathbf{u}^*$ cannot be worse than the starting point ($u_i^* \geq s_i$). The starting point can limit the bargaining domain, as is shown in Figure 8.1.

The main objective of cooperative game theory is to determine a fair solution to the game defined by $(\mathbf{s}, \mathbf{U})$. In general two different approaches can be considered. The first one assumes that there is a bargaining process. It means that the solution is achieved in several steps (in the limit it might be a continuous process). The bargaining process can begin at the starting point or from a pair of solutions which can be treated as an offer and a counter-offer (both may be Pareto optimal). The second approach ignores the bargaining process and concentrates on a solution point which is determined by a set of axioms which should provide fairness features acceptable to all players. The situation is somewhat analogous to a conflict which is resolved by an arbiter; hence the choice of name *arbitration scheme*. Observe that the arbitration scheme fits very well in the context of our network problem where the network control mechanism can be viewed as an arbiter trying to satisfy the requirements of all users or/and services. Consequently in the following we limit our considerations to arbitration schemes.

As we have indicated the cooperative game outcome should fulfill certain fairness criteria (expressed by the axioms). What are these criteria? There is no unique answer to this question. In the case of network problems the important issue might be the price of providing some fairness features and the network operators willingness to pay this price. Note that each network operator may have quite a different viewpoint on this issue depending on the economical priorities, network users, expectations, and the competitors' position. That is why we select for our study a class of arbitration schemes which provides a wide range of fairness characteristics. This class is based on three standard fairness axioms:

- *Symmetry*: If the bargaining domain is symmetric with respect to the axis $u_1 = u_2$ and the starting point is on this axis, then the solution is also on this axis.

- *Pareto optimality*: The solution is on the Pareto boundary.

- *Invariance with respect to utility transformations*: The solution for any positive affine transformation of $(U, \mathbf{s})$ denoted by $V(U), V(\mathbf{s})$ is $V(\mathbf{u}^*)$ where $\mathbf{u}^*$ is the solution for the original system.

Based on the last axiom, without losing any generality, we can linearly transform the pair $(\mathbf{s}, \mathbf{U})$ in such a way that $\max_{\mathbf{u}' \in U'}\{u'_i\} = 1$ and $s'_i = 0$ where the prime denotes the model after transformation. Henceforth this case is called normalized. Within the class defined by the axioms there are three arbitration schemes which are of particular interest due to their natural and mathematical interpretations. These are the Nash (Nash, 1950), Raiffa-Kalai-Smorodinsky (Raiffa, 1953; Kalai and Smorodinsky, 1975) (henceforth called Raiffa) and modified Thomson (Cao, 1982) solutions. The first two are uniquely defined by one additional axiom:

- Nash — *Independence of irrelevant alternatives*: If the solution for $(U_1, \mathbf{s})$ is $\mathbf{u}^*$, and $U_2 \subseteq U_1$, $\mathbf{u}^* \in U_2$, then $\mathbf{u}^*$ is also the solution for $(U_2, \mathbf{s})$. In other words this axiom states that if U_1 is reduced, in such a way that the original solution and starting point are still included, the solution of the new problem remains the same.

- Raiffa — *Monotonicity*: If $U_2 \subseteq U_1$ and $\max\{u_1 : \mathbf{u} \in U_1\} = \max\{u_1 : \mathbf{u} \in U_2\}$ and $\max\{u_2 : u \in U_2\} \leq \max\{u_2 : u \in U_1\}$ then $u_2^{2*} \leq u_2^{1*}$, where $\mathbf{u}^{j*}$ denotes the solution for $(U_j, \mathbf{s})$. In other words for any subset of U_1 the solution for the second player cannot be improved if the maximum utility of the first player is constant.

The modified Thomson solution is defined by the utilitarian rule which maximizes the sum of utilities $u'_1 + u'_2$ in the normalized case.

8.1 Game Theoretic Framework

An interesting mathematical interpretation of these arbitration schemes is given in (Cao, 1982). It is based on player preference functions defined for the normalized case as

$$v_1 = u_1' + \beta(1 - u_2') \qquad (8.1)$$
$$v_2 = u_2' + \beta(1 - u_1') \qquad (8.2)$$

where β is a weighting factor. In (Cao, 1982) it is shown that the Nash, Raiffa and the modified Thomson arbitration solutions correspond to the maximization of the product of player's preference functions with $\beta = 0, 1, -1$, respectively:

$$\mathbf{u}^* = \arg(\max_{\mathbf{u} \in \mathbf{U}}\{v_1 \cdot v_2\}) \qquad (8.3)$$

where $\arg(\max_u\{\ \})$ denotes the value of the argument $\mathbf{u}$ for which the maximum is achieved. From this formulation one can easily find that in the Raiffa case ($\beta = 1$) the preference function treats with the same weight the player's own gain and the other player's losses so the solution equalizes both utilities $u_1^* = u_2^*$. In the modified Thomson case ($\beta = -1$) both gains have the same weight; thus the solution maximizes the sum $u_1' + u_2'$. Finally in the Nash solution ($\beta = 0$) the preference function takes into account the players' own gain only so the solution is located somewhere between the Raiffa and the modified Thomson solutions.

In (Cao, 1982) it is shown that by changing β continuously from -1 to 1 one can achieve monotonically and continuously a set of solutions which relates $\beta \in [-1, 1]$ to a part of the Pareto boundary. This fact has also an appealing geometrical interpretation. Namely for each β the solution is given by the tangent point between the Pareto boundary and a hyperbola from the set of hyperbolae defined by the function $v_1 \cdot v_2 = $ const. A more detailed discussion of this feature is given in Appendix C.

The mathematical simplicity, natural interpretation, and uniformity of the preference functions' product approach makes it attractive for an investigation of the fairness issue in the context of broadband networks. In order to fully exploit these features in the following we extend this formulation to the multidimensional case. We start with extensions of the preference functions' product for the three distinguished arbitration schemes. In the case of Nash and the modified Thomson schemes the forms of the preference functions are simple and are given by

$$v_j = u_j' \qquad (8.4)$$
$$v_j = \sum_i u_i' \qquad (8.5)$$

respectively. In the case of the Raiffa scheme it can be shown that the preference function can have the form

$$v_j = u_j' + 1 - \frac{1}{n-1}\sum_{i \neq j} u_i' \qquad (8.6)$$

All three cases can be viewed as special instances of a set of preference functions defined by

$$v_j = u_j' + |\beta(n-1)| - \beta \sum_{i \neq j} u_i' \qquad (8.7)$$

for $\beta = 0, -1, 1/(n-1)$ respectively. Then a general class of solutions for multidimensional case is defined by

$$u^* = \arg\left(\max_{u \in U}\left\{\prod_j v_j\right\}\right) \tag{8.8}$$

As in the case of two players, by changing β continuously from -1 to $1/(n-1)$ one can achieve continuously a set of solutions which relates $\beta \in [-1, 1/(n-1)]$ to a part of the Pareto boundary.

8.1.2 The Pareto boundary for multi-service systems*

To make the study of the fairness concepts more illustrative we decompose the calculation of arbitration solutions into two stages. First we generate the Pareto boundary for the considered system and then the chosen solutions are computed. In this section we describe an approach to Pareto boundary generation for networks where connection state transitions can be modeled as a Markov process. We do not restrict our consideration to any particular type of network. The only presumption is that the connection admission control is based on "bandwidth" allocation (physical or logical) to connections.

In the game theoretic framework the connection classes are treated as players and the connection admission mechanism is treated as an arbiter. The natural candidate for utility units is the carried traffic so each service performance is characterized by

$$\overline{a}_j = \overline{\lambda}_j d_j \mu_j^{-1} \tag{8.9}$$

where $\overline{\lambda}_j$ is the rate of accepted connections from class j. Note that due to the *invariance with respect to utility transformations* axiom the utility defined as $u_j = \overline{\lambda}_j$ will give the same arbitration solutions.

To achieve an arbitrary point in the utility domain, one has to apply a randomized control policy. Such a policy is defined by a matrix of state-dependent probabilities, $\pi_r = [h_j^k(\mathbf{z}), \mathbf{z} \in Z, j \in J]$ where $0 \leq h_j^k(\mathbf{z}) \leq 1$. After class j connection arrival in state $\mathbf{z}$ the connection is accepted on path k with probability $h_j^k(\mathbf{z})$; otherwise it is rejected [with probability $1 - \sum_{k \in W_j} h_j^k(\mathbf{z})$]. In the case where the values of parameters $h_j^k(\mathbf{z})$ are limited to 1 or 0 the policy becomes deterministic, as described in Chapter 5.

It is well known that for a given π the performance of the system can be found from the solution of steady state equilibrium equations. This provides a model to calculate the vector of utilities as a function of the connection admission policy, $\overline{\mathbf{a}}_v = F(\pi)$. Then evaluation of a point on the Pareto boundary can be formulated as a maximization problem with constraints:

$$\max_{\pi_r}\{|\overline{\mathbf{a}}_v| = |F(\pi)|\} \tag{8.10}$$

$$\overline{\mathbf{a}}_v \cdot |\overline{\mathbf{a}}_v|^{-1} = \gamma \tag{8.11}$$

where $\overline{\mathbf{a}}_v = [\overline{a}_j]$ denotes the utility vector and the unit vector γ ($|\gamma|=1$) defines the direction in which the Pareto point is required. In general this optimization problem is non-linear but can be easily transformed into a linear problem where well known solution techniques from linear programming can be applied. Then the Pareto boundary can be generated by evaluating the Pareto points in all possible directions γ.

While the above approach is feasible, in the following we present another model which is more efficient. It is based on the deterministic policy derived from Markov decision theory as described in Section 5.2.1. In this setting each connection is characterized by a reward parameter, $r_j \in [0, \infty)$, and the objective is to find the optimal control policy, $\pi^*(\mathbf{r})$, which maximizes the average reward from the system defined as

$$\overline{R} = \sum_j \overline{\lambda}_j r_j \tag{8.12}$$

for a given vector of reward parameters, $\mathbf{r} = [r_j, j \in J]$. In the following we show that the solution of the decision problem, $\pi^*(\mathbf{r})$, possesses some features facilitating evaluation of the Pareto boundary.

Theorem 1. *Each solution of the decision problem, $\pi^*(\mathbf{r})$, is located on the Pareto boundary in the utility domain.*

Proof: The solution $\pi^*(\mathbf{r})$ provides maximum average reward from the system; thus no individual carried traffic $\overline{a}_j$ can be increased without decreasing the other's ∎

Theorem 2. *Each Pareto point is located on a hyper-plane defined by a set of n deterministic solutions of the decision problem, where n denotes the space dimension.*

Proof: Let us denote a part of the hyper-plane defined by the convex hull of n deterministic solutions as the face. It can be easily shown that any solution on a face can be achieved by a randomized policy. Let us choose the face which provides the best solution in direction γ. It is easy to show that all points, $\overline{\mathbf{a}}_v$, on this face satisfy the relation

$$\overline{R} = \sum_j \overline{a}_j r'_j \tag{8.13}$$

for certain values of $\overline{R}$ and $\mathbf{r}' = [r'_j]$ where $r'_j = r_j \mu_j / d_j$ denotes the normalized connection reward parameter. Suppose there exists a solution, $\overline{\mathbf{a}}'_v$, in direction γ better than the one on the chosen face. For this solution

$$\overline{R} < \sum_j \overline{a}'_j r'_j \tag{8.14}$$

But according to Theorem 1 this would mean that there exists a deterministic solution with

$$\overline{R}' > \sum_j \overline{a}_j r'_j \tag{8.15}$$

which contradicts the fact that the chosen face provides the best solution in direction γ ∎

Observe that based on Theorem 2 we can generate the Pareto boundary in two steps. First, by varying the values of reward parameters in the range $[0, \infty)$ the complete set of deterministic policies π_d^* can be found. It is significant that the number of these policies is finite since the number of states is also finite. Then simple linear interpolation can be used to determine the complete Pareto boundary. It should be underlined that in most cases of practical interest the best deterministic strategy can be used instead of the probabilistic one since the difference between them is usually negligible due to a large number of states.

The fairness models based on the product of the preference functions assume that the utility domain is convex, closed, and bounded. In the case of our multi-service network model the last two features are obvious. In the following we show that the first feature also holds.

Theorem 3. *The utility domain in the considered multi-service system is convex.*

Proof: Suppose there exists a concave region in the utility domain in the region corresponding to the Pareto boundary. Then one can construct a hyper-plane above this region which is tangent to the Pareto boundary in at least n points. Observe that all points on the face defined by these n solutions are feasible solutions which can be achieved by the randomized policies. Thus the Pareto boundary cannot be situated below this face. ■

To summarize, the Pareto boundary is defined by the boundary of the convex hull of the finite set of all optimal deterministic policies, $\pi^*(\mathbf{r})$. As discussed in Chapter 5 and Appendix B the deterministic policies can be found from one of three basic approaches: policy iteration, linear programming, and value iteration. The value iteration algorithm is most attractive due to its numerical simplicity, which enables the solution of relatively large systems. However, it should be stressed that, except for single-link examples and some small networks, the evaluation of optimal policy can be intractable due to the enormous cardinality of the state and policy spaces. In such cases one can resort to the simplified models described in Chapter 5.

8.2 Analysis of a Single-Link System

In this section we illustrate application of game-theoretical framework to synthesis and analysis of connection admission policies by studying single-link network examples which allow us to evaluate exact solutions. Since in many multi-link networks the CAC_{GoS} policy is at least partly decomposed into single-link problems, $\text{CAC}_{\text{GoS}}^l$, the discussed solutions and their features are also applicable to a general network case. Moreover, in ATM networks based on the virtual path design (see Section 9.3) the virtual paths can be seen as single-links from the CAC algorithm viewpoint. The traffic assumptions used in this section are consistent with the ones used in Chapter 5. In particular we assume Poissonian connection arrival rates and exponentially distributed service times. The tested examples are described in Table 8.1.

8.2 Analysis of a Single-Link System

TABLE 8.1 Description of single-link examples.

Example	d_j	$\sum a_j$	share of a_j [%]	μ_j^{-1}	L
1	1,12	32.0	50,50	1,10	50
2	1,12	33.6	90,10	1,10	50
3	1,6	130.0	50,50	1,10	120
4	1,6	110.8	90,10	1,10	120
5	1,6	110.0	10,90	1,10	120
6	1,6	84.0	50,50	1,1	120
7	1,12	15.0	50,50	1,10	50
8	1,4,10	60.0	60,30,10	1,3,10	60

As reference points we use the two traditional control objectives in telecommunication networks which are connection loss probability equalization (henceforth referred to as loss equalization) and traffic maximization. In fact these objectives can be presented in a manner similar to the Raiffa and the modified Thomson solutions, respectively. We begin with a comparison of the Raiffa solution and the loss equalization. The loss equalization solution can be obtained by maximizing the product of the following preference functions

$$v_j = (1 - B_j) + 1 - \frac{1}{n-1}\sum_{i \neq j}(1 - B_i) \qquad (8.16)$$

where B_j denotes the connection class loss probability defined as

$$B_j = 1 - \frac{\overline{a}_j}{a_j} \qquad (8.17)$$

and $a_j = \lambda_j d_j \mu_j^{-1}$ denotes the offered traffic. Let us define the normalized connection loss probability as

$$B'_j = 1 - \frac{\overline{a}_j}{\max_{u_j \in \mathbf{U}}\{\overline{a}_j\}} \qquad (8.18)$$

By using the normalized connection loss probability definition, the preference function for the Raiffa solution can be rewritten as

$$v_j = (1 - B'_j) + 1 - \frac{1}{n-1}\sum_{i \neq j}(1 - B'_i) \qquad (8.19)$$

In spite of the almost identical structure of Equations (8.16) and (8.19) and the fact that both solutions are Pareto optimal, it is easy to check that in general the *symmetry* axiom is not fulfilled in the case of loss equalization. In the following we try to illustrate what are the practical consequences of this difference.

From the preference function definitions it follows that both solutions are identical if $B_j = B'_j$ for all j. The fulfillment of this condition strongly depends on the traffic levels and the ratio of "bandwidth" required by connections to link capacity,

$\gamma_j = d_j/L$. In general the higher the overload and/or the higher the connection "bandwidth" requirement, the larger the difference between the two solutions. These relations are illustrated in Figures 8.2a and 8.2b for Examples 6 and 7 defined in Table 8.1 (the figures also include solutions for three other objectives to be discussed subsequently). The utilities are expressed in terms of carried traffic normalized by link capacity $u_j = \overline{a}_j/L$. The families of Pareto boundaries (dotted lines) are achieved by multiplying the nominal traffic from Table 8.1 by an overload factor β_j which is indicated on the figures close to the corresponding curves and axes ($a'_j = \beta_j \cdot a_j$). To make the results more readable the solutions for the given objective are joined by a line. The results show that the differences between the two solutions occur under heavy overload of narrow-band traffic in the case of small γ_2 while in the case of relatively large γ_2 the difference exists even for light traffic conditions. The interpretation of these differences is that the Raiffa solution relates losses to the maximum system throughput while the loss equalization solution relates losses to the offered traffic. In other words the loss equalization approach ignores the fact that even by providing the highest priority for a particular type of traffic it can still encounter losses if the offered traffic is higher than the system throughput. From this point of view the Raiffa solution can be seen as more fair than loss equalization. The features of the Raiffa solution can also be attractive in a dynamic situation where the system is designed for and operates at nominal conditions and overloads should not have the same priority. In this context the Raiffa solution provides a certain protection for well behaved users against heavy overloads from other users. Concerning the average traffic loss probability, $\overline{B}$, the differences between both cases are small so the Raiffa solution can be seen as an attractive alternative to loss equalization.

The preference functions for traffic maximization and the modified Thomson solutions are also very similar. In particular the traffic maximization solution can be defined by

$$v_j = \overline{a}_j + (n-1) + \sum_{i \neq j} \overline{a}_i \qquad (8.20)$$

and for the modified Thomson solution we have

$$v_j = \frac{\overline{a}_j}{\max_{u_j \in U}\{\overline{a}_j\}} + (n-1) + \sum_{i \neq j} \frac{\overline{a}_i}{\max_{u_i \in U}\{\overline{a}_i\}} \qquad (8.21)$$

It is easy to verify that both solutions are equal if $\max_{u_j \in U}\{\overline{a}_j\} = \max_{u_i \in U}\{\overline{a}_i\}$ for all i, j. In other cases the existence and magnitude of the differences depend strongly on the asymmetry of each users traffic levels. This follows from the fact that the traffic maximization and the modified Thomson solutions are in fact tangent points between the Pareto boundary and the hyper-plane normal to the vectors [1] and $[\frac{1}{\overline{a}_j}]$, respectively. Thus the more asymmetric the Pareto boundary, the more favorable the solution for users with smaller traffic level in the case of the modified Thomson solution. This feature is illustrated in Figures 8.2a and 8.2b. Note that in the case of a large overload of one type of traffic the solutions differ significantly and the grade of service of the stream with the smaller traffic level is significantly better for the modified Thomson solution than for the traffic maximization solution. In particular, under heavy overload of the narrow-band traffic in Example 6

8.2 Analysis of a Single-Link System

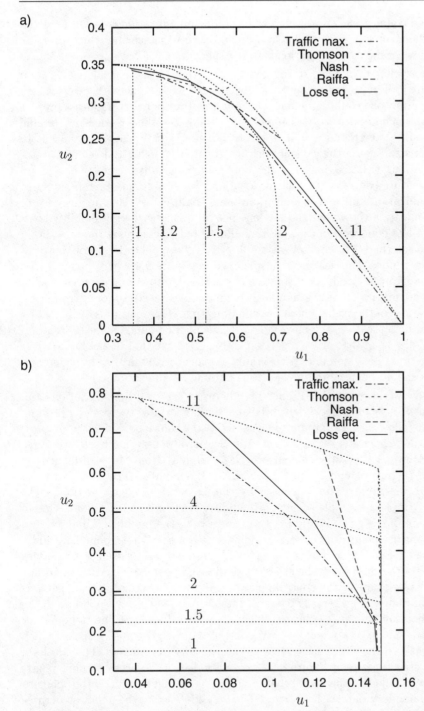

Figure 8.2. *Pareto boundaries (dotted lines) and solutions (joined by the lines indicated for each solution) for different overloads of (a) narrow-band traffic (Example 6) and (b) wide-band traffic (Example 7).*

all wide-band connections are rejected when traffic maximization is used. Thus the modified Thomson solution can be seen as an arbitration scheme which protects users with low traffic against overloads from other users.

It is interesting to note that the relative location of the considered optimal solutions on the Pareto boundary is not necessarily consistent with intuitive expectations. First, one could anticipate that the two traditional strategies represent two extremes: one of fairness (loss equalization) and the other of efficiency (traffic maximization). Nevertheless, as it is illustrated in Figures 8.2a, 8.2b, and 8.3b, in many cases the game theory solutions are located far from the set of Pareto-optimal solutions situated between the two traditional objectives. The second interesting feature concerns the relative location of the solutions corresponding to traffic maximization and loss equalization. One could expect that in the case of traffic maximization the wide-band stream encounters higher loss probability compared to the loss equalization case. But, as illustrated in Figures 8.2b and 8.3a, in many situations the traffic maximization objective gives higher loss probability to narrow-band traffic. This feature is also shown in Figure 8.4 where policy structures corresponding to traditional solutions from the Pareto boundary presented in Figure 8.3a are given. There are significantly more blocking states for narrow-band traffic in the case of the traffic maximization objective. This feature is explained by the fact that in these cases the mean holding time of wide-band connections is significantly higher than that of narrow-band traffic. Observe that at the limit (when the holding time for wide-band connections approaches infinity) only wide-band connections should be accepted in the case of traffic maximization.

Up to now we have indicated differences between the traditional solutions and the solutions from game theory which have a similar form of the preference functions. In the following we compare the features of the solutions derived from game theory amongst themselves. As indicated in Section 8.1.1 we have selected three particular solutions (modified Thomson, Nash, Raiffa) from the set defined by Equation (8.7). Surprisingly in most cases tested the modified Thomson and Nash solutions are the same. This is caused by the existence of edges on the Pareto boundary. A typical example of such a boundary is given in Figure 8.3b. One of the few exceptions is given in Figure 8.3a. Since both the modified Thomson and Nash solutions are defined by a tangent point between the Pareto boundary and a line (modified Thomson) or a hyperbole (Nash), in many cases the edge determines solutions for both objectives. The existence of these edges is explained by an abrupt change in the structure of the control policy around the edge. This phenomenon is illustrated in Figure 8.5 where structure of the optimal policy corresponding to the modified Thomson and Nash solutions, located on the edge of the Pareto boundary (Figure 8.3b), can be compared with structure of the "neighboring" solution which is located close to the edge. Let u_1^c, u_2^c denote the edge coordinate. The two level type policy structure, as in Figure 8.5a, corresponds to the Pareto boundary part with $u_1 \leq u_1^c$, while the one level type structure, as in Figure 8.5b, corresponds to the Pareto boundary part with $u_1 > u_1^c$. The big difference in the policy structures causes the slope of the Pareto boundary to be quite different in these two regions resulting in the edge.

Concerning the relation between the Raiffa solution and the Nash or modified

8.2 Analysis of a Single-Link System

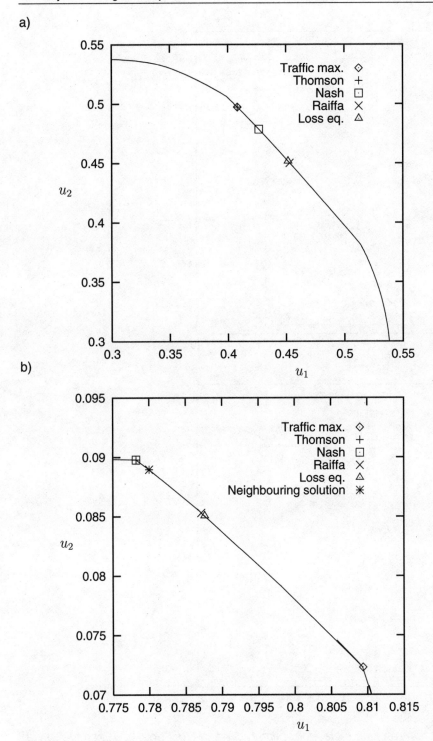

Figure 8.3. *Pareto boundary and solutions: (a) Example 3, (b) Example 4.*

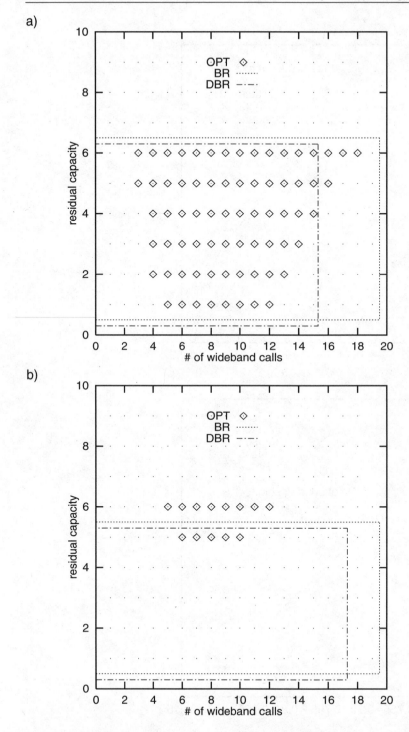

Figure 8.4. *NB connection blocking states for optimal, BR and DBR policies (Example 3): (a) traffic maximization objective, (b) loss equalization objective.*

Thomson solution, the results show that while the Raiffa solution provides equal normalized losses for any shape of the Pareto boundary, the results for the modified Thomson or Nash solution are less consistent. In particular, in Example 6 under nominal conditions or light overload, both standard and normalized losses for wide-band connections are significantly higher than the ones for narrow-band traffic, while under heavy overload of narrow-band traffic the situation is completely reversed since the scheme protects small streams. This effect results in crossing of the lines corresponding to the considered two solutions in Figure 8.2a.

As mentioned earlier some features of the modified Thomson (or Nash) and Raiffa solutions could be useful for the protection of well behaved users against overloads from other users. One limitation of such an approach is that the objective functions do not include any information about what is considered as nominal conditions. In the following we describe a more precise and effective method of handling such dynamic situations. This approach is also based on the game theoretic framework and has two stages. In the first stage only nominal conditions are considered and the solution is evaluated according to the chosen arbitration scheme with a zero starting point, $\mathbf{s} = [0]$. Then for any overload conditions a new solution is evaluated by applying an arbitration scheme with a modified starting point equal to the solution under nominal conditions, $\mathbf{s} = \mathbf{u}_n^*$. The idea behind this approach is that the users' throughput under nominal conditions should not deteriorate due to an overload generated by other users. On the contrary, the gain in the overall throughput caused by the overload should be fairly distributed amongst all users since it is an additional gain compared to the agreed nominal conditions. We consider also a more flexible approach with $\mathbf{s} = \alpha \cdot \mathbf{u}_n^*$ where $0 < \alpha < 1$. This corresponds to the philosophy that by relaxing the starting point by a small amount a large gain in overall throughput can be achieved. Thus a well behaved user may be willing to sacrifice a little bit for the sake of a global throughput increase. The illustration of this approach is given in Figures 8.6, 8.7, and 8.8. The results show that by choosing the starting point close enough to the nominal conditions solution, the GoS of the well behaved stream is protected almost ideally at the expense of the other stream's GoS. By relaxing the starting point the required trade-off between the GoS of the protected stream and the overall throughput can be achieved.

In summary, the results presented in this section show several advantages of the game theoretic framework. First of all the arbitration schemes provide better fairness features than the traditional control objectives. Moreover, the algorithm based on an adjusted starting point allows one to integrate the constraint of the guaranteed nominal throughput with the control objective. The results also indicate that the intuitive judgment that the traffic maximization and loss equalization strategies constitute two extremes corresponding to efficiency and fairness is not correct. In particular it means that in general the arbitrated solutions based on the fairness axioms cannot be approximated by a convex combination of the two traditional objectives.

200   **Optimal Operating Point — Fairness vs. Efficiency**

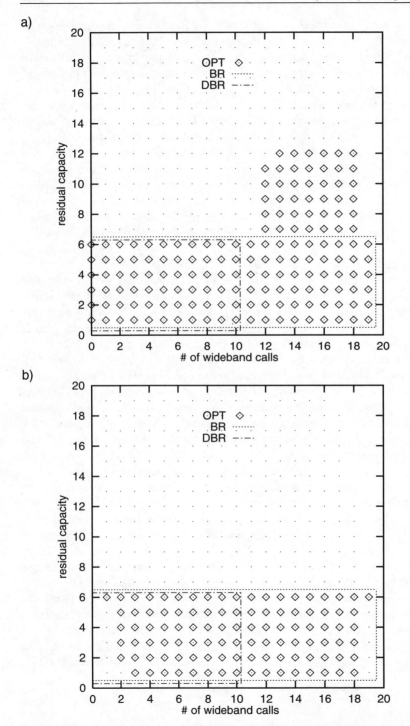

Figure 8.5. *NB connection blocking states for optimal, BR and DBR policies (Example 4): (a) Thomson and Nash solutions, (b) "neighboring" solution.*

8.2 Analysis of a Single-Link System

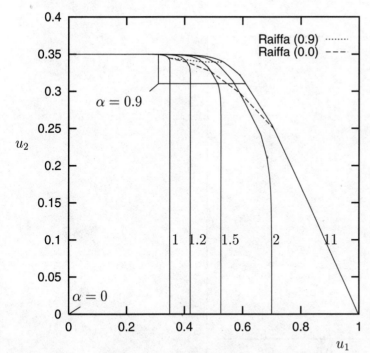

Figure 8.6. *Solutions for different starting points (Example 6).*

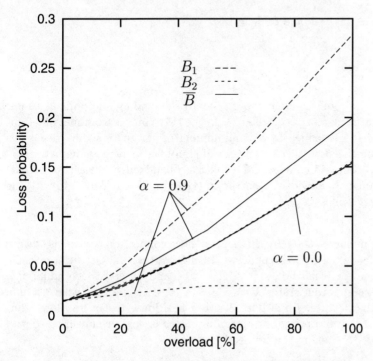

Figure 8.7. *Losses vs overload of narrow-band traffic (Example 6; $\alpha = 0.0, 0.9$).*

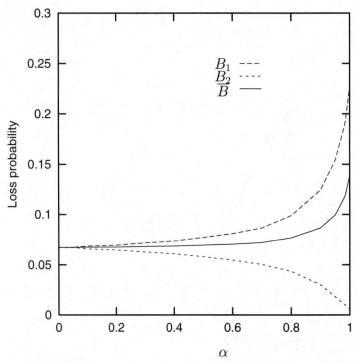

Figure 8.8. *Losses vs. parameter* α *(Example 6;* $\beta_1 = 1.5$*).*

8.2.1 Simplified CAC policies

In general the optimal policies are computationally demanding and difficult to implement due to a large memory requirement. In the following we examine some simplified policies which are defined by a small number of parameters and are easy to implement. We consider four distinct classes of simplified policies: complete sharing (CS) policy, coordinate convex (CC) policies, "bandwidth" reservation (BR) policies, and dynamic "bandwidth" reservation (DBR) policies. We begin with the definition of these schemes.

Complete sharing policy — CS. In this case a new connection arrival is always accepted if there is sufficient free capacity. In other words the set of admissible states equals to the set of all link states, $\Omega_{cs} = X$ and all states are admissible from each of the neighboring states $\mathbf{x}' = \mathbf{x} \pm \delta_j : \mathbf{x}' \in \Omega_{cs}, j \in J$ where δ_j is a vector with 1 in the position j and 0 in all other positions. It can be shown that this policy yields a Pareto optimal solution, but there is no mechanism to control the distribution of GoS among the connection classes.

Coordinate convex policies — CC. In the CC policy the set of admissible states is a closed subset of all feasible states, $\Omega_{cc} \in \Omega_{cs}$. The important characteristic of the CC policies is that all states in Ω_{cc} must be admissible from the lower neighboring states, $\mathbf{x}' = \mathbf{x} - \delta_j : \mathbf{x}' \in \Omega_{cc}, j \in J$. The attractive feature of this class is that the performance of the link is given by a product form solution (Kaufman, 1981). Let us define a subset of the CC policies where the maximum number of class j connections is limited by a threshold ($x_j \leq t_j$). In the following we limit our considerations to CC policies of the threshold type. Note that the thresholds provide a tool for controlling the GoS distribution among the services.

"Bandwidth" reservation policies — BR. Under the "bandwidth" reservation policy class j connections are rejected if the link residual capacity $C = L - \sum_i x_i d_i$ is smaller than or equal to the "bandwidth," H_j. Thus, H_j can be interpreted as the "bandwidth" reserved for other connection classes with $H_i = 0$ or $H_i < H_j$. In fact this scheme is widely used to protect direct-link-path connections against multi-link-path connections in circuit switched networks. In the case of the considered system the mechanism can be used as a tool to provide fair access for wide-band connections. Observe that in the BR scheme some states may not be accessible from the lower neighboring states so in general the product form solution is not applicable.

Dynamic "bandwidth" reservation policies — DBR. The DBR class of policies is defined in the same way as the trunk reservation policies with the exception that the "bandwidth" reservation is state dependent:

$$H_j(\mathbf{x}) = \begin{cases} \max_{i \neq j}\{h_i : x_i \leq t_i\}, & x_j > t_j \\ 0, & x_j \leq t_j \end{cases} \quad (8.22)$$

where t_j is the priority threshold and h_j denotes "bandwidth" reserved for connection class j if only this class has priority. The main idea behind this scheme is that the "bandwidth" reservation for particular type of connections is not needed if there are already many connections of this type in the system.

Observe that the BR policy is a special case of the DBR policy and that the CS policy is a particular case of all other policies.

Numerical study of simplified policies

Comparison of the simplified policies with the optimal ones for the loss equalization, traffic maximization and Raiffa objectives is presented in Tables 8.2, 8.3, and 8.4, respectively. The performance of the simplified policies is evaluated by means of the value iteration algorithm (the same as the one used for the evaluation of optimal policies). The optimal thresholds in the CC, BR, and DBR policies are found by exhaustive search. The performance is described in terms of the average traffic loss probability, $\overline{B}$, ratios of connection loss probabilities, B_j, or ratios of normalized connection loss probabilities, B'_j. The values of thresholds are also given.

The performance of the DBR policy is best among the considered simplified policies, and in many cases the results are close to optimal. The performance of

TABLE 8.2 Performance for loss equalization objective.

Example		OPT	CS	CC (t_{j-1})	BR (H_{j-1})	DBR (t_j, h_j)
1	$\overline{B}$	0.107	0.111	0.172	0.136	0.136
	B_2/B_1	1.000	17.041	0.899	1.000	1.000
	$j = 2$			(34)	(11)	(4, 11)
2	$\overline{B}$	0.065	0.030	0.134	0.076	0.076
	B_2/B_1	0.999	27.349	0.991	0.999	1.000
	$j = 2$			(20)	(11)	(3, 11)
3	$\overline{B}$	0.166	0.173	0.203	0.171	0.171
	B_2/B_1	1.000	5.691	0.957	1.000	1.000
	$j = 2$			(65)	(5)	(17, 5)
4	$\overline{B}$	0.055	0.045	0.083	0.057	0.057
	B_2/B_1	1.000	6.818	0.937	0.999	1.000
	$j = 2$			(21)	(5)	(13, 5)
8	$\overline{B}$	0.204	0.158	0.218	0.224	0.223
	$B_{\max}/B_{\min}$	1.084	9.879	2.870	1.000	1.013
	$j = 2$			(30)	(9)	(4, 3)
	$j = 3$			(9)	(9)	(2, 9)

TABLE 8.3 Performance for traffic maximization objective.

Example		OPT	CS	CC (t_{j-1})	BR (H_{j-1})	DBR (t_j, h_j)
1	$\overline{B}$	0.105	0.111	0.111	0.111	0.105
	B_2/B_1	1.405	17.041	17.072	17.046	1.148
	$j = 2$			(13)	(0)	(2, 12)
2	$\overline{B}$	0.030	0.030	0.030	0.030	0.030
	B_2/B_1	27.360	27.349	27.221	27.221	27.349
	$j = 2$			(0)	(0)	(-1, 12)
3	$\overline{B}$	0.164	0.173	0.173	0.164	0.164
	B_2/B_1	0.332	5.691	5.696	0.246	0.283
	$j = 2$			(25)	(6)	(15, 6)
4	$\overline{B}$	0.045	0.045	0.045	0.045	0.045
	B_2/B_1	6.835	6.818	6.834	6.834	6.834
	$j = 2$			(0)	(0)	(0, 0)
8	$\overline{B}$	0.158	0.158	0.158	0.158	0.158
	$B_{\max}/B_{\min}$	9.889	9.879	9.882	9.882	9.882
	$j = 2$			(0)	(0)	(-1, 4)
	$j = 3$			(0)	(3)	(-1, 10)

8.2 Analysis of a Single-Link System

TABLE 8.4 Performance for Raiffa objective.

Example		OPT	CS	CC (t_{j-1})	BR (H_{j-1})	DBR (t_j, h_j)
1	$\overline{B}$	0.105	0.111	0.163	0.136	0.182
	B'_2/B'_1	1.000	14.746	0.995	1.216	0.972
	$j = 2$			(33)	(10)	(1, 26)
2	$\overline{B}$	0.065	0.030	0.134	0.076	0.076
	B'_2/B'_1	0.998	28.129	0.991	0.999	1.000
	$j = 2$			(20)	(11)	(3, 11)
3	$\overline{B}$	0.166	0.173	0.202	0.171	0.171
	B'_2/B'_1	1.002	5.636	1.038	0.980	0.991
	$j = 2$			(64)	(5)	(14, 5)
4	$\overline{B}$	0.055	0.045	0.083	0.057	0.057
	B'_2/B'_1	1.011	8.464	1.000	1.104	0.945
	$j = 2$			(21)	(5)	(3, 6)
8	$\overline{B}$	0.204	0.158	0.218	0.224	0.223
	$B'_{\max}/B'_{\min}$	1.085	9.896	2.870	1.000	1.013
	$j = 2$			(30)	(9)	(4, 3)
	$j = 3$			(9)	(9)	(2, 9)

the BR policy is almost the same as that of the DBR policy, especially in the cases of loss equalization and traffic maximization. In the case of the Raiffa solution the additional degree of freedom in the DBR scheme provides slightly better results. Concerning the CC policy the proper threshold levels can provide relatively good results in the case of loss equalization or the Raiffa solution. However, by studying the average loss probability one can easily find that these solutions, in many cases, are not close to the Pareto boundary. With regard to the traffic maximization objective, the CC policy provides almost the same performance as the CS policy. In some cases these results are close to the optimal ones (especially for lower loss probability levels). Since the CS policy has no mechanism to control the distribution of GOS among different services, it is not surprising that in the case of loss equalization and the Raiffa solution the results are very poor (the B_j/B'_j ratio can be as large as 28 — see Example 2).

The relatively good performance of the BR and DBR policies can be explained by the fact that the structures of these policies, in many cases, are similar to the optimal ones. This is illustrated in Figures 8.4a and 8.5 for the traffic maximization and Nash solutions (the regions with blocking states for narrow-band connections are indicated by dotted and dashed lines). Moreover, with regard to the loss equalization objective, it can be easily shown that the BR policy equalizes connection congestion when $H_j = d_{\max} - 1$. Another characteristic feature of the BR policy is that the traffic maximization is achieved either when there is no reservation at all or when $H_1 = d_{\max}$. These features can facilitate the optimization of the BR policy.

Approximate evaluation of DBR parameters. Evaluation of the optimal parameters in the simplified policies requires an analytical model and can be time consuming. Further on we investigate the efficiency of two approximate models for DBR parameter evaluation. These models are used in some of the routing schemes studied in Chapters 5 and 6.

Decomposition of link problem — DBRA: This approximation is limited to two connection classes and is based on the assumptions that in the practical range of link parameters the value of h_2 should be equal to the "bandwidth" requirement, d_2, of the wide-band (WB) connections and that no "bandwidth" reservation is required for narrow-band (NB) connections ($t_1 = -1$). The main idea of the link problem decomposition is based on the assumption that since the transition rates of WB connections are expected to be significantly smaller than the ones for NB connections, the NB connections reach steady state distribution instantaneously for each state of WB connections. This feature known also as *near complete decomposability* allows one to analyze the stationary properties of the system separately for the NB Markov process (for each state of WB connections) and the WB Markov process. After applying this decomposition technique, the optimal threshold t_2 can be found by a search. This approach provides considerable reduction of the state space and substantial computational savings compared to the exact model.

Heuristic model — DBRH: In the heuristic approach it is also assumed that $h_j = d_j$ and that no "bandwidth" reservation is required for narrow-band connections, $t_1 = -1$. Concerning the thresholds, $t_j : j > 1$, they are given by the following formula:

$$t_j = [L_j/d_j] \cdot [1 + \sqrt{d_j/L_j}] \qquad (8.23)$$

where $L_j = L \cdot a_j / \sum_i a_i$ and a_j denotes the class j offered traffic. The main idea in this approach is to "allocate" the link "bandwidth" to each connection class proportionally to its offered traffic, a_j [the first factor in Equation (8.23)]. The second factor in Equation (8.23) increases the threshold for connection classes with higher traffic variability caused by the larger "bandwidth" requirement. Observe that in this setting both fairness and traffic maximization are taken into account.

The comparison of the DBR, DBRA, and DBRH policies with the optimal one is presented in Tables 8.5 and 8.6 for the traffic maximization and Raiffa objectives, respectively. The performance of the DBRA algorithm is in most cases close to that of DBR. The performance of the DBRH algorithm is, as could be expected, less accurate. Nevertheless in the case of the Raiffa solution the performance of DBRH is close to that of DBRA and much better than that of the CS policy which is also included in the tables.

8.2 Analysis of a Single-Link System

TABLE 8.5 Performance of approximated DBR policies (traffic maximization).

Example		OPT	DBR (t_j,h_j)	DBRA (t_j,h_j)	DBRH (t_j,h_j)	CS
1	$\overline{B}$	0.105	0.105	0.105	0.122	0.111
	B_2/B_1	1.405	1.148	1.148	0.367	17.041
	$j=2$		(2, 12)	(2, 12)	(3, 12)	
2	$\overline{B}$	0.030	0.030	0.030	0.074	0.030
	B_2/B_1	27.360	27.349	27.349	0.457	27.349
	$j=2$		(-1, 12)	(-1, 12)	(1, 12)	
3	$\overline{B}$	0.164	0.164	0.171	0.164	0.173
	B_2/B_1	0.332	0.283	3.279	0.434	5.691
	$j=2$		(15, 6)	(8, 6)	(13, 6)	
4	$\overline{B}$	0.045	0.045	0.045.	0.057	0.045
	B_2/B_1	6.835	6.834	6.818	0.856	6.818
	$j=2$		(0, 0)	(-1, 6)	(3, 6)	
8	$\overline{B}$	0.158	0.158		0.218	0.158
	$B_{max}B_{min}$	9.889	9.882		2.983	9.879
	$j=2$		(-1, 4)		(6, 4)	
	$j=3$		(-1, 10)		(1, 10)	

TABLE 8.6 Performance of approximated DBR policies (Raiffa objective).

Example		OPT	DBR (t_j,h_j)	DBRA (t_j,h_j)	DBRH (t_j,h_j)	CS
1	$\overline{B}$	0.105	0.182	0.105	0.122	0.111
	B'_2/B'_1	1.000	0.972	0.821	0.178	14.746
	$j=2$		(1, 26)	(2, 12)	(3, 12)	
2	$\overline{B}$	0.065	0.076	0.074	0.074	0.030
	B'_2/B'_1	0.998	1.000	0.455	0.455	28.129
	$j=2$		(3, 11)	(1, 12)	(1, 12)	
3	$\overline{B}$	0.166	0.171	0.166	0.164	0.173
	B'_2/B'_1	1.002	0.991	0.865	0.418	5.636
	$j=2$		(14, 5)	(11, 6)	(13, 6)	
4	$\overline{B}$	0.055	0.057	0.057	0.057	0.045
	B'_2/B'_1	1.011	0.945	0.945	0.945	8.464
	$j=2$		(3, 6)	(3, 6)	(3, 6)	
8	$\overline{B}$	0.204	0.223		0.218	0.158
	B'_{max}/B'_{min}	1.085	1.013		2.983	9.896
	$j=2$		(4, 3)		(6, 4)	
	$j=3$		(2, 9)		(1, 10)	

8.3 General Case

In the multi-link network case the features of solutions based on the game theoretic framework may differ from the single-link systems. In particular in most multi-link networks we have $\max_{u_j \in \mathbf{U}}\{\overline{a}_j\} = a_j$, $j \in J$ due to the large pool of network resources which can be accessed by each connection class using alternative routing. In these cases the Raiffa solution is equivalent to the loss equalization objective. One can also expect that the Pareto boundary is smooth (no edges) since the structure of network states is more convoluted and the number of these states is very large. This feature suggests that in the multi-link network case the Nash solution can be well suited to provide a balanced compromise between fairness and efficiency.

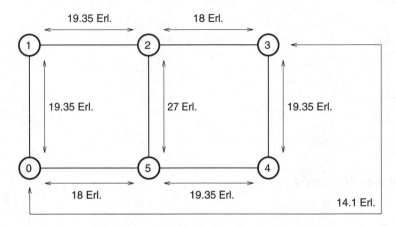

Figure 8.9. *Modification of example W6N'*.

TABLE 8.7 Connection loss probabilities [%] for different objective functions.

	Traffic maximiz.	Nash solution	Loss equaliz.	Traffic maximiz.	Nash solution	Loss equaliz.
$\overline{B}$	23	32	35			
OD pairs	NB connections			WB connections		
0-3	100	63	35	100	51	35
0-1, 1-2, 3-4, 4-5	4	27	35	27	30	35
0-5, 2-3	3	27	35	22	31	36
2-5	12	36	35	44	31	35

In the following we compare the Nash solution with the traffic maximization and loss equalization solutions on the basis of a simple example illustrated in Figure 8.9. This example is a modification of Example W6N' from Table 7.2 achieved by adding traffic offered to OD pair 0-3. The modified example can be considered as a situation after the failure of link 0-3. In this case the traffic maximization policy can result in a very high loss probability for this OD pair and the question arises as to what is the cost of providing more equitable access for this commodity.

The performance of the considered example for the three control objectives is given in Table 8.7. The results are achieved from an optimization procedure based on the Powell algorithm (Mordecai, 1976) assuming the MDPD reward maximization routing (Chapter 5). The procedure employs the network performance model described in Chapter 7. As expected, the traffic maximization operating point rejects almost all connections offered to the 0-3 OD pair. The loss equalization objective provides equitable access for all connection classes but the average loss probability is significantly increased. The Nash solution gives much better access to the connections offered to the 0-3 OD pair when compared with traffic maximization. At the same time the average loss probability of the Nash solution is smaller than that of the loss equalization, although it is still significantly higher than that of traffic maximization. These results show that, in the case of multi-link networks, the Nash solution provides a compromise between fairness and efficiency. Obviously, the final choice of the operating point depends on preferences and motivations of the network operator.

8.4 Discussion and Bibliographic Notes

While the issue of fairness has received some attention in telephone and data networks, it has usually been considered as a secondary problem. Typically control policies which use the available resources efficiently are sought. For example, the usual traffic control objectives are throughput maximization for telephone networks and average delay minimization for data networks. Once the *system optimum* is found the grades of service offered to the individual users (origin-destination pairs) are compared and some post optimization adjustments are then made if specific users do not receive an adequate grade of service.

The issue of trade-off between the efficiency and fairness has gained prominence in multi-service environments and resulted in substantial literature on connection admission in broadband networks. Although many of these works, e.g. (Gopal and Stern, 1983; Miyake, 1988; Ross and Tsang, 1989), are oriented towards throughput maximization, there are more and more studies focused on the problem of fair grade of service distribution among different services. These studies can be categorized by two different approaches. In the first one, e.g. (Oda and Watanabe, 1990; Ash et al., 1991), the GoS constraints are directly taken into account during the optimization process. In the second approach each connection class is allocated a certain weight or a reward parameter and the control objective is to minimize the network cost, e.g. (Gersht and Lee, 1988), or to maximize the reward from the network, e.g. (Kelly, 1988; Liao, Dziong, and Mason, 1989; Dziong and Mason, 1994). Thus by changing the values of weights or reward parameters different GoS allocations among the connection classes can be achieved.

Game theoretic formulations of network control problems are not new although to date they have not received widespread acceptance in the literature. In (Mason, 1972) the routing problem for telephone networks was formulated as a multi-person stochastic game where the individual players correspond to the origin-destination pairs. Stochastic learning automata were employed to obtain the non-cooperative solution (user equilibrium) for a simple circuit-switched network. The Nash equilibrium solution, where the players correspond to origin-destination pairs, and the system optimal (maximum throughput) solutions for some telephone networks were compared in (Harris, 1973) by way of simulation. The issue of user versus system optimization was discussed in (Mason and Girard, 1982). It has been noted that user optimized networks, while not always as efficient as system optimized networks, do appear to provide more equitable service to the users.

Game theoretic concepts were also employed in obtaining user optimal and system optimal routing policies for data networks (Mason, 1985). A technique first proposed in (Dafermos and Sparrow, 1969) was then employed which enabled the off-line computation of the system optimal routing policy by distributed learning automata. In (Douligeris and Mazumdar, 1987) it was demonstrated that user optimized solutions are not efficient in the Pareto sense and algorithms for computing efficient solutions were proposed. In (Mazumdar, Mason, and Douligeris, 1991) it was shown that the criterion of user powers' product corresponds to the Nash arbitration strategy of cooperative game theory. An extensive review of game theory applications to traffic control in data networks is given in (Douligeris and Kumar, 1992).

In this chapter a game theoretic framework for efficient and fair utilization of network resources in multi-rate service environment was discussed. This framework can be used to study general multi-link networks and single-link systems. The latter can correspond to CAC_{GoS} mechanisms decomposed into single-link problems (CAC^l_{GoS}) or to CAC in ATM networks based on the virtual path design. In single-link systems the arbitration solutions from co-operative game theory have several attractive features especially in overload conditions. The dynamic formulation of the framework fairly distributes an increase in throughput caused by an overload by using the nominal conditions as a reference point. Application of the game theoretic framework to study simplified link CAC^l_{GoS} policies shows that the "bandwidth" reservation and dynamic "bandwidth" reservation policies are, in many cases, near optimal in terms of efficiency and fairness and can provide a practical solution to the problem. In the general case of multi-link networks, the game theoretic framework can be used to find an optimal operating point according to desired criteria. In particular the Nash arbitration scheme provides a compromise between the two traditional objectives: traffic maximization and loss equalization.

References

Ash, G.R., Chen, J.-S., Frey, A.E., and Huang, B.D. 1991. Real-time network routing in a dynamic class-of-service network. In *Proceedings of the 13th International Teletraffic Congress*, pp. 187–194. North-Holland.

References

Cao, X. 1982. Preference functions and bargaining solutions. In *Proceedings of IEEE CDC-21*, pp. 164–171. IEEE Computer Society Press Proc.

Dafermos, S.C., and Sparrow, F.T. 1969. The traffic assignment problem for a general network. *J. N.B.S.-B, Mathematical Sci.* 73B(2).

Douligeris, C., and Mazumdar, R. 1987. On Pareto optimal flow control in integrated environments. In *Proceedings of the 25th Allerton Conf.* Univ. Illinois, Urbana.

Douligeris, C., and Kumar, L.N. 1992. A survey into the fairness problem in the networking environment. In *Proceedings of IEEE ICC'92*. IEEE Computer Society Press.

Dziong, Z., Liao, K-Q., Mason, L.G., and Tetreault, N. 1991. Bandwidth management in ATM networks. In *Proceedings of the 13th International Teletraffic Congress*, pp. 821–827. North-Holland.

Dziong, Z., and Mason, L.G. 1994. Call admission and routing in multi-service loss networks. *IEEE Transactions on Communications* 42(2/3/4):2011–2022.

Dziong, Z., and Mason, L.G. 1996. Fair-efficient call admission for broadband networks - A game theoretic framework. *IEEE/ACM Transactions on Networking* 4(1):123–136.

Foschini, G.J., Gopinath, B., and Hayes, J.F. 1981. Optimum allocation of servers to two types of competing customers. *IEEE Transactions on Communications* 29(7).

Gersht, A., and Lee, K.J. 1988. Virtual-circuit load control in fast packet-switched broadband networks. In *Proceedings of IEEE GLOBECOM'88*. IEEE Computer Society Press.

Gopal, I.S., and Stern, T.E. 1983. Optimal call blocking policies in an integrated services environment. In *Conf. Inform. Sci. Syst.*, pp. 383–388. The Johns Hopkins University.

Harris, R.J. 1973. Concepts of optimality in alternate routing networks. *A.T.R.* 7(2):3–8.

Kalai, E., and Smorodinsky, M. 1975. Other solutions to Nash's bargaining problem. *Econometrica* 43:513–518.

Kaufman, J.S. 1981. Blocking in a shared resources environment. *IEEE Transactions on Communications* 29(10):1474–1481.

Kelly, F.P. 1988. Routing in circuit switched networks: optimization, shadow prices and decentralization. *Adv. Appl. Prob.* 20:112–144.

Liao, K.-Q., Dziong, Z., and Mason, L.G. 1989. Dynamic link bandwidth allocation in an integrated services network. In *Proceedings of IEEE ICC'89*. IEEE Computer Society Press.

Luce, D., and Raiffa, H. 1957. *Games and Decisions.* New York: Wiley.

Mason, L.G. 1972. Self-optimizing allocation systems. PhD Thesis, University of Saskatchewan.

Mason, L.G., and Girard, A. 1982. Control techniques and performance models for circuit switched networks. In *Proceedings of IEEE CDC-21*. IEEE Computer Society Press.

Mason, L.G. 1985. Equilibrium flows, routing patterns and algorithms for store-and-forward networks. *Journal of Large Scale Systems* 8:187–209.

Mazumdar, R., Mason, L.G., and Douligeris, C. 1991. Fairness in network optimal flow control: Optimality of product forms. *IEEE Transactions on Communications* 39(5).

Mordecai, A. 1976. *Non-Linear Programming: Analysis and Methods.* Prentice-Hall.

Miyake, K. 1988. Optimal Trunk Reservation Control for Multi-slot Connection Circuits. *Trans. of IEICE.* November.

Nash, J. 1950. The Bargaining Problem. *Econometrica* 18.

Oda, T., and Watanabe, Y. 1990. Optimal trunk reservation for a group with multislot traffic stream. *IEEE Transactions on Communications* 38(7).

Raiffa, H. 1953. Arbitration schemes for generalized two-person Games.. In *Contributions to the Theory of Game II*, edited by H.W.Kuhn and A.W.Tucker. Princeton.

Ross, K.W., and Tsang, D.H.K. 1989. Optimal circuit access policies in an ISDN environment: A Markov decision approach. *IEEE Transactions on Communications* 37(9):934–939.

Stefanescu, A., and Stefanescu, M.W. 1984. The arbitrated solution for multiobjective convex programming. *Rev. Roum. Math. Pure. Appl.* 29:593–598.

Chapter

9

Virtual Networks as a Tool for Resource Management

THE INTEGRATION of all services into one uniform transport layer is seen as a major advantage of the ATM standard. Nevertheless, this integration also creates several new problems. In particular the broad range of services, traffic characteristics, time scales, and performance constraints, which are integrated into one transport system, causes the *resource management and traffic control* issues to become very complex and difficult. That is why, in many cases, the resource management and traffic control architecture is trying to reintroduce some kind of separation in order to make the problem manageable. The virtual network concept can ideally serve this purpose since it can provide a means for two types of separation. The first is *separation of management functions* in order to allow customization to particular needs of certain services and user groups. The second is *virtual separation of resources* in order to simplify the resource management functions and provide grade of service (GoS) guarantees for some services and user groups. In this chapter we present a coherent framework for resource management and traffic control which takes full advantage of the virtual network concept.

We start with the generic virtual network definition which can be used in many applications (Section 9.1). These applications can be divided into three categories: service oriented VN, user oriented VN and management oriented VN (Section 9.2). Several applications can coexist in the same virtual network, which implies that virtual networks can be nested. The relation between virtual networks and virtual paths is discussed in Section 9.3. In particular we show that in general the virtual path connections should not be used for resource management purposes but rather for routing and switching simplification. Resource allocation to service oriented virtual networks is a function of cell layer scheduling and flow control procedures. These issues are discussed in Section 9.4, including application of models for handling logical "bandwidth" allocations presented in Section 3.2.3. "Bandwidth" allocation enforcement on the virtual network layer can be realized logically by the connection admission algorithm because "bandwidth" allocation to connections can be enforced by UPC at UNI. A virtual private network is an exception, if the connection admission procedure is under user supervision. Possible solutions to this problem are presented in Section 9.5.

While each of the VN applications may have specific issues that need resolving, there are three generic issues common to most applications: virtual network design, virtual network design update, and design of a backup virtual network. In Section 9.6 we describe an algorithm for virtual network design which utilizes the reward maximization routing described in Chapter 5. It is based on an optimization procedure that designs the virtual network topology, allocates "bandwidth" to virtual network links, and selects routing parameters. Due to imposed economic constraints, which link the "bandwidth" cost with the CAC & routing algorithm, the solution can be achieved by a simple iterative algorithm. In general virtual network design is based on predicted traffic parameters which may be inaccurate or may be a function of time. In Section 9.7 we discuss several options for virtual network design updates based on traffic measurements. Some of these approaches can serve as an alternative implementation of the algorithm for virtual network design which uses measurements in place of an analytical model for performance evaluation. Backup virtual networks are created to enable a specified fraction of the carried connections to be restored in case of a network component's failure (e.g., a link or node). Some of the issues associated with backup virtual network design are covered in Section 9.8. More detailed discussion of this design is given in Chapter 10 since many of the problems are common with the physical network design.

9.1 Virtual Network Definition

A virtual network is defined by a set of virtual network nodes and a set of virtual network links (VNL) connecting the nodes. The virtual network is referred to by a *virtual network identifier* (VNI). In general the set of VN nodes is a subset of the physical network node set although in some cases both sets can be equal. The virtual network link defines a path (consisting of one or more physical links) between two VN nodes and is referred to by a virtual network link identifier, VNLI. In general there may be more than one VNL between two VN nodes. Nevertheless in the following we limit our considerations to a single VNL case since it is the most likely case.

A set of resources can be allocated to the virtual network. In particular this set can include "bandwidth" and a set of resource management objects. Figure 9.1 illustrates the logical "bandwidth" allocation to virtual network links along a particular VCC which requires "bandwidth" d. The connection is established on two VNLs: VNL1 and VNL2. Each of these VNLs is allocated a certain "bandwidth" (G^1 and G^2, respectively) from the physical link "bandwidths," L^1, L^2, L^3. The connection can be established if "bandwidth" d can be reserved on each of the VNLs.

9.2 Virtual Network Applications

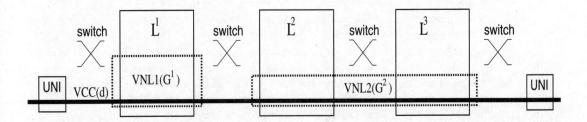

Figure 9.1. *Allocation of resources to VNL.*

Several virtual networks can co-exist in a physical network (PN). They can constitute independent entities, but in some cases a particular virtual network can be nested in its parent virtual network or its resources can be divided into several subsets, each of them being nested in a different parent virtual network. These possibilities are illustrated in Figure 9.2 where the virtual networks are represented as sets. Interrelations among particular applications of virtual networks are discussed in the next section.

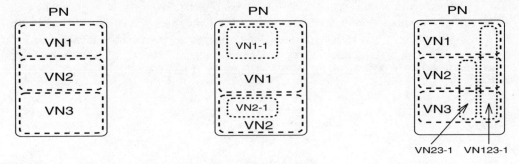

Figure 9.2. *Examples of generic relations between VNs.*

9.2 Virtual Network Applications

In general VN applications can be divided into three categories: service, user, and management oriented. Service oriented virtual networks are created to separate management functions specialized for different services (e.g., real-time vs. data services) and/or to simplify the QoS management (each QoS class is served by a separate virtual network). Allocation of "bandwidth" to service oriented VNs aims at providing sufficient GoS and fairness on the connection layer for different services. Moreover, "bandwidth" allocation to QoS virtual networks can increase resource utilization and simplify "bandwidth" allocation to connections, as will be discussed in Section 9.4. In most cases the set of service oriented VN nodes is equal to the set of physical network nodes.

User oriented virtual networks are created for groups of users who have specific requirements (e.g., guaranteed throughput, customized control algorithms, resource management under user control, increased security and reliability, "group" tariff, etc). The two most likely applications are private networks and multi-point connections. Creation of virtual networks for multi-point connections is especially attractive in the case of many-to-many connections with strong negative correlation between the signals transmitted in different multi-pont VC's. This follows from the fact that in this case such a VN can provide very good "bandwidth" utilization (see Section 6.1). In most cases the set of user oriented VN nodes will include only a subset of the physical network nodes.

Management oriented virtual networks are created to facilitate some of the management functions (not associated with a particular service or user group). The first application is connected with fault management and is called a backup virtual network. The "bandwidth" allocated to a backup VN should ensure that in case of a network component failure (e.g., a link or node), a specified fraction of connections affected by the failure can be restored in the backup VN. The second possible application is aimed at simplification of the "bandwidth" reservation procedure during connection setup. In particular, if all connections are routed via end-to-end VNLs which connect directly UNIs of the source and destination, the connection admission procedure at the origin UNI can reserve "bandwidth" required by a new connection based on local information only. Thus there is no need for "bandwidth" reservation in the transit nodes. This application is in fact equivalent to using end-to-end virtual paths for resource management. In Section 9.3 we indicate that this approach has several drawbacks and should be avoided if possible.

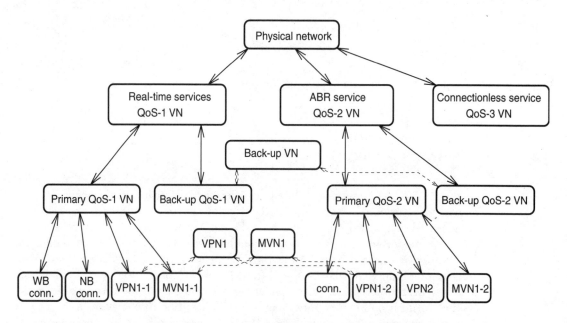

Figure 9.3. *Interrelations between different applications of virtual networks.*

Interrelation between different applications of VNs, from the "bandwidth" allocation viewpoint, is illustrated in Figure 9.3. This interrelation has a dominant hierarchical structure. The highest layer of this hierarchy is defined by the service oriented VNs associated with different QoS constraints. The allocation of "bandwidth" to QoS VNs is a function of ATM link scheduling mechanisms and can be done in several ways, as discussed in Section 9.4. Each of the QoS virtual networks can be divided into two parts, which are referred to as primary and backup VNs. A primary QoS VN is used for carrying connections demanding corresponding QoS features. The "bandwidth" allocated to a backup QoS VN is reserved for a partial restoration of connections, established in the corresponding primary QoS VN, which are affected by a network component failure. The "bandwidth" allocated to the primary QoS VN can be divided further into service and user oriented VNs. For example, due to a very large difference in "bandwidth" requirements of different services, it may be practical to create separate VNs for wide-band and narrow-band connections. This approach provides a simple tool for providing different services with fair access to the network resources. Moreover, the CAC & routing protocols can be simplified. Other potential virtual networks nested in a primary QoS VN are VPNs and VNs associated with multi-point connections (MVNs).

Apart from the top-down hierarchical structure of the virtual network interrelations, resulting from the QoS constraints, there are also potential horizontal associations. A particular user oriented VNs, handling multimedia services, can require different QoS "bandwidth" for different components of multimedia connections at the same time. In this case the user oriented VN can be composed of several subnetworks, each of them nested in different primary QoS VN. While "bandwidth" allocation to each of these subnetworks has to be negotiated separately with the resource manager of the corresponding QoS VN, there are several service management functions which require coordination between the subnetworks (e.g., CAC & routing of multimedia connections). From this perspective the set of subnetworks is treated as one virtual network. Analogously, all backup QoS VNs can be viewed as one backup VN since the design and management of each backup QoS VN has many common elements.

9.3 Relation Between Virtual Network and Virtual Path Concepts

In general virtual network links could correspond directly to virtual path connections. In such a case the virtual network link identifier is equal to the virtual path identifier, which means that the virtual paths are also used for resource management. On the other hand, the original function of the virtual path concept is to simplify network routing addressing and switching (see Chapter 1). In other words a virtual path defines a *connection routing path* (CRP) on which the connection setup can be significantly simplified by using only cross-connect switches. In the following we analyze whether expansion of the virtual path functionality, by adding resource management functions, is beneficial.

Let us compare three generic cases of the relation between the virtual network links and virtual paths.

- VP_{ee} design: VPC=CRP=VNL, resource management by VPC. In this case the virtual paths define first the optimal connection routing paths. Then the "bandwidth" is allocated to the virtual paths according to the traffic matrix. To simplify presentation we assume that the virtual path address range is sufficient to create end-to-end virtual paths for all routing paths between all OD pairs. The main motivation for considering this case is simplification of the CAC procedure which can be realized in UNI without communication with any of the transit nodes. An example of VP_ee structure is given in Figure 2.4.

- VP_{nn} design: VPC=VNL, resource management by VNL. Here the virtual paths are also equal to virtual network links, but the corresponding virtual network is designed to minimize the VN "bandwidth" cost for given traffic matrix and GoS constraints. In this case a connection routing path can be composed of several virtual paths. An example of VP_nn structure is given in Figure 2.4.

- VN design: VPC=CRP, resource management by VNL. As in VP_{ee} design the virtual paths define the optimal connection routing paths. However, resource management is realized by means of the virtual network, as in VP_{nn} design. Thus the functions of VPC and VNL are separated. In particular a virtual path can pass through several VNLs and several virtual paths can be carried on one VNL.

In the following we discuss important features of these cases from the viewpoint of resource utilization, adaptation to traffic changes, bandwidth reservation, updating of routing tables, and scalability.

Resource utilization. The resource utilization of VP_{ee} design can be significantly inferior to that of VP_{nn} design and VN design, especially for large networks. This follows from the fact that in VP_{ee} design each physical link can be used by many end-to-end virtual paths from the same virtual network. Thus the physical link "bandwidth" is divided into a large number of parts allocated to the virtual paths which are seen as independent links. This results in poor resource utilization due to restricted statistical multiplexing. Firstly, contrary to VP_{nn} design and VN design there is no statistical multiplexing on the connection layer between connections using different virtual paths on the particular physical link. This reduces the resource utilization since, for real-time connection streams with a given GoS constraint, the smaller the link "bandwidth" the smaller the "bandwidth" utilization. Secondly, assuming that the CAC decision is based solely on local information in the new connection UNI, the gain from statistical multiplexing on the cell layer between connections using different virtual paths on a particular physical link cannot be realized. Thus the equivalent bandwidth of real-time connections with variable bit rate is larger than in VP_{nn} design and VN design. Thirdly, the gain from alternative routing cannot be realized due to fixed allocation of the "bandwidth" to all

9.3 Relation Between Virtual Network and Virtual Path Concepts

virtual paths between a particular OD pair. These factors cause the "bandwidth" required in VP_{ee} design to be up to several times larger than in VP_{nn} design and VN design. An additional problem can arise when the "bandwidth" allocation to virtual paths is smaller than the "bandwidth" required by a wide-band connection. This problem is illustrated in Figure 9.4 where an example based on a network with three QoS virtual networks is presented. Assuming a symmetrical traffic matrix, in VP_{ee} design each virtual path is allocated 2.7 Mbps. Thus, even in this small network example, the "bandwidth" allocated to a virtual path, based on the traffic distribution, is not sufficient to carry a good quality video signal with the average rate of let say 4 Mbps. As a result the "bandwidth" allocation to virtual paths would have to be changed at the moment of wide-band connection arrival.

Routing paths (CRP) on link 2-5:

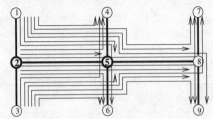

Example:

three QoS virtual networks,
150 Mbps links,
symmetrical traffic matrix.

Virtual network links used by routing paths on VNL 2-5:

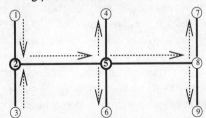

Results:

VP_{ee} design : VPC=CRP, BM by VPC

2.7 Mbps per VPC

VN design : VPC=CRP, BM by VNL

50.0 Mbps per VNL

Figure 9.4. *Analysis of possible VPC applications within the VN framework.*

Adaptation to traffic changes. In VP_{nn} design and VN design the adaptation of VN "bandwidth" allocation to traffic changes is significantly simpler in comparison with VP_{ee} design. This follows from the fact that, due to alternative routing, the adaptation in VP_{nn} design and VN design is, in most cases, necessary only when the overall traffic offered to the VN exceeds the design level. In contrast in VP_{ee} design the adaptation is necessary whenever the traffic offered to a particular OD pair exceeds the design level, although the other virtual paths of this VN can be underutilized at the same time. Moreover, the adaptation in VP_{nn} design and VN design can be done in a distributed manner in the VNL resource managers (see Section 9.7), while in VP_{ee} design the process requires communication between several nodes or has to be centralized.

"Bandwidth" reservation in the connection setup procedure. In VP_{ee} design "bandwidth" reservation for a new point-to-point connection can be realized at UNI since the end-to-end virtual path state is known locally. On the other hand, in VP_{nn} design and VN design admission of a new connection requires "bandwidth" reservation in the VNL's resource managers. Thus if the routing path consists of more than one VNL, the reservation involves communication between the origin node and the transit nodes of the routing path in the VN. The cost of this communication, in terms of processing capacity and time delay, depends on the employed signaling protocol. The traditional approach, derived from a circuit switched environment, is based on meta-signaling, e.g., B-ISUP standard (ITU-T Rec. Q.2761-Q.2764, 1993), which provides a general protocol for communication between the network control units. Its generality causes relatively high complexity since it has to cover the complete protocol stack and the transmitted information field has to include the complete connection and function identification. In large networks this type of protocol can require significant processing capacity and can cause noticeable delays in the CAC procedure, see e.g. (Veeraraghavan et al., 1995). An attractive alternative can be implemented by means of the resource management cells (called CAC cells in this application — see Section 4.1) which are already used for "bandwidth" reservation for CTP services (e.g., ABR, ABT — see Section 2.1). In this case the protocol is significantly simplified since the header of the CAC cell directly identifies the function and switch output port in which the "bandwidth" reservation is required (via VPI corresponding to the chosen routing path). The cost of this resource management cell functionality extension is negligible due to the fact that the functions of CAC cells and RM cells for CTP services are very similar, and the generation rate of CAC cells is few orders of magnitude smaller than that of RM cells used for CTP services. "Bandwidth" reservation by means of CAC cells is, most effective when the connection routing paths are equal to virtual paths as in VN design. In VP_{nn} design this approach is not attractive since one must first establish the virtual channel connection along the routing path which requires communication with some of the transit nodes by means of meta-signaling. Note that in the case of multi-point connections, VP_{ee} design also requires communication with transit nodes. This results from the fact that it is not practical to pre-establish multi-cast virtual paths with reserved "bandwidth" due to the enormous number of such paths. On the other hand, using a set of point-to-point VPs for realizing multi-point connections would further decrease resource utilization in VP_{ee} design (see performance of the LLP strategy in Tables 6.4 and 6.5).

Updating routing tables. In VP_{ee} design and VN design the routing paths are associated with end-to-end virtual paths. This allows the use of cross-connect switching which does not require updating of routing tables in switches at the connection setup time. In VP_{nn} design the routing paths can be composed of more than one virtual path; therefore the routing tables have to be updated at the VN transit nodes in order to establish a VC connection. This feature increases the switches' cost and complexity of the connection setup procedure.

Scalability. The most noticeable scalability problem is in VP_{ee} design. Namely, for large and not well connected networks the number of paths carried on a physical link can be so large that, besides very poor resource utilization, the "bandwidth" allocation to these paths would have to be changed with almost every new wideband connection arrival, thus defeating the main advantage of this option.

TABLE 9.1 Comparison of different VPC applications.

	VP_{ee} design	VP_{nn} design	VN design
resource utilization	$--$	$++$	$++$
adaptation to traffic changes	$--$	$++$	$++$
"bandwidth" reservation	$++$	$--$	$+$
updating routing tables	$++$	$--$	$++$
scalability	$--$	$+$	$+$

$++$ very good, $+$ good, $-$ bad, $--$ very bad

An abbreviated comparison of the considered generic cases, resulting from the above discussion, is given in Table 9.1. The comparison shows that using VPCs for resource management involves some contradictions. First, when used for connection setup simplification (VP_{ee} design), the "bandwidth" utilization drops significantly due to the restrictions in statistical multiplexing (on the connection and cell layers) between the virtual paths. Second, when used to optimize resource utilization (VP_{nn} design), the connection setup becomes complex since the routing tables in the nodes between VPCs constituting the routing path have to be updated. VN design avoids these inherent contradictions by separation of the resource management function from VPC implementation. In this case the virtual paths are used to simplify routing and switching only, while the virtual networks are used to optimize resource utilization. This approach is illustrated in Figure 9.5 where a possible implementation of the VNL identifier is also shown. In such a case VNLI is defined by a part of the VPI field with the highest weight. Alternatively VNLI can be a general function of VPI or it may be unrelated directly to VPI.

It should be noted that under certain technical constraints, VP_{ee} design can be a viable solution. In particular this may be the case if there is "bandwidth" abundance, or there is no efficient signaling protocol, or the CAC_{GoS} control procedure is outside the network manager's jurisdiction (e.g., VPN — see Section 9.6).

9.4 Resource Allocation to QoS Virtual Networks

Resource allocation to QoS virtual networks is a function of the applied scheduling mechanism and can be done in several ways. In the following we discuss three generic options. Consider an example with two distinct QoS classes in the network. Each of these classes is managed by its QoS virtual network. To provide high resource utilization a buffer is allocated to each of these virtual networks. Assume

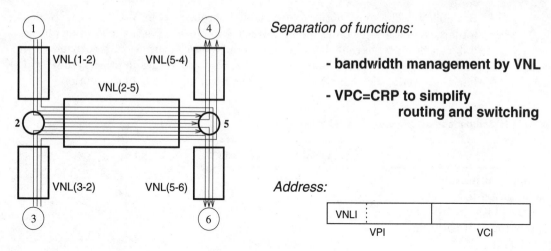

Figure 9.5. *Decomposition of VN and VP functions.*

that the higher the VN index, the more stringent the QoS requirements. The service priorities are implemented by a scheduler which can be realized as a non-preemptive multi-priority system or a fair queuing system (see Chapter 1).

Three basic approaches to "bandwidth" allocation to virtual networks are presented in Figure 9.6. To illustrate the main features of these approaches, a schematic representation of the exact admissible regions (continuous line) and linear admissible regions corresponding to each of the approaches (dotted lines) is shown in Figure 9.7. These admissible regions correspond to a non-preemptive multi-priority system and statistical QoS constraints (see Section 3.2.3). Nevertheless similar features are expected for other scheduling algorithms which provide different QoS classes.

Complete sharing of the "bandwidth" (Figures 9.6a and 9.7a): Here all virtual networks share the same pool of link "bandwidth." Thus full statistical multiplexing on the connection layer, between virtual networks, is achieved which can provide good average resource utilization. On the other hand, if linear admissible regions are required to simplify the CAC procedure, the resource utilization may be reduced due to the gap between the exact and linear admissible regions. Moreover, this approach requires additional tools for providing fair access for services associated with different QoS VNs.

Limited sharing of the "bandwidth" (Figures 9.6b and 9.7b): In this case the "bandwidth" allocated to a particular priority can also be used by lower priority connections. This feature gives limited statistical multiplexing between the virtual networks on the connection layer. In this case the linear admissible regions are efficient if the "bandwidth" allocated to higher priority VNs provides required GoS (small gap between the exact admissible region and the inclined part of the linear region in Figure 9.7b). To provide access fairness, tools for protecting high priority traffic access against low priority traffic are required. A particular implementation of such a tool is the flow control mechanism for CTP services (see Chapter 2). In this case the high priority real-time connections can always be accepted within the "bandwidth" allocated to this service, regardless of the CTP connection state. The CTP connections requiring the minimum "bandwidth" allocation are accepted if the sum of the allocated minimum "bandwidths" does not exceed the "bandwidth" reserved for exclusive use of CTP services. The free "bandwidth" of the real-time VN can be used by the CTP service under the flow control mechanism which regulates the CTP connection rates between the minimum and maximum rates in such a way that the QoS constraints are met.

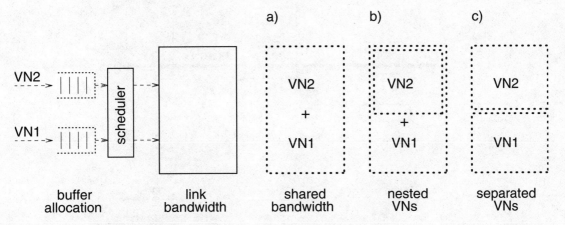

Figure 9.6. *Resource allocation alternatives in service oriented VN.*

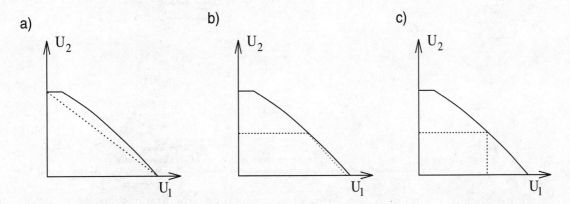

Figure 9.7. *Admissible regions for homogeneous on-off sources (U_1, U_2 — connection class throughput).*

"Bandwidth" separation (Figure 9.6c and 9.7c): In this case there is no statistical multiplexing among the virtual networks on the connection layer. The linear admissible region is efficient only when the ratio of the VN traffic levels is similar to the designed operating point. Thus, to achieve high resource utilization, the "bandwidth" allocation should adapt to the changes in the traffic matrix (this issue is discussed in Section 9.7). The "bandwidth" separation provides a simple tool for controlling GoS fairness among VNs.

The choice of a particular resource allocation scheme depends on the considered services and the design objectives.

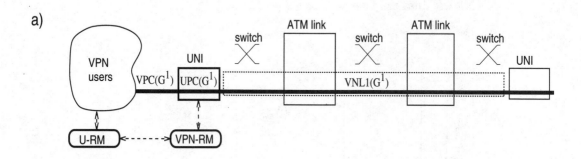

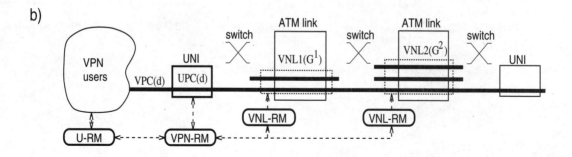

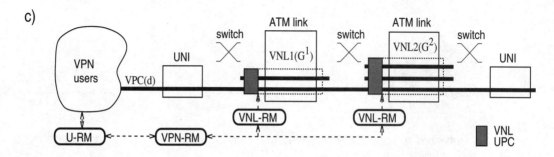

Figure 9.8. *"Bandwidth" enforcement in virtual private networks.*

9.5 "Bandwidth" Enforcement in Virtual Private Networks

In virtual private networks, the connection admission procedure and "bandwidth" allocation to VCCs can be under user responsibility. In this case the virtual private network resource manager should provide enforcement of "bandwidth" allocation to VNLs in a manner that is independent from the user's actions. Three basic possibilities are illustrated in Figure 9.8. The simplest solution is when the virtual network links are realized as end-to-end VPC (VP_{ee} design from Section 9.3). In this case the "bandwidth" allocated to VNL can be enforced by a user parameter control (UPC) algorithm applied to VPC at UNI as shown in Figure 9.8a.

When the functions of VPN links and end-to-end VPCs are separated (VN design), the enforcement can be realized assuming that communication exists between the VPN resource manager (VPN-RM) and the user resource manager (U-RM). In this case the end-to-end VPC is treated by VPN-RM as a connection with "bandwidth" enforced by UPC. Whenever U-RM wants to increase or decrease "bandwidth" reserved for a particular VPC, it asks the VPN-RM to change the UPC parameters. The request is realized if the sum of "bandwidth" required by VPCs does not exceed the VNL capacities. While this scheme, shown in Figure 9.8b, can be seen as an adaptive version of VP_{ee} design, there are two important differences. First, the overall "bandwidth" allocated to VPN links remains intact. Second, the statistical multiplexing on the cell layer among the VPCs using the same VNL can be taken into account when "bandwidth" used by VNL is evaluated.

Another alternative is to enforce "bandwidth" allocated to VNL directly at the switch output port at the origin of this VNL. This option can easily be realized if scheduling in the switch output ports is based on fair queuing (see Chapter 1). Otherwise the VNL UPC mechanism could be installed in a special server inserted in a loop joining a designated output port with a designated input port of the switch under consideration. In this case all connections using this VNL have to be switched via the loop. This architecture is similar to the one for connectionless services illustrated in Figure 1.12.

9.6 Virtual Network Design

The process of virtual network design is illustrated in Figure 9.9. The key element of the VN design method is an optimization procedure which gives the VN topology, allocation of "bandwidth" to VNLs and routing algorithm parameters. The optimization procedure is supplied with the offered traffic matrix, GoS constraints, "bandwidth" cost functions, route setup cost, "bandwidth" reservation cost during the setup procedure, and routing policy. This optimization stage results in a VN demand for resource allocation which is considered by a higher level resource manager (physical network or virtual network in case of nested virtual networks). The demand can be accepted if there are enough free resources or rejected in case of resource shortage. In some cases of resource shortage the demand can be realized partially. In such a case the optimization routine can be applied one more time to take into account the resource limitations.

The optimization procedure depends on many factors, including the applied CAC & routing mechanism, required accuracy, and design time constraint which can vary from days (new physical network) to seconds (multi-point VN). In general optimization of virtual networks is simpler than optimization of physical networks. This is caused by the fact that in the case of virtual networks the assumptions of linear "bandwidth" cost and the treatment of the VNL capacity as a continuous variable correspond well to reality. This is not the case for physical networks where there is usually an initial cost of the link installation and the link "bandwidth" can only be increased in some modules which can be quite large. Nevertheless in many cases the linear cost and "bandwidth" continuity assumptions are also used to find an initial solution for the modular problem of resource allocation on the physical network layer. That is why many of the dimensioning procedures for physical networks, discussed in Chapter 10, can also be used for optimization of virtual networks. In the next section we describe an optimization procedure for design of virtual networks with the reward maximization routing presented in Chapter 5. The main advantage of this model is the introduction of economic constraints which significantly simplify the optimization procedure. This approach can also serve for optimization of networks with other routing strategies since the connection flow distribution resulting from reward maximization strategy can be approximated by a simplified routing strategy in the actual network. The model also provides a simple framework for adaptation of virtual networks to variations in the offered traffic matrix. This framework is described in Section 9.7.

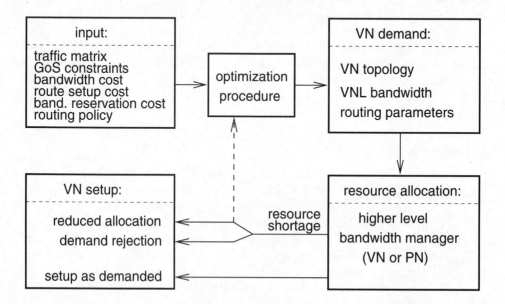

Figure 9.9. *Virtual network design.*

9.6.1 Optimization procedure for reward maximization routing*

To simplify presentation, in the following we consider only the "bandwidth" cost. Nevertheless the route setup and "bandwidth" reservation costs can be easily incorporated in the procedure by reducing reward parameters of connections using multilink paths. The linear cost function for VNL "bandwidth" is given by

$$c^s(G^s) = G^s \cdot c^{s'} \cdot \gamma^s \qquad (9.1)$$

where G^s denotes "bandwidth" allocated to the sth VNL, $c^{s'}$ is the cost of the "bandwidth" unit, and γ^s denotes the normalization factor. The normalization factor is introduced to simplify description of the optimization model. In this case a connection requires the same "bandwidth" allocation on all VNLs. Since in reality the connection equivalent bandwidth may depend strongly on the physical link speed, the normalization factor takes into account this effect. For example, if there are links with two speeds (e.g., 150 and 600 Mb/s) resulting in the expected average "bandwidth" requirement which differ by a factor of 1.5, we can have $\gamma^s = 1$ for a VNL consisting of one high speed link or $\gamma^s = 1.5$ for a VNL consisting of one low speed link. In the case of a path consisting of high speed and low speed links with respective costs c'_l and c'_h, we have $\gamma^s = (1.5c'_l + c'_h)/c^{s'}$ where $c^{s'} = c'_l + c'_h$. It is obvious that once the solution is found the "bandwidth" allocated to the VNLs has to be transformed into "bandwidth" required on each physical link, according to the definition of normalization factors. Observe, that for virtual networks the cost of "bandwidth" unit, $c^{s'}$, can be defined by a real cost of resources or by a dynamic cost which can be expressed as an average shadow price. This average shadow price defines the influence of increasing (or decreasing) the VNL "bandwidth" allocation by one unit on the average reward from all link s connections excluding the considered VN.

If the CAC & routing algorithm was determined in advance, our problem would be formulated as a traditional dimensioning problem with the objective to find

$$\min_{\mathbf{G}} \left\{ c = \sum^s c^s(G^s) \right\} \qquad (9.2)$$

subject to inequality connection loss probability constraints

$$B_j(\mathbf{G}) \leq B_j^c, \quad j \in J \qquad (9.3)$$

where $\mathbf{G} = [G^s]$ is the VN "bandwidth" allocation vector, $B_j(\mathbf{G})$ denotes class j connection loss probability, and B_j^c denotes the loss constraint. The inequality loss constraints are used due to the fact that once the CAC & routing algorithm is determined, it may be not feasible to achieve a solution with equality loss constraints or such a solution may be not cost efficient if the constraint functions are not independent.

As illustrated in Chapter 5, in the case of CAC & routing based on state-dependent link shadow prices, the connection class loss probabilities can be controlled almost continuously and independently by means of the connection reward parameters. Based on these features combined with the convexity of the Pareto boundary (**Theorem 3** in Section 8.1.2) it can be easily shown that the minimum network cost is achieved when the connection class loss probabilities equal the constraints. Thus by applying this CAC & routing strategy and treating the reward parameters as optimization variables, the optimization procedure can be formulated with equality loss constraints

$$\min_{\mathbf{G},r}\left\{c = \sum^{s} c^s(G^s)\right\} \quad (9.4)$$
$$B_j(\mathbf{G},r) = B_j^c, \quad j \in J$$

One could try to solve this problem by a general descent method. This approach has some disadvantages. The first is a requirement of an accurate model for the function derivatives. The second is the fact that gradient methods can find a local minimum which is not optimal or the procedure can be stuck in a flat region which does not include the solution (more information on solutions of the general dimensioning problem is given in Section 10.2). To make the solution simpler and more robust we introduce the link efficiency factor defined as the ratio of the average reward from the VNL to the VNL cost

$$h^s(\mathbf{G},r) = \frac{\overline{R}^s}{c^s} \quad (9.5)$$

Now the dimensioning problem (P2) can be defined with additional link efficiency constraint, h^c:

$$\min_{\mathbf{G},r}\left\{c = \sum^{s} c^s(G^s)\right\} \quad (9.6)$$
$$B_j(\mathbf{G},r) = B_j^c, \quad j \in J$$
$$h^s(\mathbf{G},r) = h^c, \quad s \in \{s : G^s > 0\}$$

Note that in this formulation the routing decision automatically takes into account the costs of VNLs constituting the considered paths (the larger the VNL cost, the larger the shadow price). This feature can also be usefull in designing tariff by taking into account the values of reward parameters which reflect the average cost of connections within a class. Concerning the optimal solution it can be easily shown that in general the equality constraints can be met for more than one network connectivity configuration. In the next section we show that the optimal solution is achieved for a configuration with maximum connectivity (for which a solution exists) and that in many practical cases it can be proved that there is only one solution for a given connectivity. Based on these features a solution which is optimal or close to the optimal can be found from a simple iterative algorithm. To simplify presentation we first describe this algorithm (OPT1) for networks with single-rate connections (one connection class per OD pair):

1. choose initial **r** and **G**,

2. evaluate network performance from the analytical model (Section 3.2)

3. if $\max_j | B_j^c - B_j | < \epsilon$ and $\max_s | h^c - h^s | < \epsilon$ stop the calculations,

4. evaluate new values of G^s for $s \in \mathbf{J}_d \setminus \mathbf{J}_l$ (where $\mathbf{J}_d$ denotes all OD pairs with direct-link-path and $\mathbf{J}_l$ denotes OD pairs with direct-link-paths carrying a significant share of multilink connections, e.g., more than 40%; or with a small share of the OD pair traffic carried on the direct-link-path) defined by

$$G^s \leftarrow G^s + \frac{B_s^c - B_s}{\frac{\partial B_s}{\partial G^s}} \qquad (9.7)$$

5. evaluate new value of r_j for $j \in (\mathbf{J}_a \cup \mathbf{J}_l)$ (where $\mathbf{J}_a$ denotes OD pairs without direct-link-path) given by

$$r_j \leftarrow r_j + \frac{B_j^c - B_j}{\frac{\partial B_j}{\partial r_j}} \qquad (9.8)$$

6. evaluate new value of G^s for $s \in \mathbf{J}_l$ given by

$$G^s \leftarrow G^s + \frac{h^c - h^s}{\frac{\partial h^s}{\partial G^s}} \qquad (9.9)$$

7. evaluate new value of r_j for $j \in \mathbf{J}_d \setminus \mathbf{J}_l$ given by

$$r_j \leftarrow r_j + \frac{h^c - h^j}{\frac{\partial h^j}{\partial r_j}} \qquad (9.10)$$

8. if $h^j > h^c$ for $j \in \mathbf{J}_a$ (the value of h^j is evaluated from a dummy link with assigned incremental capacity, ∂G^s) move the jth OD pair from $\mathbf{J}_a$ to $\mathbf{J}_d$,

9. go to step 2.

It is important that all derivatives $\partial B_s/\partial G^s$, $\partial B_j/\partial r_j$, $\partial h^s/\partial G^s$, $\partial h^j/\partial r_j$ can be replaced by very rough approximations since the direction of the search is determined by the sign of the difference between the variable and the constraint.

Now let us consider the general case where each OD pair can be offered several connection classes with different "bandwidth" requirements. To simplify the presentation we still assume that each pair of nodes is offered one class of connections but this class can consist of several subclasses which can differ in loss constraints, "bandwidth" requirements and mean holding times. Let $B_{j,i}^c$ and $r_{j,i}$ denote the loss constraint and reward parameter for subclass i of class j, respectively. Note that the connection class loss probabilities and their constraints can be defined as weighted averages over corresponding subclass variables. Thus, the solution can be achieved by extending the iterative procedure for the single-rate connection case as follows. The procedure remains basically the same with the exception of adjusting the connection reward parameters. Namely, since there is no connection class reward parameter to be adjusted, the equivalent effect is achieved by multiplying all connection subclass reward parameters, included in the considered connection class, by the same factor. This approach provides a solution for the connection class loss constraints and link efficiency constraints. The subclass loss constraints are met by adjusting the ratio of subclass reward parameters while keeping the average reward parameter of the connection class unchanged. This is done by adding an additional step in the original procedure. Since in this scheme the subclass loss probabilities are independent and monotonic functions of the corresponding reward parameters, we can meet all subclass loss constraints and this solution is unique.

9.6.2 Solution analysis*

We break up the analysis of the solutions satisfying constraints of Equation (9.6) into three parts. First we consider the issue of solution uniqueness in a fully connected network. Then the case of limited connectivity is addressed. Finally we show which connectivity provides the best solution. As in the previous section, to simplify presentation we assume only one connection class per OD pair with understanding that this class can consist of several subclasses.

Fully connected network. In the considered optimization problem, Equation (9.6), we have the same number of variables and constraint functions. Thus if the functions are independent (Jacobian not equal to zero in the domain) and monotonic the solution for the constraint set is unique. Let us introduce the assumption (**A.9.1**) that there exists a solution where for each OD pair connection class the traffic carried on the direct-link-path, $\bar{a}_j^j$, is significantly larger than any other stream carried on this link, $\bar{a}_j^j \gg \bar{a}_i^j : i \in J \setminus \{j\}$, and that it constitutes significant majority of this class's overall carried traffic $\bar{a}_j^j \gg \bar{a}_j - \bar{a}_j^j$. Then we have

$$\left|\frac{\partial B_j}{\partial G^j}\right| \gg \left|\frac{\partial B_k}{\partial G^j}\right|, \quad k \neq j \tag{9.11}$$

$$\left|\frac{\partial B_j}{\partial r_j}\right| \gg \left|\frac{\partial B_k}{\partial r_j}\right|, \quad k \neq j \tag{9.12}$$

$$\left|\frac{\partial h^j}{\partial G^j}\right| \gg \left|\frac{\partial h^k}{\partial G^j}\right|, \quad k \neq j \tag{9.13}$$

$$\left|\frac{\partial h^j}{\partial r_j}\right| \gg \left|\frac{\partial h^k}{\partial r_j}\right|, \quad k \neq j \tag{9.14}$$

Consequently each pair of functions, h^j, B_j, is independent from the other pairs. The independence of the loss and link efficiency functions for the same OD pair follows from the fact that we have

$$\frac{\partial B_j}{\partial G^j} < 0, \quad \frac{\partial B_j}{\partial r_j} < 0, \quad \frac{\partial h^j}{\partial G^j} < 0, \quad \frac{\partial h^j}{\partial r_j} > 0 \tag{9.15}$$

Thus by appropriately choosing values of both variables (G^j, r_j) we can change the value of each function in any direction, keeping the other function unchanged and

9.6 Virtual Network Design

the projection $(h^j, B_j) \Rightarrow (r_j, G^j)$ is unequivocal. Finally since all functions are monotonic the solution for the constraint set is unique. These features together with the fact that

$$\left|\frac{\partial h^j}{\partial r_j} r_j\right| \gg \left|\frac{\partial h^j}{\partial G^j} G^j\right| \qquad (9.16)$$

imply that the OPT1 algorithm provides an optimal solution to the dimensioning problem, Equation (9.6), if the assumptions are fulfilled.

Limited connectivity. Even in the case of incomplete connectivity, the dimensioning problem still has the same number of variables and functions. Let us assume that for each connection class without a direct-link-path, its total carried traffic is distributed over several multi-link paths in such a way that the traffic carried on any particular path is minor in relation to the total carried traffic. Then, under the assumption **A.9.1**, for connection class j without a direct-link-path we have

$$\left|\frac{\partial B_j}{\partial r_j}\right| \gg \left|\frac{\partial B_k}{\partial r_j}\right|, \quad k \neq j \qquad (9.17)$$

$$\left|\frac{\partial h^k}{\partial r_k}\right| \gg \left|\frac{\partial h^k}{\partial r_j}\right|, \quad k \neq j \qquad (9.18)$$

$$\left|\frac{\partial B_k}{\partial G^k}\right| \gg \left|\frac{\partial B_j}{\partial G^k}\right|, \quad k \neq j \qquad (9.19)$$

Therefore the loss function for the jth OD pair is independent from the other functions and, due to the applied routing, its dependence on r_j is monotonic. These features imply that, for given network connectivity, if the solution for the constraint set exists it is unique and can be found by the OPT1 algorithm.

Optimal connectivity In general different solutions can be obtained for many different connectivity patterns. This results from the fact that having a solution for a given connectivity, another solution can be arrived at for any subset of links from the given connectivity, providing that all OD pairs have alternative paths. The question is how to decide which connectivity is best. The concept of link efficiency factors, $h^s(\mathbf{G}, r)$, makes determination of the optimal connectivity very simple. First, notice that the loss constraints can be satisfied for any connectivity (providing that all OD pairs have alternative paths) due to the reward maximization routing. Also, it can be easily shown that, for given reward parameters, the larger the link capacity, the smaller the link efficiency. Then consider a solution for a given limited connectivity. If for any link with $G^s = 0$ (no connection) we have

$$\frac{\partial \overline{R}^s}{\partial c^s} > h^c \qquad (9.20)$$

the link should be reinstated, $G^s > 0$. The argument supporting this decision is simple. Equation (9.20) shows that the link under consideration can provide more reward per "bandwidth" unit cost than the existing solution. Thus a new solution with the reinstated link will lower the network cost by reducing the resource allocation to other links until all link efficiency parameters are equal. This argument implies that the optimal solution is achieved for a configuration with maximum connectivity (for which a solution exists). The OPT1 algorithm provides optimal connectivity by verifying condition 9.20 in step 8.

9.6.3 Numerical study

In this section we present numerical results illustrating convergence of the optimization procedure. The derivatives in the procedure were replaced by very rough approximations and a damping factor was introduced to avoid oscillations. The performance evaluation model described in Section 7.2 is used for evaluation of the loss probabilities and average link rewards. Three network examples used for this study are described in Table 9.2. The two first examples have five nodes and single-rate connection classes. The third example has four nodes and two connection subclasses with different "bandwidth" requirements. The potential full connectivity structures for all examples are illustrated in Figure 9.10.

TABLE 9.2 Description of networks.

		OD pair	1	2	3	4	5	6	7	8	9	10
Example 1 $B_j^c = 2\%$		$c^{s'}$	1	1	1	1	1.5	1.5	1.5	1.5	3	3
		λ_j	41	41	41	41	41	41	41	41	41	41
Example 2 $B_j^c = 0.65\%$		$c^{s'}$	1	1	1	1	1.5	1.5	1.5	1.5	2	2
		λ_j	20	40	30	60	10	20	40	60	10	70
Example 3 $B_{j,1}^c = 1\%, d_1 = 1$ $B_{j,2}^c = 1\%, d_2 = 3$		$c^{s'}$	1	∞	1	1	1	1				
		$\lambda_{j,1}$	15	15	15	15	15	15				
		$\lambda_{j,2}$	5	5	5	5	5	5				

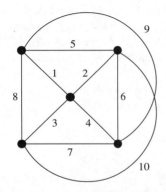
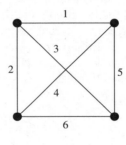

Figure 9.10. *Network structure.*

The convergence of the procedure is illustrated in Figures 9.11, 9.12, and 9.13, where the performance variables are given as a function of the iteration number. Since in Examples 1 and 3 there are three groups of symmetrical OD pairs, only plots corresponding to groups are given. The results indicate a very good convergence despite rough heuristics used for the derivative approximations. It is also important that in the first example the most expensive links were eliminated in a few iterations. In contrast, an optimization program based on the gradient method, derived for the network dimensioning problem, was not able to eliminate these links.

9.7 Virtual Network Update

The issue of virtual network update is illustrated in Figure 9.14. The objective of the VN update is to correct the original VN design and to adapt the VN design to changes in traffic profiles. The structure of the VN update procedure is similar to the design procedure. The difference is that the algorithm is fed by measurements which are associated with the offered traffic matrix, GoS metrics, and "bandwidth" utilization metrics. This information is used to update the VN topology, allocation of "bandwidth" to VNLs, and routing algorithm parameters. As in the case of VN design demand, the update demand can be accepted, rejected, or realized partially.

The OPT1 procedure created for design of VNs can be easily adapted for the purpose of virtual network updating. Two approaches could be considered. In the first the procedure is used in its entirety. The only difference is that the offered traffic matrix is estimated based on measurements at the origin nodes. This option is viable if the analytical performance model can be executed in a reasonable time. Another possibility is to replace the analytical model results with estimates based on measurements which would provide the required parameters for each iteration of the optimization procedure. Here the convergence time is a function of the time required for parameter estimation. While the average link rewards and traffic distribution are easy to estimate in a relatively short period, estimation of connection loss probabilities in short intervals may be more difficult due to a small number of important events (blocked connections). To improve the estimation quality one may approximate the connection loss probabilities using the link blocking probabilities evaluated from a simplified link model, fed by measurements of the traffic carried on links.

In general the adaptation procedure can be implemented in a distributed manner since most of the actions concerning a particular OD pair and its direct VNL can be taken based on measurements in the origin node. This approach has some attractive features. First, it can react faster to sudden changes in the traffic matrix distribution because the algorithm is closer to the information sources. Second, in many cases the adaptation is needed only for a few OD pairs at a time (local problem). Thus the adaptation process is simplified. The local adaptation is initiated once the GoS of a particular connection class deteriorates. If the overall traffic loss level in the virtual network indicates that there is spare capacity in the network,

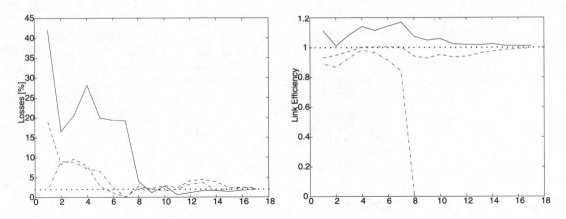

Figure 9.11. *Constraint functions versus iteration number (Example 1).*

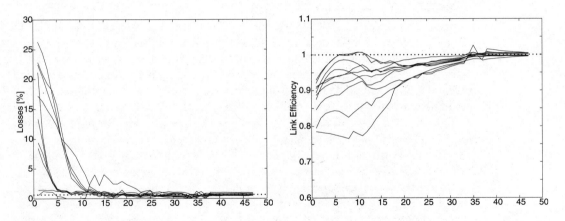

Figure 9.12. *Constraint functions versus iteration number (Example 2).*

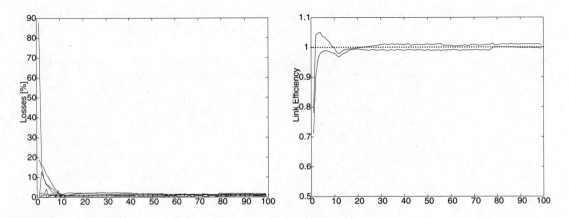

Figure 9.13. *Constraint functions versus iteration number (Example 3).*

9.7 Virtual Network Update

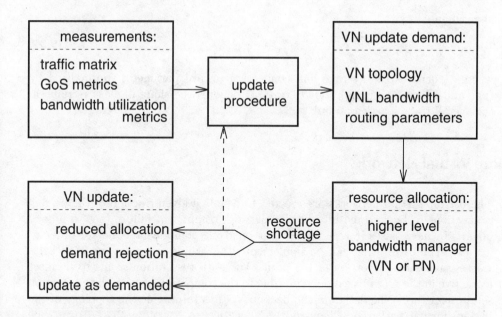

Figure 9.14. *Virtual network update.*

one can try to adapt the local CAC & routing parameters to improve the GoS. Otherwise the "bandwidth" allocation to the VNLs has to be updated. This can be done based on link marginal costs, $c_j^{s'}$, of carrying additional traffic

$$c_j^{s'} = \frac{\partial c^s}{\partial \overline{a}_j^s} \qquad (9.21)$$

Assuming that the marginal costs are given, the update algorithm should allocate more "bandwidth" to the bottleneck VNLs on the path for which the marginal cost of the update is minimal.

In the reward maximization model, evaluation of the link marginal costs is very simple. This follows from the fact that the average link shadow price, for the connection class with the minimal "bandwidth" unit requirement and unit mean holding time, can be expressed as the difference between the average rewards from the links with capacity G^s and $G^s - d_j$ (assuming the same offered traffic):

$$\overline{p}_1^s = \overline{R}^s(G^s) - \overline{R}^s(G^s - 1) \qquad (9.22)$$

Therefore by extending the system function definitions to real values of link capacities by linear interpolation, we get the following approximation

$$\frac{\partial \overline{R}^s(G^s)}{\partial G^s} = \overline{p}_1^s \qquad (9.23)$$

This result has been proven for single-rate networks with load sharing routing strategy in (Kelly, 1988). By using Equation (9.23), the marginal cost can be approximated by

$$c_j^{s'} = \frac{c^{s'} \overline{r}^s}{\overline{p}_1^s \beta_j^s} \qquad (9.24)$$

where $\overline{r}^s$ denotes the average link connection reward parameter (normalized by time and bandwidth units) and $\beta_j^s = \overline{a}_j^s / \overline{a}^s$ denotes the share of class j traffic in the overall traffic carried on link s.

9.8 Backup Virtual Network

The backup virtual networks are created to enable a given fraction of the carried connections to be restored in case of a network component failure (e.g., a link or node). In most cases it is assumed that only one component is out of order at a time. The restoration can be implemented on a link, fragment, or path basis where fragment is a portion of the path. The path restoration seems to be most attractive in the ATM based networks due to the compatibility with the connection setup procedure. The design of the backup VN has to provide the topology of the backup virtual paths and "bandwidth" allocation to backup VNLs. To provide network reliability cost-effectively, an important objective of the backup VN design is to minimize the "bandwidth" allocation to backup VNLs while satisfying the restoration requirements. The backup VN design is quite different from the design of the primary VN. In particular the "bandwidth" allocated to the backup VNL is not equal to the sum of "bandwidths" required by all backup VPs using this VNL since only a part of these VPs will be activated by a link or node failure (see Figure 9.15).

The backup VN design is strongly associated with the design of the physical network; therefore more detailed discussion of the backup network design is given in Section 10.4. It should be noted that the backup virtual networks are needed only when there is no automatic restoration mechanism in the underlying transport layer (see Section 10.1).

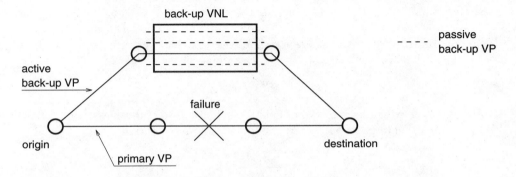

Figure 9.15. *Example of primary VP, backup VP, and backup VNL.*

9.9 Discussion and Bibliographic Notes

The notion of virtual networks is not new and was used in several publications, e.g. (Mason, 1987; Mason et al., 1990; Walters and Ahmed, 1992; Atkinson, Anido, and Bradlow, 1992; Dziong, Liao, and Mason, 1993; Dziong, Montanuy, and Mason, 1994; Wernik, Pretty, and Smith, 1993; Fotedar et al., 1995; Dupuy, Nilsson, and Inoue, 1995; Gupta and Domenico, 1995; Ji, Hui, and Karasan, 1996), for particular applications such as virtual private networks or virtual networks associated with different qualities of service. In this chapter we generalized the virtual network concept by showing that a common generic definition of virtual networks can be applied to all potential applications.

There is also an extensive literature on using virtual paths for resource management, e.g. (Addie and Warfield, 1988; Ohta, Sato, and Tokizawa, 1988; Burgin, 1989; Sato, Ohta, and Tokizawa, 1990), which can also be seen as a particular application of the virtual network concept for connection setup simplification. Some of the drawbacks of this approach were indicated in (Dziong et al., 1991; Crocetti et al., 1994). In this chapter we showed that these drawbacks can be avoided by separating the virtual network link and virtual path functions and applying an efficient signaling protocol based on the resource management cells. In this case virtual paths are used mainly for routing and switching simplification, as was the original reason for their introduction. The function of virtual network link is to control the resource allocation. Under this scenario several virtual paths can be carried on one virtual network link. A similar concept was presented in (Guillemin and Hamchaoui, 1996) for virtual private networks, under the name of virtual trunk.

To take full advantage of the virtual network concept several generic and application oriented problems have to be resolved. In this chapter we presented a rather general framework concentrating on critical issues. A particular design depends strongly on switch output port schedulers and the CAC & routing and flow control algorithms. Naturally the design and management of virtual networks have many common features with the design and management of physical networks. For this reason the material presented in Chapter 10 includes sections which are also relevant to virtual network design and management.

References

ITU-T Recommendation Q.2761-Q.2764. 1993. B-ISDN User Part. Draft. Geneva.

Addie, R.G., and Warfield, R.E. 1988. Bandwidth switching and new network architectures. In *Proceedings of the 12th International Teletraffic Congress*, pp. 2.3iiA.1.1–7. North-Holland.

Anderson, J., Doshi, B., Dravida, S. and Harshavardhana, P. 1994. Fast restoration of ATM networks. *IEEE Journal on Selected Areas in Communication*, 12(1):128–138.

Aneroussis, N.A., and Lazar, A.A. 1996. Virtual path control for ATM networks with call level quality of service guarantees. In *Proceedings of IEEE INFOCOM'96*, IEEE Computer Society Press.

Atkinson, D.J., Anido, G.J., and Bradlow, H.S. 1992. B-ISDN Traffic Characterization and Management. In *Proceedings of the 7th Australian Teletraffic Research Seminar*. pp. 457–466. Mannum, South Australia.

Burgin, J. 1989. Broadband ISDN Resource management. In *Proceedings of the 6th ITC Specialist Seminar*, Adelaide, Australia.

Crocetti, P., Fratta, L., Gallassi, G., and Gerla, M. 1994. ATM Virtual private networks: Alternatives and performance comparisons. In *Proceedings of IEEE ICC'94*, pp. 608–612. IEEE Computer Society Press.

Dupuy, F., Nilsson, G., and Inoue, Y. 1995. The TINA Consortium: Towards networking telecommunications information services. In *Proceedings of ISS'95*, Vol. 2, pp. 207–211. Berlin: VDE-VERLAG GMBH.

Dziong, Z. 1988. Dimensioning of nonhierarhical networks with state-dependent routing based on reward maximization – New Features. INRS-Telecommunications report.

Dziong, Z., and Liao, K.-Q. 1989. Reward maximization as a common basis for routing, management and planning in ISDN. In *Proceedings of Networks'89*. Spain.

Dziong, Z., Liao, K-Q., Mason, L.G., and Tetreault, N. 1991. Bandwidth management in ATM networks. In *Proceedings of the 13th International Teletraffic Congress*, pp. 821–827. North-Holland.

Dziong, Z., and Mason, L.G. 1992. An analysis of near optimal call admission and routing model for multi-service loss networks. In *Proceedings of IEEE INFOCOM'92*, pp. 141–152. IEEE Computer Society Press.

Dziong, Z., Liao, K-Q., and Mason, L.G. 1993. Effective bandwidth allocation and buffer dimensioning in ATM based networks with priorities. *Computer Networks and ISDN-Systems*, 25:1065–78.

Dziong, Z., Mignault, J., and Rosenberg, C. 1993. Blocking evaluation for networks with reward maximization routing. In *Proceedings of IEEE INFOCOM'93*, pp. 593–601. IEEE Computer Society Press.

Dziong, Z., and Mason, L.G. 1994. Call admission and routing in multi-service loss networks. *IEEE Transactions on Communications* 42(2/3/4):2011–2022.

Dziong, Z., Montanuy, O., and Mason, L.G. 1994. Adaptive bandwidth management in ATM networks. *International Journal of Communication Systems (John Wiley & Sons)*, 7:295–306.

Dziong, Z., Xiong, Y., and Mason, L.G. 1996. Virtual network concept and its applications for resource management in ATM Based Networks. In *Proceedings of Broadband Communications'96, An International IFIP-IEEE Conference on Broadband Communications*, pp. 223–234. Chapman & Hall.

Dziong, Z., Zhang, J., and Mason, L.G. 1996. Virtual network design – An economic approach. In *Proceedings of The 10th ITC Specialist's Seminar on "CONTROL IN COMMUNICATIONS"*, pp. 75–86. Lund, Sweden, September.

Dziong, Z., Juda, M., and Mason, L.G. 1996. A framework for bandwidth management in ATM networks – Aggregate equivalent bandwidth estimation approach. *IEEE/ACM Transactions on Networking*, in press (considered for February 1997 issue).

References

Fotedar, S., Gerla, M., Crocetti, P. and Fratta, L. 1995. ATM virtual private networks. *Communications of the ACM,* 38(2):101–109.

Guillemin, F., and Hamchaoui, I. 1996. Some traffic aspects in virtual private networks over ATM. In *Proceedings of Broadband Communications'96, An International IFIP-IEEE Conference on Broadband Communications,* pp. 235–246. Chapman & Hall.

Gupta, A., and Domenico, F. 1995. Resource Partitioning for Real-Time Communication. *IEEE/ACM Transactions on Networking,* 3(5):501–506.

Ji, H., Hui J.Y., and Karasan, E. 1996. GoS-based pricing and resource allocation for multimedia broadband networks. In *Proceedings of IEEE INFOCOM'96,* IEEE Computer Society Press.

Kawamura, R., Sato, K-I, and Tokizawa, I. 1994. Self-healing ATM networks based on virtual path concept. *IEEE Journal on Selected Areas in Communication,* 12(1):120–127.

Kelly, F.P. 1988. Routing in circuit switched networks: Optimization, shadow prices and decentralization. *Adv. Appl. Prob.* 20:112–144.

Mason, L.G., and Gu, X.D. 1986. Learning automata models for adaptive flow control in packet switching networks. In *Adaptive and Learning Systems – Theory and Applications,* Narendra K.S., pp. 213-227. New-York: Plenum Press.

Mason, L.G. 1987. Virtual network services and architectures", INRS-Telecommunications report 87–24.

Mason, L.G., Dziong, Z., Liao, K.-Q., and Tetreault, N. 1990. Control architectures and procedures for B-ISDN. In *Proceedings of the 7th ITC Specialist Seminar,* Morristown.

Medhi, D. 1995. Multi-hour, multi-traffic class network design for virtual path-based dynamically reconfigurable wide-area ATM networks. *IEEE/ACM Transactions on Networking,* 3(6):809–818.

Murakami, K., and Kim, H. 1995. Joint optimization of capacity and flow assignment for self-healing ATM networks. In *Proceedings of IEEE ICC'94,* pp. 216–220. IEEE Computer Society Press.

Ohta, S., Sato, K., and Tokizawa, I. 1988. A dynamically controllable ATM transport network based on virtual path concept. In *Proceedings of IEEE GLOBECOM'88.* IEEE Computer Society Press.

Sallberg, K., and Stavenow, B. 1994. ATM Traffic Management at the Initial Deployment of B-ISDN. *Erricsonn Review* 71(4):150–159.

Sato, K., and Tokizawa, I. 1990. Flexible asynchronous transfer mode networks utilizing virtual paths. In *Proceedings of IEEE ICC'90.* pp. 831–838. IEEE Computer Society Press.

Walters, M.S., and Ahmed, N. 1992. Broadband virtual private networks and their evolution. In *Proceedings of ISS'92.*

Wernik, M., Pretty, R., and Smith, D. 1993. Evolution of broadband network services – A North American perspective. In *Proceedings of IEEE ICC'93,* pp. 68-74. IEEE Computer Society Press.

Veeraraghavan, M., La Porta, T.F., and Lai, W.S. 1995. An alternative approach to call/connection control in broadband switching systems. In *Proceedings of the 1st IEEE International Workshop on Broadband Switching Systems,* pp. 319–332. Poznan, Poland.

Yamanaka, N., Oki, E., Pitcho, F., and Sato, H. 1995. In *Proceedings of ISS'95,* pp. 195–199. Berlin: VDE-VERLAG GMBH.

Chapter 10

Physical Resource Allocation to ATM Networks

IN ALL previous chapters we dealt with resource management issues within the considered ATM network assuming that a pool of physical resources is allocated to this network. In this chapter we concentrate on managing this allocation. This issue can be divided into three generic functions: resource allocation design, adaptation of resource allocation to traffic matrix changes, and protection against network component failures. While the objectives of these functions are similar to the objectives of equivalent functions defined for virtual networks in Chapter 9, realization of these functions can be quite different. This follows from the fact that the transport layer, which provides transmission facilities for the ATM layer, can be realized in many ways characterized by different constraints and functionality. In particular, usually the bandwidth can be allocated to ATM links only in certain modules. The time needed to allocate additional modules can vary from seconds (if the transport layer is based on a digital hierarchy system with cross-connect capabilities) to weeks (if new installation is required). Moreover, the transport layer can have its own survivability mechanisms which influence the failure protection mechanisms on the ATM network layer. We discuss important features of different transport layer options in Section 10.1, including digital hierarchy systems (PDH, SDH, SONET). The initial allocation of resources from the transport layer to the ATM network is described in Section 10.2. We consider several options including multi-hour traffic design. Adaptation to traffic changes is described in Section 10.3. Depending on the underlying transport layer, two options are considered: fast adaptation and slow adaptation. The survivability issues are treated in Section 10.4. First we discuss possible failure scenarios and restoration mechanisms. Then the objectives of the connection and GoS restorations are described followed by generic design formulations for self-healing networks.

10.1 ATM Network Layer Versus Transport Layer

10.1.1 Dedicated transport layer

Here we consider a transport layer dedicated to the ATM network, which means that the transmission facilities are owned by the ATM network operator or are leased on a long term basis. We assume that in this case the ATM link capacity expansion can require significant time, on the order of days or weeks. Concerning the reliability issue, three generic options, illustrated in Figure 10.1, can be considered. When the ATM link is realized on one physical cable (or concatenation of cables), as shown in Figure 10.1a, the protection against link failures has to be provided on the ATM network layer. On the other end of the spectrum, the link can be fully protected on the transport layer by installing a hot stand-by link (Figure 10.1b). In the case of failure, the stand-by link can take over the connections from the primary link with minimal disruption of the cell streams. If the stand-by link is realized on a path that is physically disjoint from the primary link, failure of both links at the same time is unlikely and the ATM network layer does not need to cope with the link failure issue since the possible minor disruptions of the cell stream can be handled by the higher layer protocols. Obviously this excellent survivability feature is achieved at a high cost of duplication of the required network bandwidth. An attractive compromise between the cost and reliability of the transport layer can be achieved by realizing the link on two (or more) physically disjoint paths (Figure 10.1c). In this case failure of one transmission facility reduces the ATM link capacity only by a fraction (by half or less). Thus, the ATM layer can handle link failures without reserving a large amount of "bandwidth" for the affected connection restoration.

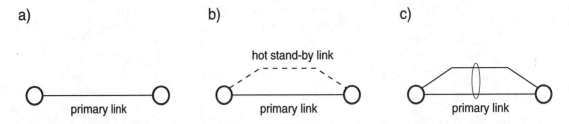

Figure 10.1. *Possible dedicated ATM link realizations.*

Note that in the case of a dedicated transport layer, the link bandwidth modules can be chosen from a wide spectrum of transmission link standards in both synchronous and asynchronous modes used in circuit switched and packet switched environments. Nevertheless due to presumably high installation cost one can expect that rather large modules will be used to minimize the number of network design updates.

10.1.2 ATM over the digital hierarchy (PDH, SDH, SONET)

The digital hierarchy (DH) is a dominant medium providing the transport layer for many different types of networks. Originally developed for voice communication over copper cables, *plesiochronous digital hierarchy* (PDH), it is being replaced by newer systems designed for optical cables, *synchronous digital hierarchy* (SDH), known as SONET (Synchronous Optical NETwork) in North America. We start with a description of the main features of these systems. There are three basic components of digital hierarchy systems: terminal multiplexers and demultiplexers (TM), add-drop multiplexers (ADM), and digital cross-connect switches (DCS).

These elements are connected by means of digital links and an example of their interrelation is depicted in Figure 10.2. The underlying transfer mode is based on hierarchical time division multiplexing. In particular it means that the digital channel on layer n of the hierarchy consists of k_n digital channels from layer $n-1$ and that each of these channels is associated with a predetermined synchronous time slot.

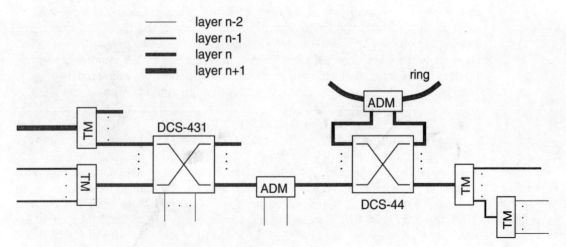

Figure 10.2. *Example of digital hierarchy configuration.*

As indicated in Figure 10.2 the function of TMs is to multiplex channels from the lower layers into one channel of the higher layer. The basic function of a digital cross-connect switch (DCS) on layer n is to switch channels on layer $n-1$ between the nth layer links connected to DCS. Since the layer $n-1$ channels can consist of several lower layer channels, DCS is effectively switching bundles of channels which correspond to a certain bandwidth associated with these channels. That is why the name cross-connect is used in this case. Optionally DCS on layer n can also switch channels on lower layers than $n-1$ including direct connection of lower layer links to DCS (e.g., DCS-431 in Figure 10.2). The add-drop multiplexer is a device "tapped" in the middle of the nth layer link. Its function is to recover and/or insert a signal from/into specified lower layer channels. This component

TABLE 10.1 Transmission rates in DH systems.

Plesiochronous digital hierarchy			Synchronous digital hierarchy	
Europe [Mbps]	North America [Mbps]	Japan [Mbps]	SONET SDH	[Mbps]
E1 : 2	DS1 : 1.5	1.5		
E2 : 8	DS2 : 6.3	6.3		
E3 : 34	DS3 : 45	32	STS-1 /OC1 :	52
E4 : 140	274	100	**STS-3,**/OC3 **STM-1** :	**155**
		400	STS-9, /OC9 STM-3 :	466
			STS-12/OC12**STM-4** :	**622**
			STS-18/OC18 STM-6 :	933
			STS-24/OC24 STM-8 :	1244
			STS-48/ OC48 STM-16 :	2488

can be seen as a specific switching device limited to one transmission line only.

It is important to underline the difference between the digital cross-connect switch and the ATM cross-connect switch (see Chapter 1). Namely, the ATM cross-connect switch does not switch the bandwidth but rather bundles of signals. In this case the bandwidth switching function can only be emulated by logical "bandwidth" allocation to virtual paths and this function is optional.

The rounded off transmission rates on different layers of digital hierarchy are given in Table 10.1 for PDH, SDH, and SONET systems. Note that these rates are not integer multiplies of the lower layer rates. This feature follows from the fact that a particular transmission rate is a sum of the payload rate and the overhead rate. The overhead rate is associated with transmission channel management functions such as synchronization and communication between the digital hierarchy network elements. For example, the principal layer for ATM transport is STM-1 in SDH (STS-3 in SONET). In this case the transmission rate is 155.520 Mbps, of which 149.760 Mbps is dedicated to payload. In Table 10.1 the basic rates in SONET and SDH systems are given in boldface.

Digital hierarchy systems and their elements can have many applications. In the following we discuss a general application where the transport network based on a digital hierarchy system provides transmission facilities for many different users and networks including the ATM network under consideration. In this case there are two network layers from the ATM network's viewpoint, illustrated in Figure 10.3. In the ATM layer, the links between the ATM nodes are seen as direct transmission connections with a certain bandwidth allocated from the digital hierarchy network. In the digital hierarchy layer, the same links are seen as paths which can traverse several TMs, DCSs, and ADMs.

There are several important consequences of this architecture from the ATM network resource management viewpoint. First, assuming a sufficient number of interfaces between the ATM nodes and the DH layer, as well as spare capacity in the DH network, one can realize fast adaptation of the ATM link bandwidth allocation to meet a variable demand. Obviously this adaptation can be realized only in modules available in the considered DH. Second, usually the DH network has its own dynamic restoration algorithms for recovery from transmission element failures. This feature implies that the ATM links affected by a failure in the transmission facilities can be at least partially restored by the DH layer. Thus the failure is less critical for the ATM network survivability and the ATM network design can reserve less capacity for connection restoration strategy if needed. On the other hand, the same feature requires some kind of cooperation between the restoration strategies on the ATM and DH layers in order to avoid conflicting actions. The simplest approach would be to activate the restoration mechanism in sequence, starting from the DH layer. Nevertheless, the parallel execution of both algorithms in a complementary way can speed up the restoration process and provide better resource utilization at the expense of increased complexity. Also, it is important to note that in the case of digital cross-connect systems the failure of one transmission facility can affect more than one ATM link as indicated in Figure 10.3. Such an event is less likely when dedicated transmission facilities are used, since in this case the network designer has control over the links' physical configuration.

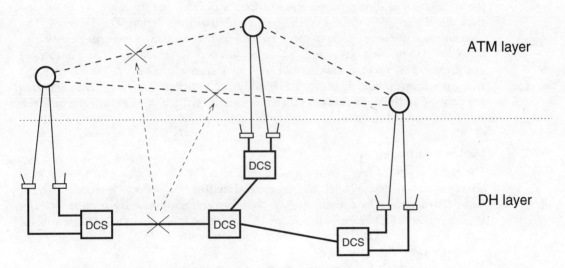

Figure 10.3. *Interaction of ATM and DH layers.*

10.2 Dimensioning of ATM Networks

In this section we analyze several important issues of the ATM network design. First we assume a continuous link "bandwidth" allocation capability and describe several options for network dimensioning. Then we discuss methods for taking into account link "bandwidth" modularity, including approaches based on solutions for the continuous case. Finally we focus on the *multi-hour* case where traffic patterns change in a predictable way during certain cyclic periods (e.g., day, week).

10.2.1 Continuous link bandwidth*

The issue of ATM network dimensioning is strongly connected with the design of QoS virtual networks discussed in Chapter 9. For example, when designs of the QoS VNs are given, the "bandwidth" required by an ATM link can easily be calculated from the sum of the "bandwidth" allocations required by the VNLs using this ATM link. If the QoS virtual networks are not used for resource management ("bandwidth" is shared among all QoS VN — see Figure 9.6a), then the issue of ATM network dimensioning becomes very similar to the design of virtual networks. Thus the general network design methods can be applied for both the ATM and virtual networks. In the following we discuss several generic methodologies for network dimensioning except for the optimization procedure for networks with reward maximization routing which was already described and analyzed in Section 9.6 in the context of VN design. We concentrate on a variety of optimization procedure formulations and discuss the main features of different alternatives. It should be underlined that we do not describe complete algorithms but rather concepts. Implementation of these concepts can be a tedious task and often requires resorting to simplified models or even rough approximations. Since these issues and numerical evaluation of different options could well fill a separate book, we do not attempt to cover them in this chapter. To simplify presentation, the obvious constraints are omitted in the optimization procedure definitions (e.g., constraints defining the variables' natural domain).

Problem formulation

For the purpose of network dimensioning the ATM network can be treated as a circuit-switched loss network serving several multi-rate connection classes, each of them characterized by arrival rate λ_j, "bandwidth" requirement d_j, mean holding time μ_j^{-1}, and OD node pair. Note that data connections with a flow control mechanism (e.g., ABR) can also be taken into account in this model by using minimum rate allocation as d_j.

The objective of the generic dimensioning problem is to find the minimum network cost defined as the sum of link costs

$$\min\left\{c = \sum_{s \in S} c^s\right\} \tag{10.1}$$

subject to inequality loss probability constraints over all connection classes

$$B_j \leq B_j^c, \quad j \in J \tag{10.2}$$

for given CAC & routing strategy. Note that although the CAC & routing strategy is specified, its optimal parameters may be a function of network flow distributions and link dimensions which are not known in advance (e.g., optimal path sequence

10.2 Dimensioning of ATM Networks

for sequential routing). Thus, in general, the CAC & routing parameters, Υ, should also be treated as optimization variables.

Before sketching different network dimensioning procedures we describe in brief the relation between network performance and network dimensioning models. In general the network dimensioning procedures require network performance models since the network design should meet the connection loss probability constraints. As discussed in Chapter 7, the performance models, in most cases, are based on the statistical link independence assumption resulting in decomposition of the network model into a set of link loading and link performance functions making up the following fixed-point equations:

$$\mathbf{A}^s = f_l(\mathbf{\Pi}, \pi), \quad s \in S \tag{10.3}$$
$$\Pi^s = f_p(\mathbf{A}^s, L^s), \quad s \in S \tag{10.4}$$

where addition of L^s compared to Equation (7.6) emphasizes that the link capacities are given. As a consequence most of the dimensioning procedures also assume statistical link independence resulting in a set of link performance functions coupled by means of link loading functions and possibly link dimensioning functions if the link capacities are not optimization variables. The link dimensioning function is usually defined by the inverse of the link performance function

$$L^s = f_d(\mathbf{A}^s, \Pi^s), \quad s \in S \tag{10.5}$$

In the following the link dimensioning function will be represented by $L^s(\mathcal{X})$ where $\mathcal{X}$ denotes the optimization variables.

Link capacities as optimization variables

A natural approach to the dimensioning problem is to use the link capacities and CAC & routing parameters as the optimization variables. In this case the problem is formulated as

$$\min_{\mathbf{L},\Upsilon}\left\{c = \sum_{s\in S} c^s(L^s)\right\} \tag{10.6}$$
$$B_j(\mathbf{L}, \Upsilon) \leq B_j^c, \quad j \in J$$

where $\mathbf{L} = [L^s, s \in S]$ is the link "bandwidth" allocation vector. Application of general optimization methods to this problem is practical only when the continuous gradients of the constraint functions with respect to the CAC & routing parameters can be obtained. This is feasible only when the CAC & routing strategy is defined by a set of continuous parameters, as in the case of load sharing and reward maximization strategies. In fact, in these cases the inequality constraints can be replaced by equality constraints which can simplify the problem solution as shown in Section 9.6 for the reward maximization routing.

When the routing parameters are not continuous variables (e.g., sequential routing), the dimensioning problem can be simplified by dividing it into dimensioning and routing subproblems. This is done by applying an iterative approach where in each step the two subproblems are solved in sequence. First the dimensioning problem is solved for given routing parameters. Then the CAC & routing parameters are optimized for given link capacities in order to reduce the network cost in the next iteration. There is no guarantee of convergence, and if convergence occurs, it may be not at the optimum. Nevertheless, in practice, this approach generally works well. In this case the optimization of the CAC & routing parameters is simplified and can be based on some approximations specific to the applied routing strategy. Obviously this approach can be applied to all types of CAC & routing strategies including the ones with continuous parameters.

The dimensioning subproblem is defined by

$$\min_{\mathbf{L}} \left\{ c = \sum_{s \in S} c^s(L^s) \right\} \qquad (10.7)$$
$$B_j(\mathbf{L}) \leq B_j^c, \quad j \in J$$

This is a nonlinear multi-variable constrained minimization problem which one can try to solve by means of a general-purpose algorithm [see e.g. (Luenberger, 1984)], assuming that the gradients of the objective and constraint functions, with respect to the optimization variables, can be calculated. While the gradient of the objective function is usually given in an explicit form, the gradients of the constraint functions can be evaluated analytically only under certain approximations and for simple CAC & routing schemes, see e.g. (Girard, 1990). Thus in most cases one has to approximate the gradients from finite differences based on an analytical model for performance evaluation. Since the required precise analytical model is usually quite complex (fixed-point equations), this approach is feasible for small networks only.

Link performance characteristics as optimization variables

A significant reduction of the dimensioning subproblem complexity can be achieved by using the link performance characteristics, Π^s, as optimization variables. In this case the problem is formulated as follows:

$$\min_{\mathbf{\Pi}} \left\{ c = \sum_{s \in S} c^s\left(L^s(\mathbf{A}^s(\mathbf{\Pi}, \pi), \Pi^s)\right) \right\} \qquad (10.8)$$
$$B_j(\mathbf{\Pi}) \leq B_j^c, \quad j \in J$$

The central feature of this approach is that, as a result of having link performance characteristics, the connection class loss probabilities and link offered traffic parameters can be evaluated without the necessity for an iterative solution of the fixed-point equations. Thus, the evaluation of the constraint function gradients is much simpler. While the gradient of the objective function is more complex than in formulation (10.7), it is still based on independent link-dimensioning models. Implementation of this formulation is relatively simple when a single parameter link performance characteristic is sufficient. This is the case in single-rate networks where link blocking probabilities, b^s (probability that link s is in blocking state), can be used as optimization variables. Such an approach was proposed in (Pióro, 1983; Pióro and Wallstroem, 1985) where Rosen's method (Rosen, 1960) is suggested as an optimization procedure, and both one and two moment models for evaluation of connection loss probabilities are compared. In the case of multi-rate networks one can consider approximations where link traffic loss probabilities, $\overline{B}^s = 1 - \overline{a}^s/a^s$, are used as optimization variables assuming that link connection class loss probabilities B_j^s can be derived approximately from $B_j^s = f_b(\overline{B}^s, \mathbf{A}^s)$ where $\mathbf{A}^s = [A_j^s, j \in J^s]$, see e.g. (Labourdette and Hart, 1990). In the following considerations $\overline{B}^s$ will be used in this context.

Flow distribution as optimization variables

Another formulation of the optimization problem can be arrived at using the flow distribution among the alternative paths as optimization variables. Let $\overline{a}^s$ and $\overline{a}_j^k$ denote the traffic carried on link s and connection class j traffic carried on path k, respectively. The network flow distribution is given by $\overline{\mathbf{a}} = [\overline{a}_j^k, j \in J, k \in W_j]$. In an analogous fashion the connection class flow distribution, $\overline{\mathbf{a}}_j = [\overline{a}_j^k, k \in W_j]$, and link flow distribution, $\overline{\mathbf{a}}^s = [\overline{a}_j^k, \overline{a}_j^k \in \mathcal{F}_s]$, can be defined, where $\mathcal{F}_s$ is the set of flows using link s. The flow formulation can be described by

$$\min_{\overline{\mathbf{a}}} \left\{ c = \sum_{s \in S} c^s(L^s(\overline{a}^s)) \right\} \quad (10.9)$$

$$\sum_{\overline{a}_j^k \in \mathcal{F}_s} \overline{a}_j^k = \overline{a}^s, \quad s \in S$$

$$\sum_{k \in W_j} \overline{a}_j^k \geq (1 - B_j^c) a_j, \quad j \in J$$

$$\overline{\mathbf{a}}_j \in \Omega_j(a_j, \pi), \quad j \in J$$

where $\overline{a}^s$ is total traffic carried on link s, and the routing constraints are defined by the domains of the connection class flow distributions, $\Omega_j(a_j, \pi)$, which can be obtained under the applied CAC & routing policy π. This formulation has the structure of a non-linear multi-commodity flow problem with attractive linear connection loss constraints. Nevertheless the linearization of the loss constraints is achieved at the expense of increased complexity of the objective function.

The additional complexity arises from the fact that the link dimensioning function, $L^s(\overline{a}^s)$, is not independent from the rest of the network parameters as was the case in formulation (10.8). In fact, if treated independently the problem is undetermined since in general the link carried traffic function

$$\overline{a}^s = f_c(\mathbf{A}^s, \overline{B}^s), \quad s \in S \quad (10.10)$$

can give the same $\overline{a}^s$ for many different pairs $\mathbf{A}^s, \overline{B}^s$ which would correspond to different link capacities. One can try solving the problem by using the link loading function in Equation (10.10), resulting in

$$\overline{a}^s = f_c(\mathbf{A}^s(\overline{\mathbf{B}}, \pi), \overline{B}^s), \quad s \in S \quad (10.11)$$

where $\overline{\mathbf{B}} = [\overline{B}^s, s \in S]$. Then Equation (10.11) can be solved for $\overline{B}^s$. Having the link performance characteristics and the link offered traffic levels from the link loading function, the link capacities can be found from the link dimensioning function (10.5). Unfortunately, in general, it is not only difficult to solve this nonlinear problem, but it is also difficult to define conditions under which there is one or more solutions. Moreover, the complexity of the objective function leads to difficulties in evaluation of the objective function gradients required in the optimization procedure. Despite all the complexity the flow formulation is considered an attractive approach for large network dimensioning if some approximations or heuristics are used to overcome the mentioned obstacles. A good example of such an approach is the model for single-rate networks described in (Berry, 1971) which corresponds well to formulation (10.9). It is based on a flow-deviation method derived from the Frank-Wolfe formulation (Frank and Wolfe, 1956).

There are also many other dimensioning methods which are based on certain modifications of formulation (10.9). All of them exploit the linear form of the loss constraints which, with some approximations, permits the solution of large network problems. In the following we describe three basic options which take advantage of this feature.

Option 1 — routing problem separation. To avoid complexity associated with the routing constraints, the problem can be solved first without routing constraints. Then the routing algorithm is synthesized with the objective to match the designed flow distribution as closely as possible.

Option 2 — problem linearization. Note that even with the removal of the routing constraints the problem is still non-linear due to the form of the objective function (non-linear relation between the link cost and link traffic). One way to simplify the solution is to linearize the problem. For example, by assuming that the link traffic loss probabilities, $\overline{\mathbf{B}}$, are constant one can solve the problem iteratively. In each iteration the flow distribution is evaluated assuming a linear objective function where the cost of carrying a unit of traffic is equal to the link's marginal cost, $\partial c^s / \partial \overline{a}^s$. This corresponds to optimization in the hyperplane tangent to the point defined by $\overline{\mathbf{B}}$ and is defined by the following linear program:

$$\min_{\overline{\mathbf{a}}} \left\{ c = \sum_{s \in S} \frac{\partial c^s}{\partial \overline{a}^s} \overline{a}^s \right\} \quad (10.12)$$

$$\sum_{\overline{a}_j^k \in \mathcal{F}_s} \overline{a}_j^k = \overline{a}^s, \quad s \in S$$

$$\sum_{k \in W_j} \overline{a}_j^k \geq (1 - B_j^c) a_j, \quad j \in J$$

$$\overline{a}_j^k \leq (1 - B_j^k(\overline{\mathbf{B}})) a_j, \quad j \in J, \ k \in W_j$$

Note that because the link traffic loss probabilities are given, additional linear constraints limiting the maximum flow on a path are introduced. This feature facilitates matching the CAC & routing parameters to the designed flow distribution.

10.2 Dimensioning of ATM Networks

The values of marginal costs for the next iteration are evaluated from the new values of $\overline{a}^s$ and a link model, $\overline{B}^s = f^s(\overline{a}^s/(1-\overline{B}^s), L^s)$, which also serves for link dimensioning. The algorithm is illustrated in Figure 10.4. The main difficulty of this formulation is an *a priori* evaluation of the link traffic loss probabilities which can be done only approximately (e.g., based on experience). In a heuristic approach proposed in (Knepley, 1973), which is similar to the considered formulation, the link blocking probabilities are related to the average network traffic loss probability constraint, $b^s = 2\overline{B}^c$.

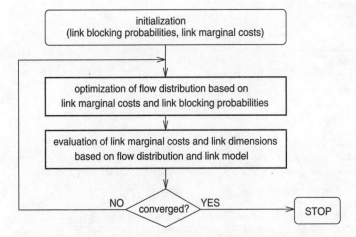

Figure 10.4. *Iterative procedure for given link traffic loss probabilities.*

Note that when end-to-end virtual paths are used for resource management (VP_{ee} design from Section 9.3), the traffic flow associated with a particular path can be represented by its "bandwidth." Thus, assuming a linear link "bandwidth" cost function, the flow formulation becomes linear and can be easily solved in one step since both the non-linearity of the objective function and routing constraints are removed.

Option 3 — conjunction with link traffic loss probability variables. The main difficulty in the general case of *Option 2* is the *a priori* evaluation of the link traffic loss probabilities. This problem can be removed by treating these values as optimization variables. In this case the optimization problem is defined as

$$\min_{\overline{a},B} \left\{ c = \sum_{s \in S} c^s \left(L^s(\overline{a}^s, \overline{B}^s) \right) \right\} \qquad (10.13)$$

$$\sum_{\overline{a}_j^k \in \mathcal{F}_s} \overline{a}_j^k = \overline{a}^s, \qquad s \in S$$

$$\sum_{k \in W_j} \overline{a}_j^k \geq (1 - B_j^c) a_j, \qquad j \in J$$

$$\overline{a}_j^k \leq (1 - B_j^k(\overline{\mathbf{B}}))a_j, \qquad j \in J, \quad k \in W_j$$

While the general solution of this problem is still quite difficult, an approximate solution can be obtained by extending the iterative procedure from *Option 2*. In this case each iteration consists of three steps as illustrated in Figure 10.5.

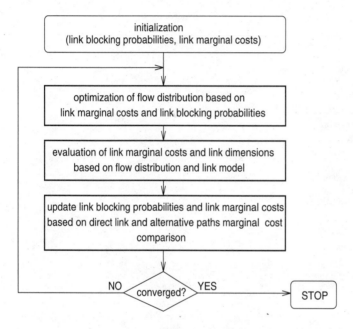

Figure 10.5. *Iterative procedure with link traffic loss probabilities as optimization variables.*

First, the flow distribution is optimized based on the given link traffic loss probabilities, marginal link costs, and routing feasibility constraints [problem (10.12)]. Consider, for example, the case of sequential routing. In this case the sequence of alternative paths for a given connection class can be easily defined by ordering the feasible paths according to their marginal costs approximated by the sum of link marginal costs on the path. Then the number of alternative paths and the flow on each path can be determined from the connection loss constraint since the path and total connection class loss probabilities can be computed from the given link traffic loss probabilities. Then, based on the new flow distribution, new values of the marginal link costs and link capacities are evaluated from the link model. Finally, the link traffic loss probabilities and link marginal costs are updated based on a marginal cost analysis. To explain this last step let us consider the optimality conditions from the perspective of class j traffic flow allocation to the direct-link-path, $\overline{a}_j^d$, versus allocation to all other alternative paths $\overline{a}_j^a$. From the first order optimality conditions (Kuhn-Tucker) it follows that the optimal allocation can be achieved when the marginal costs of carrying class j traffic on the direct-link-path,

$\partial c/\partial \overline{a}_j^d$, and all alternative paths, $\partial c/\partial \overline{a}_j^a$, are equal

$$\frac{\partial c}{\partial \overline{a}_j^d} = \frac{\partial c}{\partial \overline{a}_j^a} \qquad (10.14)$$

By substituting the left side of this equality by the marginal cost of the direct link $\partial c^d/\partial \overline{a}_j^d$ we obtain a condition which can be used to update this link's traffic loss probability. Namely, assuming that the marginal cost of all alternative paths is known, we can evaluate, based on a link model, a new direct link capacity and corresponding B^d which satisfy the condition (10.14).

The described iterative procedure takes full advantage of both the flow formulation and link traffic loss probabilities treated as optimization variables. In particular the flow distribution phase is linear and the link traffic loss probability update and dimensioning phases are based on a link model only. This simplicity allows one to consider very large networks. More details on this type of solution can be found in (Ash, Cardwell, and Murray, 1981) where an algorithm for a multi-hour problem in single-rate networks is presented (the multi-hour problem will be discussed in Section 10.2.3).

10.2.2 Link bandwidth modularity

Bandwidth modularity introduces additional complexity to the already difficult problem. Besides transforming the issue into an integer optimization problem, the cost of different size modules may not be proportional to the bandwidth. In this case the exact relation between the link cost and its "bandwidth" can be quite complex. These features limit application of integer programming packages for network dimensioning to very small networks. In most practical cases the solution is first approximated by an optimization model for continuous link "bandwidth" with a linearized link "bandwidth" cost function. Then an "intelligent" rounding off can be applied in order to provide the final design, see e.g. (Pióro, 1989; Pióro and Tomaszewski, 1989). This approach seems to be especially reasonable in modern multi-service networks where the prediction of the offered traffic can be very approximate, and usually the traffic demand grows very fast, so the network design has to be updated anyway in relatively short time intervals.

Another issue associated with the link modularity is that in some cases an ATM link capacity allocation is realized as a sum of several modules which have separate interfaces with the node (i.e., separate switch input and output ports). In this case a particular connection should be carried on one module only, to provide the cell sequence integrity. This feature influences the connection's "bandwidth" allocation which is not a function of the link capacity but rather a function of the used module capacity. Moreover, the same feature can have slight influence on the link "bandwidth" utilization since in some instances a wide-band connection can be rejected due to the lack of adequate free capacity in any of the modules, although the sum of available capacity on all modules can be sufficient to carry this connection. Nevertheless in most cases this effect is not significant, see e.g. (Lutton, 1984), and can be ignored in the network design procedures.

10.2.3 Multi-hour case*

Up to now we have assumed that the offered traffic matrix **a** is fixed, which corresponds to the case where all connection classes have the maximum average arrival rates in the same time period called the *busy hour*. Nevertheless, in many cases these maximum values can be associated with different time periods. This may be caused by different time zones which are covered by the network, or by migration of the population from work to homes and vice versa, or by certain habits of the network users which can use different services in different time periods. In such cases the network design has to provide required GoS constraints in every *hour* which is characterized by its own traffic matrix $\mathbf{a}^t$ and possibly its own connection loss constraints, $B_j^{c,t}$, where t is the time period index.

In general there are two different approaches to the multi-hour problem. The first assumes that the underlying transport network has fast adaptation capability based on the digital hierarchy cross-connect systems. In this case the network can be designed separately for each hour (based on the methods discussed in Sections 10.2.1 and 10.2.2). Then the "bandwidth" required in each particular hour is provided by rearranging the transport layer. Alternatively, the transport layer can be rearranged based on an adaptive mechanism where the demand for the "bandwidth" is predicted from traffic measurements. This option is covered in the next section.

The second approach to the multi-hour problem is used for the cases where the transport layer has only slow adaptation capabilities. Here the design is based on the same resource allocation for all *hours*, while the CAC & routing algorithm parameters can be *hour* dependent in order to take advantage of the changing flow distributions. In this case the natural problem formulation is given by

$$\min_{\mathbf{L},\Upsilon}\left\{c = \sum_{s\in S} c^s(L^s)\right\} \quad (10.15)$$
$$B_j^t(\mathbf{L},\Upsilon^t) \leq B_j^{c,t}, \quad j \in J, \ t \in T$$

In a manner similar to the original formulation of the single hour problem, Equation (10.6), one can try to solve this multi-hour problem by a general non-linear programming method. Unfortunately, the significantly increased number of the constraints, together with all of the difficulties listed for problem (10.6), limit application of this approach to very small networks. The most attractive option seems to be the formulation using both the flow distribution and the link traffic loss probabilities as the optimization variables (analogously to *Option 3* from Section 10.2.1). Although direct solution of this problem is still difficult, an efficient approximate solution can be achieved by an iterative algorithm consisting of three blocks similar to the single-hour case (see Figure 10.5). This concept was introduced in (Ash, Cardwell, and Murray, 1981) where the detailed description of the algorithm and numerical examples are given for single-rate networks. In this case the flow distribution is evaluated from the following linear program formulation:

$$\min_{\overline{\mathbf{a}}}\left\{c = \sum_{s \in S} \frac{\partial c^s}{\partial \overline{a}^s} \overline{a}^s\right\} \quad (10.16)$$

$$\max_{t}\left\{\sum_{\overline{a}_j^{k,t} \in \mathcal{F}_s} \overline{a}_j^{k,t}\right\} = \overline{a}^s, \quad s \in S$$

$$\sum_{k \in W_j} \overline{a}_j^{k,t} \geq (1 - B_j^{c,t})a_j^t, \quad j \in J, \; t \in T$$

$$\overline{a}_j^{k,t} \leq \left(1 - B_j^{k,t}(\overline{\mathbf{B}}^t)\right)a_j^t, \quad j \in J, \; k \in W_j, \; t \in T$$

Note that this is an extension of the single hour formulation of *Option 2* from Section 10.2.1. In (Ash, Cardwell, and Murray, 1981) an alternative formulation for the flow distribution optimization is also considered where the CAC & routing constraints are taken into account directly in the optimization process. This is done by defining the routing domains $\Omega_j^t(a_j, \pi)$ as a linear combination of finite number of fixed flow distributions which are evaluated based on the link traffic loss probabilities and the routing strategy constraints. Then the optimal combination of these flow distributions (for each hour) minimizing the linearized network cost is found by a linear program.

Observe that in the case of VP_{ee} design the multi-hour formulation is significantly simplified since the connection loss constraints are separated from the optimization process. Moreover, under the assumption of linear link "bandwidth" cost function, the problem becomes linear.

10.3 Adaptation to Traffic Changes

The issue of adaptation of the ATM network design to changes in the offered traffic matrix has many facets and can be grouped into several subproblems depending on the considered time scale and resource availability.

The long term adaptation strategies can basically use the same tools as the ones used for the original network design except that the offered traffic matrix forecast can be evaluated based on traffic and performance measurements in the network. These strategies can be divided into reactive and predictive categories. The reactive strategies add additional resources when the current resource allocation is not sufficient to sustain the required GoS constraints. In this case the adaptation is based on the current traffic matrix measurement with a heuristic margin for further growth of the traffic. In the case of predictive strategies, the measurements are used to predict the traffic growth so the network design is updated before the GoS is compromised. The latter option also allows for planning the updating periods.

Short term adaptation strategies are possible when the ATM network transport layer is based on a digital hierarchy with cross-connect capabilities. Here, one can also apply reactive and predictive strategies. In particular the predictive strategies can be applied to adapt to highly predictable traffic variations, e.g., caused by a difference in time zones covered by the network, or by the work to home movements and vice versa, or by the changes in tariff at predetermined day time instances. The reactive strategies are based on the current traffic measurements and should react to unexpected changes in the traffic patterns, e.g., caused by special events, local disasters, etc.

As in the case of virtual network update (see Section 9.7), the physical network design update algorithms can be implemented in a centralized or distributed fashion. The centralized option fits well to the long term adaptation strategies since the design update time is not critical while the more efficient solution can bring visible savings over long periods. On the other hand, the distributed option corresponds well to the short term reactive adaptation schemes where the reaction time is sometimes critical and the decision should be evaluated locally. The marginal link costs based on the average shadow prices can be very useful in such local adaptation schemes. Thus the approach described in Section 9.7 for virtual networks is also applicable to the physical network layer.

10.4 Survivability Issues

Network survivability is a very diverse and complex issue which involves many network functions and components in the area of network planning, protocols, and resource management in several time scales. To cope with this complexity we divide the discussion of survivability issues into four parts: failure scenarios, restoration mechanisms, connection and GoS service restoration objectives, and design of self-healing networks.

10.4.1 Failure scenarios

In general the reliability of modern network components is very high and in most cases the assumption that there is only one element failure at a time is very reasonable. Obviously this does not include natural disasters such as earthquakes or human made disasters like wars. Nevertheless, the civilian networks, which are the focus of this book, are usually not designed for such environments. This leaves us with the possible failure of one transmission facility (defined as a separate physical system) or one network node at a time. Due to a potentially large impact of a node failure on network performance (e.g., cutoff of many network subscribers), the nodes are usually well protected against global failure by using hot stand-by elements (obviously a local natural disaster such as fire can still succeed in disabling the node). For this reason most of the failure protection efforts and mechanisms focus on failures of transmission facilities caused by electronic equipment failures or human-made local disasters such as a cable that is cut. Nevertheless, in some cases a node failure can be still treated as a possibility although with a smaller probability than that of a transmission facility.

10.4 Survivability Issues

A transmission facility failure can affect the considered ATM network layer in many ways depending on the realization of the transport layer associated with the network. As discussed in Section 10.1.1, in the case of a dedicated transport layer a transmission facility failure can be seen on the ATM layer as a total or a partial loss of transmission capacity between two ATM nodes. Obviously if there is full automatic protection of the transmission facility (see Figure 10.1b) the ATM layer is not affected except for possible minor disruptions of the cell streams which can be handled by the higher layer protocols.

When the transport layer is based on a digital hierarchy, the effects of a transmission facility failure from the ATM layer viewpoint are more complex. Such a failure can be considered as having two phases. In the immediate failure aftermath, assuming that there is no automatic protection of this facility, two basic scenarios are possible. When only one ATM link used the failed facility, the effect is analogous to the one with a dedicated transport layer: total or partial loss of the transmission capacity by this link. If more than one ATM link used the failed facility, all of these links are affected by a partial or total loss of their capacity (see Figure 10.3).

Due to the self-healing capabilities of the digital hierarchy networks based on cross-connect functionality, the lost capacity of the ATM links affected by the failures can be totally or partially restored in the second phase of the failure state. There are two important features of this restoration. First, the degree of restoration may be unknown in advance because it can depend on the distribution of spare capacity in the DH network which can be time dependent. Second, the restoration time is not negligible. In the case of centralized restoration algorithms, which are currently implemented, the time is in the range of minutes which is a significant time from the ATM layer viewpoint. In the case of a distributed restoration mechanism this time can be reduced to the order of seconds with the objective to have it limited to 2 seconds which allows affected connections to be sustained. More details on these issues can be found in (Wu, 1995; Sosnosky, 1994).

In summary, the failure scenarios seen on the ATM layer can range from a single link failure (total or partial) to many synchronized link failures (total or partial) with total or partial recovery time varying from negligible (automatic restoration of transmission facilities) via order of seconds or minutes (digital hierarchy) to hours and days (dedicated transmission facilities). Note that this description covers also ATM node failures which can be interpreted as a failure of all links connected to the failed node.

10.4.2 Restoration mechanisms

In general, failure protection on the ATM layer can be based on a link, path, or path fragment restoration, as illustrated in Figure 10.6. Although link restoration seems to be most natural, it has several disadvantages. In particular, it requires in general more resources for the same level of restoration, see e.g. (Xiong and Mason, 1997). Also, in the case of partial "bandwidth" restoration, the restoration algorithm has to communicate with the origin nodes to decide which connections should be restored; thus the restoration time is increased. On the other hand, the path restoration has several important advantages. Besides being more efficient from the resource utilization viewpoint due to more distributed demand (many OD pairs are affected by a link failure), it can be well integrated with the connection setup procedure in the connection origin nodes. This feature can speed up the connection restoration process, and takes full advantage of the CAC & routing algorithms which can significantly enhance the self-healing capabilities. It is also important that if virtual paths are used as the routing paths (VP_{ee} design and VN design from Section 9.3) the process is further simplified since the ATM switch tables do not have to be modified if the restoration paths are pre-established. For these reasons in the following we focus on the path restoration approach.

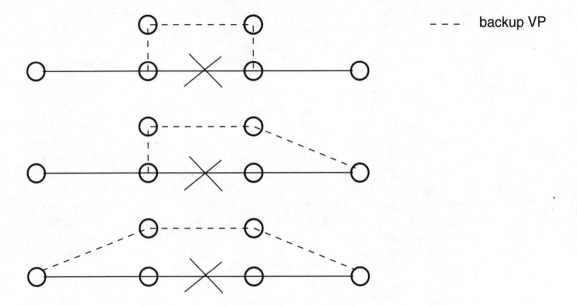

Figure 10.6. *Link, fragment, and path restoration concepts.*

The restoration process can be divided into two parts: *connection restoration* and *GoS restoration*. The connection restoration refers to the connections which were using the failed element and the objective of the connection restoration mechanism is to restore these connections on the restoration paths. The connection restoration process should be relatively fast, less than two seconds (Wu, 1995), so the affected connections are not permanently disrupted. The GoS restoration is associated with new connections which arrive during long lasting failures (a few minutes and more). In this case the GoS of connection classes which are directly affected by the failure can be deteriorated significantly for a long period of time, up to a complete cutoff. Thus the objective of the GoS restoration in these periods is to improve the connection classes' GoS level.

Let us first consider the connection restoration. Note that the restoration paths can be pre-established so that they can be used immediately after the failure is recognized. If sufficient "bandwidth" is also preallocated (deterministic case), all affected connections with restoration priority can be switched to the restoration path by simply changing its VP address. This type of restoration can be done without any interruption of the connections if special protocols are applied, see e.g. (Ohta and Ueda, 1993). Note that this *hitless* path switching on the ATM network layer is more attractive than the ATM link restoration on the digital hierarchy network layer.

If the restoration path has no "bandwidth" reserved in advance (statistical case), the restoration algorithm can ask resource managers of the links constituting the restoration path for "bandwidth" by using the connection setup algorithm. Obviously this request should have a priority. In the case of preemptive priority, an additional mechanism would be required to disconnect some connections or reduce "bandwidth" allocated to other service oriented VNs. Observe that if alternative routing is installed in the network, the restoration paths can be integrated with the alternative routing paths. The only requirement is that the set of alternative paths includes some paths disjoint from the others so that they can be used as restoration paths. This feature also indicates that the connection restoration can be done on several paths and in the limit it can be done on VC by VC basis. In this case the CAC & routing algorithm can constitute an integral part of the connection restoration mechanism.

The key mechanism of the GoS restoration is the CAC & routing algorithm. If the restoration path has reserved capacity (deterministic case), the failed path is simply replaced by the restoration path in the routing table. If the "bandwidth" is not reserved (statistical case), the failed path is removed from the table and the CAC & routing algorithm adjusts its parameters to provide the required distribution of GoS among different connection classes.

Note that the mechanisms of the connection and GoS restorations are interrelated due to the fact that they operate, at least partially, in the same "bandwidth" space. Since the "bandwidths" allocated to primary and backup virtual networks can overlap and each part of the "bandwidth" can be designated for an exclusive or shared usage with or without priority, there are several possible interrelations between these VNs which are illustrated in Figure 10.7. The choice of the particular option depends on the available restoration mechanisms and the restoration objectives which are described in the next section.

10.4.3 Connection and GoS restoration objectives

Here we specify in more detail the objectives of the connection and GoS restoration mechanisms on the ATM network layer when the duration of the failure affecting the transport layer is long enough (e.g., more than 2 seconds) to permanently disrupt connections on the affected paths. Observe that in the multi-service and multi-user environment, these objectives are in general service, user, and network operator dependent.

Let us consider first the connection restoration. In general, depending on the available capacity, the network can restore all or a fraction of existing connections.

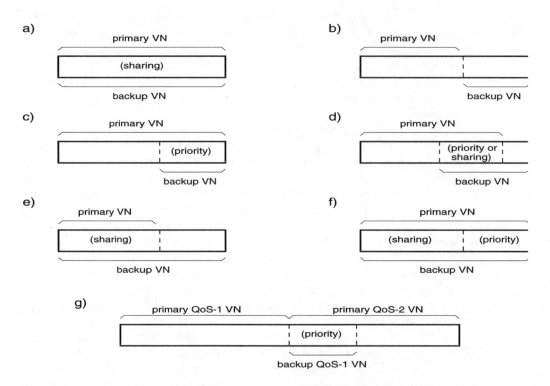

Figure 10.7. *Different relations between primary and backup virtual networks.*

In the case of a partial restoration the selection of connections to be restored can be a function of service (e.g., priority for real-time connections) or a contract with the user (first connections with restoration guarantee). The degree of restoration can be defined differently for the deterministic and statistical cases. In the deterministic case, the degree is a function of the "bandwidth" reserved for VPs in the backup VNs. The reservation level can be fixed or state-dependent (a function of "bandwidth" required by accepted connections with restoration guarantees). In the statistical case (the "bandwidth" is shared between the primary and backup VNs — Figure 10.7a) without preemptive priority, the "bandwidth" available for restoration depends on the current state of the network. Thus only the average degree of restoration can be specified in advance, being a function of the average free capacity of the links constituting the restoration paths.

In the GoS restoration case the degree of restoration also depends on the available capacity. In general the GoS can be restored totally (to the level before failure) or partially (somewhere between the level before the failure and the case where no action is taken). The degree of restoration, defined by the connection loss probabilities, can be also a function of the service priority. The level of GoS restoration is controlled by different mechanisms in the deterministic and statistical cases. In the first case (e.g., Figure 10.7b), the level of GoS restoration is defined principally by the amount of "bandwidth" allocated to the restoration paths in the backup VN.

This amount can be fixed or can be a function of the connection class's average traffic. In the statistical case (Figure 10.7a) the GoS restoration level is a function of the "bandwidth" allocated to the network and the CAC & routing algorithm.

Note that in general the connection and GoS restoration objectives should be synchronized in the sense that the "bandwidth" reserved for the connection restoration is at the same time reserved for the GoS restorations and vice versa. Nevertheless the connection restoration can be limited by the protocol capabilities. Thus in some cases the network can offer the GoS restoration without the connection restoration (all existing connections affected by the failure are lost). Also observe that the statistical and deterministic options can coexist. For example, the affected connection classes can first use the "bandwidth" allocated to the backup VN, and if this "bandwidth" is not sufficient, they can share the "bandwidth" with the primary VN (Figure 10.7e).

Up to now we have considered the restoration objectives from the affected connection classes viewpoint. Nevertheless, the other connection classes can also be influenced by the failure. This influence depends on the applied restoration philosophy. One possible approach is to design the network in such a way that fulfillment of the required restoration degree does not influence the other connection classes' performance. For example, this is the case when the affected paths can be fully restored in the backup virtual network with reserved capacity. An alternative, economic approach assumes that, under the failure conditions, the GoS of unaffected connection classes can be reduced in order to restore connections and GoS of the affected connection classes. In the deterministic case this option can be realized by overlapping the backup virtual network with the primary network (Figure 10.7c,d). Here under nominal conditions the primary network has full or partial access to the "bandwidth" allocated to the backup VN. Under failure conditions the affected connection classes can have priority (possibly preemptive) in the backup VN space over the primary connections.

Note that if several QoS VNs exist in one network, the "bandwidth" allocated to high priority QoS backup VNs can be reserved in the space of low priority QoS VNs (Figure 10.7g). For example, the real-time service VN can have its backup VN in the space of ABR and UBR service VNs. Due to priority on the cell level provided by the scheduling mechanism, the real-time connections affected by a failure can be immediately switched on the restoration paths. The reduction of the "bandwidth" available for the data services will be recognized by the flow control mechanisms which will reduce the data source rates. This example indicates that in a multi-service environment the self-healing capabilities can be achieved with minimal (or without any) additional resources.

10.4.4 Design of self-healing networks*

In general the problem of self-healing network design is quite complex due to the large number of possible failure states, $e \in E$. The most promising formulations are the ones which treat the flow distributions as optimization variables due to the linear form of the GoS constraints. Under the continuous link "bandwidth" assumption the general formulation of the statistical option (primary and backup VNs share the "bandwidth" — Figure 10.7a) for GoS restoration can be expressed as

$$\min_{\overline{\mathbf{a}}} \left\{ c = \sum_{s \in S} c^s \left(L^s(\overline{a}^s) \right) \right\} \quad (10.17)$$

$$\max_e \left\{ \sum_{\overline{a}_j^{k,e} \in \mathcal{F}_s} \overline{a}_j^{k,e} \right\} = \overline{a}^s, \quad s \in S$$

$$\sum_{k \in W_j} \overline{a}_j^{k,e} \geq (1 - B_j^{c,e}) a_j, \quad j \in J, \ e \in E$$

$$\overline{\mathbf{a}}_j^e \in \Omega_j^e(a_j, \pi), \quad j \in J, \ e \in E$$

Note a similarity to the multi-hour problem which indicates that the method based on an iterative approach, where the link traffic loss probabilities are also treated as optimization variables, could be used for an approximate problem solution (see Section 10.2.3 and Figure 10.5). Nevertheless in the case of self-healing design the complexity of the problem is increased due to the fact that in general the number of failure states can be significantly larger than the number of hours in the multi-hour problem.

Let us now consider an extended general formulation which covers both the statistical and deterministic cases. The design consists of two virtual networks: primary VN with link "bandwidth" allocation represented by G_p^s and backup VN with link "bandwidth" allocation denoted as G_b^s. Let $u_j^{k,e}$ denote the "bandwidth" allocated to restoration path k serving connection class j in state e. By treating this "bandwidth" as a flow we can define the backup network flow distribution as $\mathbf{u} = [u_j^{k,e}, j \in J, k \in W_j, e \in E]$. The deterministic degree of restoration is defined by the "bandwidth", $d_j^e = \sum_{k \in W_j} u_j^{k,e}$, which should be allocated to the restoration virtual paths of connection class j in state e. In general the required degree of restoration can depend on the particular failure likelihood (the smaller the likelihood, the smaller the degree of restoration). Having the required deterministic degree of restoration, the generic design problem can be formulated as follows:

$$\min_{\overline{\mathbf{a}},\mathbf{u}} \left\{ c = \sum_{s \in S} c^s \left(G_p^s(\overline{a}^s) + G_b^s \right) \right\} \quad (10.18)$$

$$\max_e \left\{ \sum_{\overline{a}_j^{k,e} \in \mathcal{F}_s} \overline{a}_j^{k,e} \right\} = \overline{a}^s, \quad s \in S$$

$$\max_e \left\{ \sum_{u_j^{k,e} \in \mathcal{F}_s} u_j^{k,e} \right\} = G_b^{\prime s}, \quad s \in S$$

$$\sum_{k \in W_j} (\overline{a}_j^{k,e}) \geq (1 - B_j^{c,e}) a_j, \quad j \in J, \ e \in E$$

$$\sum_{k \in W_j} u_j^{k,e} \geq d_j^e, \quad j \in J, \ e \in E$$

10.4 Survivability Issues

$$\overline{\mathbf{a}}_j^e \in \Omega_j^e(a_j, \pi), \quad j \in J, \; e \in E$$

Note that this formulation takes into account both the deterministic constraints for connection and GoS restoration, d_j^e, and the statistical constraints for GoS restoration, $B_j^{c,e}$. This approach allows for several alternative uses of the backup VN "bandwidth" under the failure condition. In particular, the backup VN "bandwidth" can be reserved only for the affected connection classes (Figure 10.7b) or, once the affected connections are restored in the backup VN, all connection classes can have access to the backup network. In other words, the connection restoration corresponds to the situation depicted in Figure 10.7b, while the GoS restoration corresponds to Figure 10.7c (or Figure 10.7f). The second option can provide good GoS restoration at low cost when the required degree of deterministic connection restoration is low.

The general solution to problem (10.18) is even more difficult than that of problem (10.17) due to the increased number of constraints. The exception is the case where the VP$_{ee}$ design (Section 9.3) is used for the primary network, and the virtual paths not affected by the failure are not reconfigured. To keep the same notation, let us assume for this case that the flow a_j represents the "bandwidth" required by connection class j to provide the GoS constraint, and $\overline{a}_j^k$ denotes "bandwidth" allocated to path k for connection class j in the nominal state (no failure). As before, d_j^e is the "bandwidth" which should be allocated to the restoration virtual paths of connection class j in state e. Under the linear link cost assumption the VP$_{ee}$ design can be obtained by solving the following linear program:

$$\min_{\overline{\mathbf{a}},\mathbf{u}} \left\{ c = \sum_{s \in S} (G_p^s + G_b^s) c^{s'} \right\} \qquad (10.19)$$

$$\sum_{\overline{a}_j^k \in \mathcal{F}_s} \overline{a}_j^k = G_p^s, \quad s \in S$$

$$\max_e \left\{ \sum_{u_j^{k,e} \in \mathcal{F}_s} u_j^{k,e} - \overline{a}_j^k \delta_j^{k,e} \right\} = G_b^s, \quad s \in S$$

$$\sum_{k \in W_j} \overline{a}_j^k = a_j, \quad j \in J$$

$$\sum_{k \in W_j} u_j^{k,e} \geq d_j^e, \quad j \in J, \; e \in E$$

where $\delta_j^{k,e} = 1$ if path k (used by class j) is affected by the failure in state e, and $\delta_j^{k,e} = 0$ otherwise.

The problem (10.18) can be significantly simplified by applying a decomposition into separate subproblems. First the primary virtual network is designed using methods presented in Section 10.2, with a provision that the set of routing paths has to provide restoration paths for each possible failure. Then based on the obtained flow distribution the network operator decides the degree of restoration for each possible network failure state, d_j^e. Finally the backup VN is designed by solving the following linear problem:

$$\min_{\mathbf{u}} \left\{ c = \sum_{s \in S} c^s(G^s) \right\} \quad (10.20)$$

$$\max_{e} \left\{ \sum_{u_j^{k,e} \in \mathcal{F}_s} u_j^{k,e} \right\} = G^s, \quad s \in S$$

$$\sum_{k \in W_j} u_j^{k,e} = d_j^e, \quad j \in J, \ e \in E$$

Obviously the relative simplicity of this formulation is achieved at the expense of some loss in efficiency since the design of flow distribution in the primary VN design does not take into account the flow distribution in the backup VN. In the examples presented in (Xiong and Mason, 1997) for VP_{ee} design, the increase of the decomposed design cost varied from one to nine percent.

The assumption of continuous link "bandwidth" in the presented formulations for the self-healing network design is reasonable for a particular QoS backup virtual network. In the case of physical network design with modular increments of the link capacity, the continuous design can serve as an input to the rounding off procedures which were discussed in Section 10.2.2. In a fashion analogous to the primary network design, the "bandwidth" allocated to the backup VNs can be updated based on the measurements of flow distribution in the network and the number of accepted connections with restoration guarantees.

10.5 Discussion and Bibliographic Notes

In this chapter we discussed physical resource allocation to ATM networks. In particular we focused on the allocation of resources from the transport layer to the ATM layer. The transport layer is treated here as a monolithic layer, which is reasonable from the ATM layer viewpoint. Nevertheless the transport layer itself can be divided further into several sub-layers. A generic layered architecture of the network resources is described in (Lubacz and Tomaszewski, 1996). This model identifies six hierarchically ordered layers and is applicable to both STM and ATM networks including the digital hierarchy systems. In Section 10.1.2 the digital hierarchy systems (SONET, SDH) were discussed in the context of providing the transmission resources for the ATM layer. More details on the digital hierarchy standards can be found in (Ballart and Ching, 1989; Shirakawa, Maki, and Miura, 1991).

Since in general the resource management in ATM networks can be expressed as a bandwidth constrained problem, the network dimensioning models derived for the circuit-switched environment can be easily extended to the VN based ATM network design. Most of the design formulations described in Section 10.2 fall into this category. Solutions to these design formulations are usually quite complex and involve several approximations. Readers interested in implementation of these models are referred to (Girard, 1990; Pióro, 1989) where a thorough description of the models and alternatives for single-rate networks is given. An important contribution to ATM network design is presented in (Mitra, Morrison, and Ramakrishnan, 1996) where the multi-rate formulation is solved by an optimization procedure using implied costs for gradient evaluation. This concept is similar to the formulation presented in Section 9.7 where the average shadow prices are used to update link capacities. In (Mitra, Morrison, and Ramakrishnan, 1996) asymptotic techniques are used extensively to reduce the complexity of the link models. This feature is of considerable importance in network design problems where the link models may be evaluated repeatedly many thousand times consuming a significant portion of the calculation time. The VP based design of ATM networks was not treated extensively in this chapter since it is relatively simple. In the literature one can find several models covering this issue, e.g. (Medhi, 1995; Miyao, 1992).

The crucial element of the ATM network design is provision of self-healing capabilities in case of a failure. The network survivability issues are quite complex due to many elements involved such as protocols, resource management, QoS and GoS metrics, different time scales, etc. To cope with this complexity we divided the problem into the connection and GoS restoration subproblems indicating that the degree of restoration can be specified with statistical or deterministic guarantees. Then we focused on the design of self-healing networks indicating several options of applying the backup virtual network concept to achieve cost-efficient solutions. In the literature a lot of attention has been paid to the VP based design of self-healing networks, e.g. (Anderson et al., 1994; Murakami and Kim, 1995; Vanderstraeten et al., 1996). Although this approach is relatively simple in implementation and design, it should be underlined that it requires significantly more resources compared to the VN based design. For example, the resources required for the VP based primary network design are, on average, sufficient for the primary and backup virtual networks in the VN based design with 100% restoration guarantees (Dziong, Xiong, and Mason, 1996b), while in the VP based design the same level of protection would require a 50-70% increase in the network resource cost (Xiong and Mason, 1997).

The large complexity of the ATM network design formulations, especially with self-healing capabilities, limits application of general formal optimization methods. For this reason many design models are based on some heuristics and approximations. In most cases these heuristics are oriented towards a specific problem. One of the interesting exceptions is *Simulated Allocation* (Pióro and Szcześniak, 1996). This approach, which can be paraphrased as an intelligent stochastic wandering in the solution domain, can be applied to many different and difficult network optimization problems including modular designs. An example of its application to VP based design of ATM networks is described in (Pióro and Gajowniczek, 1995).

The issue of the physical resource allocation to ATM networks is large enough to fill a separate book. For this reason we focused in this chapter on the basic issues which are relevant to the framework for resource management in ATM networks presented in this book. There are several other important issues which are not covered in this chapter. For example, the ATM node location design can influence the resource allocation to ATM networks. In general this issue is difficult to formalize since it depends on many management, political, and geographical factors, and in most cases some intuitive heuristics are used. Another important factor in the physical resource allocation procedures is the design of the tariff which can influence strongly the offered traffic matrix distribution. This powerful tool can be used for optimization of the network design and adaptation of the traffic matrix to available resources. One of the main difficulties in considering this issue is the unknown precise relation between the demand and the tariff (elasticity function). Different applications of the tariff in the ATM network design are discussed in e.g. (Kelly, 1994; Ji, Hui, and Karasan, 1996; Bencheikh, Girard, and Liau, 1996). In this chapter we focused on the meshed ATM network topology. While this approach is resource efficient in medium and large networks, a ring topology can be attractive in the case of small networks (e.g., LAN), due to inherent self-healing capabilities, or when the bandwidth utilization is not critical; see e.g. (May et al., 1995).

References

Anderson, J., Doshi, B. T., Dravida, S., and Harshavardhana, P. 1994. Fast restoration of ATM networks. *IEEE Journal on Selected Areas in Communication,* 12(1):128–138.

Ash, G.R., Cardwell, R.H., and Murray, R.P. 1981. Design and optimization of networks with dynamic routing. *Bell System Technical Journal* 60:1787–1820.

Ballart, R., and Ching, Y-C. 1989. SONET: Now it's the standard optical network. *IEEE Communications Magazine,* 27(3):8–15.

Bencheikh, A., Girard, A., and Liau, B. 1990. Pricing and design of a corporate B-ISDN network in a competitive environment. In *Proceedings of Informs 96.*

Berry, L.T.M. 1971. A mathematical model for optimizing telephone networks. Ph.D. Thesis. University of Adelaide.

Dziong, Z., Xiong, Y., and Mason, L.G. 1996a. Virtual network concept and its applications for resource management in ATM based networks. In *Proceedings of Broadband Communications'96, An International IFIP-IEEE Conference on Broadband Communications,* pp. 223–234. Chapman & Hall.

Dziong, Z., Xiong, Y., and Mason, L.G. 1996b. VN based design versus VP based design for ATM networks. INRS-Telecommunications report.

Frank, M., and Wolfe, P. 1956. An algorithm for quadratic programming. *Naval Research Logistics Quarterly,* 3:95–110.

Gavish, B., Trudeau, P., Dror, M., Gendreau, M. and Mason, L.G. 1989. Fiberoptic circuit network design under reliability constraints. *IEEE J. Select. Areas Commun.,* 7(8):1181–1187.

Girard, A. 1990. *Routing and Dimensioning in Circuit-Switched Networks.* Addison-Wesley.

Girard, A., and Sansó, B. 1997. Multicommodity flow models, failure propagation and reliable network design. *IEEE/ACM Transactions on Networking,* submitted.

Herzberg, M., Bye, S.J., and Utano, A. 1995. The hop-limit approach for sparecapacity assignment in survivable networks. *IEEE/ACM Transaction on Networking*, 3(6):775–784.

Ji, H., Hui, J.Y., and Karasan, E. 1996. GoS-based pricing and resource allocation for multimedia broadband networks. In *Proceedings of IEEE INFOCOM'96*, IEEE Computer Society Press.

Kawamura, R., Sato, K.-I., and Tokizawa, I. 1994. Self-healing ATM networks based on virtual path concept. *IEEE J. Select. Areas Commun.*, 12(1):120–127.

Kelly, F. 1994. Tariffs and effective bandwidths in multiservice networks. In *Proceedings of the 14th International Teletraffic Congress*, pp. 401–410. North-Holland.

Knepley, J.E. 1973. Minimum cost design for circuit switched networks. Report No. AD-A014 101, Defence Communications Agency Systems, Engineering Facility, Reston, Virginia.

Labourdette, J.F.P., and Hart, G.W. 1990. Link access blocking in very large multimedia networks. *Computer Communication Review*, 20(4):108–117.

Lubacz, J., and Tomaszewski, A. 1996. Generic architecture and core network design. In *Methods for the performance evaluation and design of broadband multiservice networks*. The COST 242 Final Report. Springer Verlag.

Luenberger, D.G. 1984. *Linear and Nonlinear Programming*. Addison-Wesley.

Lutton, J-L. 1984. Traffic performance of multi-slot call routing strategies in an integrated services digital network. In *Proceedings of ISS '84*.

May, K.P., Semal, P., Du, Y., and Herrmann, C. 1995. A fast restoration system for ATM-ring-based LANs. *IEEE Communications Magazine*, 33(9):90–98.

McDonald, J., 1994. Public network integrity — Avoiding a crisis in trust. *IEEE J. Select. Areas. Commun.*, 12(1):5–12.

Medhi, D. 1995. Multi-hour, multi-traffic class network design for virtual pathbased dynamically reconfigurable wide-area ATM networks. *IEEE/ACM Transactions on Networking*, 3(6):809–818.

Mitra, D., Morrison, J.A., and Ramakrishnan, K.G. 1996. ATM network design and optimization: a multirate loss network framework. *IEEE/ACM Transactions on Networking*, 4(4):531–543.

Miyao, Y. 1992. A dimensioning scheme in ATM networks. In *Proceedings of Networks'92*, pp.171–176.

Murakami, K., and Kim, H. 1995. Joint optimization of capacity and flow assignment for self-healing ATM networks. In *Proceedings of IEEE ICC'95*, pp. 216–220. IEEE Computer Society Press.

Ohta, H., and Ueda, H. 1993. Hitless line protection switching method for ATM networks. In *Proceedings of IEEE ICC '93*, pp. 272–276. IEEE Computer Society Press.

Pióro, M.P. 1983. A uniform approach to the analysis and optimization of circuit switched communication networks. In *Proceedings of the 10th International Teletraffic Congress*, pp. 4.3A:1–7. North-Holland.

Pióro, M.P., and Wallstroem, B. 1985. Multihour optimization and of non-hirarchical circuit-switched communication networks with sequential routing. In *Proceedings of the 11th International Teletraffic Congress*. North-Holland.

Pióro, M.P. 1989. *Design Methods for Non-Hierarchical Circuit Switched Networks with Advanced Routing.* Warsaw University of Technology Press.

Pióro, M.P., and Tomaszewski, A. 1991. Modular engineering of telephone networks with dynamic routing. In *Proceedings of the 13th International Teletraffic Congress*, pp. 219–224. North-Holland.

Pióro, M.P., and Gajowniczek, P. 1995. Stochastic allocation of virtual paths to ATM links. In *Performance Modelling and Evaluation of ATM Networks*, pp. 135–146. Chapman & Hall.

Pióro, M.P., and Szcześniak, M. 1996. Robust design problems in telecommunication networks. In *Proceedings of The 10th ITC Specialist's Seminar on "CONTROL IN COMMUNICATIONS"*, pp. 359–372. Lund, Sweden, September.

Rosen, J.B. 1960. The gradient projection method for non-linear programming. Part I. Linear Constraints. *JSIAM* 8(1).

Sato, K.-I., Ohta, S., and Tokizawa, I. 1990. Broad-band ATM network architecture based on virtual paths. *IEEE Transactions on Communications*, 38(8):1212–1222.

Shirakawa, H., Maki, K., and Miura, H. 1991. Japan's network evolution relies on SDH-based systems. *IEEE LTS* 2(4):14–18.

Sosnosky, J. 1994. Service applications for SONET DCS distributed restoration. *IEEE J. Select. Areas Commun.*, 12(1):59–68.

Wu, T.-H. 1992. *Fiber Network Service Survivability.* Artech House.

Wu, T.-H. 1995. Emerging technologies for fiber network survivability. *IEEE Communications Magazine*, 33(2):58–74.

Vanderstraeten, H., Chantrain, D., Ocakoglu, G., Nederlof, L., and Van Hauwermeiren, L. 1996. Integration of distributed restoration procedures in the control architecture of ATM cross-connects. In *Proceedings of Broadband Communications'96, An International IFIP-IEEE Conference on Broadband Communications,* pp. 549–560. Chapman & Hall.

Zaakoun, S. 1996. Dimensionement multi-matrice et tarifs optimaux. Master's thesis (supervisor: prof. A. Girard). INRS-Telecommunications.

Xiong, Y., and Mason, L.G. 1997. Restoration strategies and spare capacity requirements in self-healing ATM networks. Accepted for publication. In *Proceedings of IEEE INFOCOM '97.* IEEE Computer Society Press.

Appendix A

Kalman Filter

OPTIMAL estimation theory has a very broad range of applications which vary from estimation of river flows to satellite orbit estimation and nuclear reactor parameter identification. In this appendix we present an informal description of the Kalman filter, which is one of the basic tools stemming from estimation theory. We begin with a short definition of the optimal estimation domain to indicate the role of the Kalman filter.

According to (Gelb, 1974) *"an optimal estimator is a computational algorithm that processes measurements to deduce a minimum error (in accordance with some stated criterion of optimality) estimate of the state of a system by utilizing: knowledge of system and measurement dynamics, assumed statistics of system noises and measurement errors, and initial condition information."* The key feature of this formulation is that all measurements and the knowledge about the system are used to evaluate the estimate and that the estimation error is minimized in a well defined statistical sense. There are three main estimation problem classes: filtering, prediction, and smoothing.

The filtering problem corresponds to the cases where an estimate is needed at the moment of the last measurement. In the case of prediction the estimate is required for an instant after the last measurement. When the time of required estimate is between the first and last measurement the problem falls into the smoothing category. As the name indicates, the Kalman filter provides an optimal estimate for the last measurement instant. It is one of the basic filtering techniques which is applicable for estimation of the state of a linear system. It is also a good example of optimal estimators' capabilities and limitations.

The general structure of the filtering process is given in Figure A.1. Here it is assumed that a linear system model and models characterizing system and measurement errors are available. In this case the Kalman filter processes the measurements to provide an optimal estimate of the system state which minimizes the mean square estimation error. It should be underlined that while the Kalman filter provides the information how to process the measured data, it does not indicate the optimal mesurement schedule. Before describing the Kalman filter in more detail, we will introduce some basic concepts and models used in the filtering procedure.

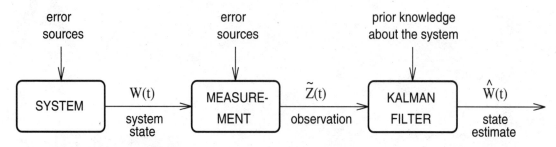

Figure A.1. *General structure of the filtering process.*

A.1 Linear System Model

In general the dynamics of a linear system can be described in either the frequency domain or the time domain. In the following we use the time domain, which is more convenient from mathematical and notational viewpoints. Also, such a description is more natural, which results in a better understanding of the system's behavior.

Let $W(t) = [W^1(t), W^2(t), ..., W^n(t)]^T$ denote the system state vector described by n parameters which are a function of time t. The dynamics of this system can be described by the following first-order differential equation:

$$\dot{W}(t) = F(t)W(t) + G(t)\mathbf{e}(t) + L(t)\mathbf{c}(t) \tag{A.1}$$

where $\mathbf{e}(t)$ is a random forcing function, $\mathbf{c}(t)$ is a control (deterministic) function, and $F(t)$, $G(t)$, $L(t)$ are matrices defining the dynamics of the system. The differential equation determines the system's subsequent behavior assuming that the state vector at a certain point in time and a description of the forcing and control functions are given. A block representation of the linear system dynamics is shown in Figure A.2.

Transition matrix

Let us consider a system without the forcing and control functions:

$$\dot{W}(t) = F(t)W(t) \tag{A.2}$$

For this system one can define the transition matrix $\Phi(t, t_0)$ which defines the system state at a time t based on the knowledge of the state at t_0:

$$W(t) = \Phi(t, t_0)W(t_0) \tag{A.3}$$

A.1 Linear System Model

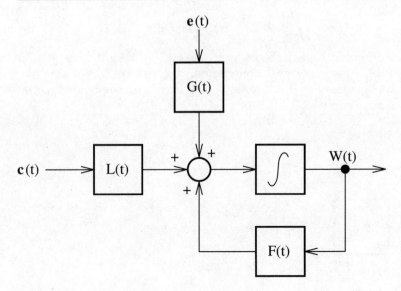

Figure A.2. *Linear system dynamics.*

Obviously the transition matrix is a function of matrix $F(t)$. In particular the following general relations can be derived:

$$\frac{d}{dt}\Phi(t, t_0) = F(t)\Phi(t, t_0) \tag{A.4}$$

$$|\Phi(t, t_0)| = \exp\left[\int_{t_0}^{t} \text{trace}[F(\tau)]d\tau\right] \tag{A.5}$$

In the case of stationary systems the transition matrix only depends on the difference $t - t_0$ and the F matrix is time invariant. This leads to the following definition of the transition matrix:

$$\Phi(t - t_0) = e^{(t - t_0)F} \tag{A.6}$$

Discrete representation

Up to now we considered a continuous time model. Nevertheless, in many cases only discrete points in time, t_k, $k = 1, 2, ...$, are of interest. In this case the system dynamics can be described by the following difference equation:

$$W_{k+1} = \Phi_k W_k + \Gamma_k \mathbf{e}_k + \Lambda_k \mathbf{c}_k \tag{A.7}$$

where

$$\Phi_k = \Phi(t_{k+1}, t_k) \tag{A.8}$$

$$\Gamma_k \mathbf{e}_k = \int_{t_k}^{t_{k+1}} \Phi(t_{k+1}, \tau) G(\tau) \mathbf{e}(\tau) d\tau \tag{A.9}$$

$$\Lambda_k \mathbf{c}_k = \int_{t_k}^{t_{k+1}} \Phi(t_{k+1}, \tau) L(\tau) \mathbf{c}(\tau) d\tau \tag{A.10}$$

A block representation of the discrete system dynamics is given in Figure A.3. In the remainder of this appendix we will consider only discrete systems. Obviously, most of the models and features to follow have their corresponding representation in the continuous time domain (Gelb, 1974).

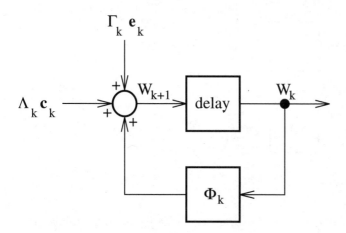

Figure A.3. *Discrete system dynamics.*

A.1.1 Observability and controllability

To discuss observability we introduce the concept of measurements, $\tilde{Z}_k$, $k = 1, 2, \ldots$. The measurements are assumed to be linearly related to the system state:

$$\tilde{Z}_k = H_k W_k + \mathbf{u}_k \tag{A.11}$$

where H_k is the observation matrix and $\mathbf{u}_k$ is the measurement noise. A system is observable if it is possible to determine $W_1, \ldots, W_k$ based on corresponding measurements in a noise free environment. A precise observability condition expressed in terms of matrices Φ and H can be found in (Gelb, 1974).

The issue of controllability is related to the ability of achieving an arbitrary state, in a given number of steps, in a deterministic (noise free) linear dynamic system. In particular a system is controllable in time t_k if for any arbitrary pair of states, W_1, W_k, there is a control, $\mathbf{c}_1, \ldots, \mathbf{c}_k$ which can drive the system from state W_1 to state W_k. A precise controllability condition expressed in terms of matrixes Φ and Λ can be found in (Gelb, 1974). In the remainder of this appendix we do not consider systems with control and we assume $\mathbf{c}_k = 0$.

Introduction of the measurement model completes the basic linear system model description for estimation purpose which is given by

$$W_{k+1} = \Phi_k W_k + \Gamma_k \mathbf{e}_k \tag{A.12}$$

$$Z_k = H_k W_k + \mathbf{u}_k \qquad (A.13)$$

It should be mentioned that this model is not unique in the sense that for given system input and output values, there are many different sets of Φ_k, Γ_k, and H_k which will give the same input-output behavior which corresponds to the choice of a coordinate system (time).

A.1.2 Covariance matrix

The concept of covariance matrix is important in estimation error analysis. Let us begin with the definition of cross-covariance matrix C of two vectors, A and B, whose elements are random variables:

$$C(A, B) = E[(A - E[A])(B - E[B])^T] = E[AB^T] - E[A]\,E[B^T] \qquad (A.14)$$

If $A = B$, the covariance matrix C defines the second central moments of the vector elements. In particular the matrix diagonal consists of vector elements' variances while other matrix elements are covariances of two vector's elements identified by the matrix element indices.

In this appendix we consider systems whose forcing functions $\mathbf{e}_k$ are vectors of random variables. It is assumed that any two values of forcing function, $\mathbf{e}_k$, $\mathbf{e}_{k-i}$, $i = 1, 2, ...$, are uncorrelated, which means that the forcing function generates a white sequence. Observe that once the forcing function is a random variable, the system state is also a random variable. To simplify presentation it is also assumed that the forcing function is unbiased (zero ensemble average values). This does not restrict the generality of the presented models since the bias can be easily removed by subtraction.

Let us define the error in the estimate of the system state as a difference between the estimated value $\hat{W}_k$ and the actual value W_k:

$$\Delta_k = \hat{W}_k - W_k \qquad (A.15)$$

Then the estimation error covariance matrix is defined as

$$\mathcal{P}_k = E[\Delta_k \Delta_k^T] \qquad (A.16)$$

The covariance matrix $\mathcal{P}_k$ expresses the statistical measure of estimation uncertainty.

A covariance matrix is also used for description of the uncorrelated random sequence $\Gamma_k \mathbf{e}_k$. Here we have

$$E[(\Gamma_k \mathbf{e}_k)(\Gamma_k \mathbf{e}_k)^T] = \Gamma_k \mathcal{Q}_k \Gamma_k^T \qquad (A.17)$$

where $\mathcal{Q}_k$ is the covariance matrix of the white sequence.

Estimation error propagation

Based on the transition matrix Φ_k one can define the estimate of the predictable portion of the next state as

$$\hat{W}_{k+1}^e = \Phi_k \hat{W}_k \tag{A.18}$$

Henceforth $\hat{W}_k^e$ is called state estimate extrapolation. By subtracting Equation (A.12) from Equation (A.18) we get

$$\Delta_{k+1} = \Phi_k \Delta_k - \Gamma_k \mathbf{e}_k \tag{A.19}$$

This equation can be used to derive a relation for extrapolation of the error covariance matrix from time t_k to t_{k+1}

$$\mathcal{P}_{k+1}^e = \Phi_k \mathcal{P}_k \Phi_k^T + \Gamma_k \mathcal{Q}_k \Gamma_k^T \tag{A.20}$$

This result indicates that in some cases the error covariance can become unbounded if there are no state measurements.

A.2 Discrete Kalman Filter

In this section we consider a linear discrete system whose dynamics are given by

$$W_{k+1} = \Phi_k W_k + \mathbf{e}_k \tag{A.21}$$

where system state W_k is a n-dimensional vector and $\mathbf{e}_k$ is a white sequence vector with zero mean and covariance matrix $\mathcal{Q}_k$. The system measurements are defined by

$$\tilde{Z}_k = H_k W_k + \mathbf{u}_k \tag{A.22}$$

where measurement $\tilde{Z}_k$ is an l-dimensional vector and $\mathbf{u}_k$ is a white sequence vector with zero mean and covariance matrix $\hat{Y}_k$.

Let us define an optimal, unbiased, and consistent estimator. Here optimality is defined as minimization of the mean square estimation error which corresponds to minimization of

$$J_k = E[\Delta_k^T I \Delta_k] = \text{trace}[\mathcal{P}_k] \tag{A.23}$$

where I is identity matrix. An unbiased estimate is defined as the one whose expectation is equal to the expectation of the actual state. A consistent estimate converges to the actual value with the increase in the number of measurements.

The Kalman filter provides an optimal, unbiased, consistent estimate which can be expressed in the linear and recursive form

$$\hat{W}_k = \mathcal{K}_k' \hat{W}_k^e + \mathcal{K}_k \tilde{Z}_k \tag{A.24}$$

where $\mathcal{K}_k'$ and $\mathcal{K}_k$ are weighting matrices. It can be shown (Gelb, 1974) that in order to have the estimate unbiased the following condition must hold:

$$\mathcal{K}_k' = I - \mathcal{K}_k H_k \tag{A.25}$$

A.2 Discrete Kalman Filter

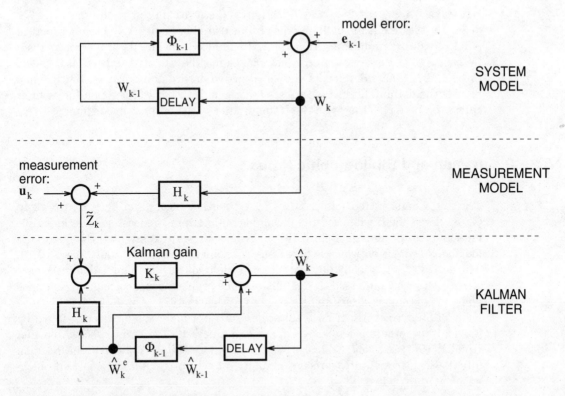

Figure A.4. *System model and Kalman filter.*

Using this relation in Equation (A.24) gives the state estimate update

$$\hat{W}_k = \hat{W}_k^e + \mathcal{K}_k[\tilde{Z}_k - H_k \hat{W}_k^e] \tag{A.26}$$

Based on this relation one can derive the error covariance matrix update

$$\mathcal{P}_k = (I - \mathcal{K}_k H_k)\mathcal{P}_k^e(I - \mathcal{K}_k H_k)^T + \mathcal{K}_k \tilde{Y}_k \mathcal{K}_k^T \tag{A.27}$$

The optimum value of $\mathcal{K}_k$ can be found from minimization of expression A.23 which corresponds to minimization of the length of the estimation error vector. This can be done by evaluating the partial derivative of J_k with respect to $\mathcal{K}_k$ and solving it for zero value. Based on the general relation for the partial derivative of the trace of the product of two matrices the solution gives

$$\mathcal{K}_k = \mathcal{P}_k^e H_k^T [H_k \mathcal{P}_k^e H_k^T + \tilde{Y}_k]^{-1} \tag{A.28}$$

which defines the Kalman gain matrix. Using this matrix in Equation (A.27) defines, after some transformations, the optimized value of the updated error covariance matrix

$$\mathcal{P}_k = (I - \mathcal{K}_k H_k)\mathcal{P}_k^e \tag{A.29}$$

Equations (A.26), (A.28), and (A.29), together with initial conditions, $W_0, \mathcal{P}_0$, define the discrete Kalman filter which is illustrated in Figure A.4. From a practical point of view it is important that the Kalman filter provides its own error analysis by means of the estimation error covariance matrix, $\mathcal{P}_k$. It has been also shown (Weiss, 1970) that, despite its simple recursive nature and linearity, the Kalman filter is the optimal filter if $\mathbf{e}_k$, $\mathbf{u}_k$ are Gaussian (in other words a non-linear filter cannot be better). Otherwise the Kalman filter is the optimal linear filter.

A.3 Discussion and Bibliographic Notes

The first significant contribution to the estimation theory can be traced back to Gauss (circa 1800) who used the technique of deterministic least-squares in simple measurement problems (Mehra, 1970). Fisher (circa 1910) invented the maximum likelihood estimation which is based on probability density function (Weiss, 1970). The design of statistically optimal filters in the frequency domain is due to Wiener (circa 1940) who addressed the continuous time problem using correlation functions and the continuous filter impulse response (Mehra, 1971; Abramson, 1968). The Kalman filter, an optimal linear filter designed in time domain, was developed by Kalman and others; see e.g. (Kalman and Bucy, 1961; Uttam, 1971; Aoki and Huddla, 1967; Tse and Athans, 1970). It is interesting to note that the Kalman filter basically constitutes a recursive solution to the original least-squares problem formulated by Gauss.

In this appendix the Kalman filter presentation follows in principle a description of this technique given in (Gelb, 1974) which provides a simple and interesting picture of the central issues underlying estimation theory and practice. Moreover, it fits very well into the estimation problem treated in Chapter 4. There are many other works dealing with both estimation theory in general and the Kalman filter in particular, e.g. (Papoulis, 1991; Nahi, 1969; Proakis, 1989).

References

Abramson, P.D. Jr. 1968. Simultaneous estimation of the state and noise statistics in linear dynamic systems. Ph.D. Thesis, M.I.T., TE-25.

Aoki, M., and Huddla, J.R. 1967. Estimation of state vector of a linear stochastic system with a constrained estimator. *IEEE Transactions on Automatic Control*, AC-12(4).

Gelb, A. 1974. *Applied Optimal Estimation*. The M.I.T. Press.

Kalman, R.E., and Bucy, R.S. 1961. *Trans. ASME Ser. D.J. Basic Eng.*, 83:95–107. December.

Kalman, R.E. 1962. Fundamental study of adaptive control systems. Technical Report No.ASD-TR-61-27, Vol.I, Flight Control Laboratory, Wright-Patterson Air Force Base, Ohio.

References

Mehra, R.K. 1970. On the identification of variances and adaptive Kalman filtering. *IEEE Transactions on Automatic Control*, AC-15(2):175–184.

Mehra, R.K. 1971. On-line identification of linear dynamic systems with applications to Kalman filtering. *IEEE Transactions on Automatic Control*, AC-16(1):12–21.

Nahi, N.E. 1969. *Estimation Theory and Applications*. Wiley.

Papoulis, A. 1991. *Probability, Random Variables, and Stochastic Processes*. 3rd ed. McGraw-Hill.

Proakis, J.G. 1989. *Digital Communications*. 2nd ed. McGraw-Hill.

Tse, E., and Athans, M. 1970. Optimal minimal-order observer estimators for discrete linear time-varying systems. *IEEE Transactions on Automatic Control*, AC-15(4).

Uttam, B.J. 1971. On the stability of time-varying systems using an observer for feedback control. *International Journal of Control*, December.

Weiss I.M. 1970. A survey of discrete Kalman-Bucy filtering with unknown noise covariances. In *Proceedings of AIAA Guidance, Control and Flight Mechanics Conference*, Paper No.70–955, Santa Barbara, Calfornia.

Appendix B

Markov Decision Theory

MARKOV decision theory has many potential applications over a wide range of topics such as inventory control, computer science, maintenance, resource allocation, etc. Surprisingly it has not permeated well the field of communication network control and management where the issue of resource allocation, is of great importance and where Markov processes have been used for a long time for system modeling. One possible explanation of this fact is that Markov decision theory is not yet familiar to many engineers and researchers working in this domain. In this appendix we try to provide a short description of basic concepts and models of Markov decision theory. Before taking up the main subject we give a brief review of underlying concepts of Markov processes.

B.1 Markov Processes

Markov processes (also called Markov chains) are based on two fundamental concepts: states and state transitions. A state is treated as a random variable which describes some properties of the system: for example, the number of passengers in a bus. A state transition describes a change in the system state at a given time instance. In our example the state transitions occur at the bus stops. One can classify Markov processes into discrete-time and continuous-time categories. In many cases the system can be modeled in either of these categories. If one is interested only in the state sequence, it is more convenient to use the discrete-time description. This would be the case in our bus example if the time between the stops can be assumed to be predictable (fixed). On the other hand, if the bus would stop at the passengers' requests the time between subsequent stops could be of interest. In this case it would be better to treat the time between state transitions also as a random variable, which corresponds to a continuous-time process description.

Discrete-time Markov processes

Let $\{z_t, t = 0, 1, ...\}$ denote a sequence of system states being random variables with a discrete state space Z. The index t is interpreted as a discrete-time description. The state sequence forms a discrete Markov process if the future probabilistic behavior of the system depends only on the current state and is not influenced by the sequence of previous states. This *Markovian* property can be defined formally as a condition that

$$P\{z_{t+1} = l_{t+1} | z_0 = l_0, ..., z_t = l_t\} = P\{z_{t+1} = l_{t+1} | z_t = l_t\}, \quad t = 0, 1, ... \quad (B.1)$$

for all possible values of $l_0, ..., l_{t+1}$.

Let us define time-homogeneous one step transition probabilities

$$p_{lk} = P\{z_{t+1} = k | z_t = l\}, \quad l, k \in Z \quad (B.2)$$

which are independent of the time t. In the following we consider Markov processes which are defined by the probability distribution of the initial state z_0 and the one-step transition probabilities p_{lk} ($\sum_{k \in Z} p_{lk} = 1$). Moreover, we will assume that the considered Markov processes have a regeneration state z^r, which means that for any initial state z_0 the system will arrive in state z^r in a finite number of time steps.

A Markov process is *periodic* if the state space can be divided into N subsets ($N \geq 2$) which are always visited in the same sequence. If this is not the case the Markov process is called *aperiodic*. Let us define the n-step ($n = 1, 2, ...$) transition probabilities as

$$p_{lk}^{(n)} = P\{z_{t=n} = k | z_0 = l\}, \quad l, k \in Z \quad (B.3)$$

For an aperiodic Markov processes we can define *steady-state* probabilities Q_k as

$$Q_k = \lim_{n \to \infty} p_{lk}^{(n)}, \quad l, k \in Z \quad (B.4)$$

The steady-state probabilities fulfill the following system of linear equations:

$$Q_k = \sum_{l \in Z} p_{lk} Q_l, \quad k \in Z \quad (B.5)$$

$$\sum_{k \in Z} Q_k = 1 \quad (B.6)$$

Equations (B.5) are called the *equilibrium* or *balance* equations and the equation (B.6) is called the *normalizing* equation.

Continuous-time Markov processes

Let $\{z(t), t \geq 0\}$ denote a continuous-time stochastic process with a discrete-state space Z. This process is a continuous-time Markov process if:

- The time of staying in state l is exponentially distributed with mean $\tau_l = 1/\nu_l$, and is independent from the trajectory of the system before entering state l. The rates ν_l, $l \in Z$ are bounded.

- The process moves from state l to state k with transition probability p_{lk}, $k \neq l$ independently of the state l duration. The transition probabilities fulfill

$$\sum_{k \neq l} p_{lk} = 1, \quad l \in Z \tag{B.7}$$

The central concept in the analysis of the continuous-time Markov processes is that of *transition rate*. First note that based on the exponential distribution characteristics, see e.g. (Tijms, 1986), we have

$$P\{z(t + \Delta t) = l | z(t) = l\} = 1 - \nu_l \Delta t + o(\Delta t) \tag{B.8}$$

where $o(\Delta t)$ is negligible small compared with Δt when $\Delta t \to 0$. Therefore $\nu_l \Delta t + o(\Delta t)$ corresponds to the probability that the process moves to another state in Δt time units. Since the next state is a function of the transition probabilities p_{lk} we have

$$P\{z(t + \Delta t) = k | z(t) = l\} = \lambda_{lk} \Delta t + o(\Delta t), \quad k \neq l \tag{B.9}$$

where

$$\lambda_{lk} = \nu_l p_{lk} \quad k \neq l \tag{B.10}$$

denotes transition rates of the continuous-time Markov process. Observe that based on Equation (B.7) the relation between the state sojourn-time rates and the transition rates is given by

$$\nu_l = \sum_{k \neq l} \lambda_{lk}, \quad l \in Z \tag{B.11}$$

The continuous-time Markov process is completely defined by the transition rates λ_{lk}, $l, k \in Z$, $k \neq l$. Note that from Equation (B.9) it follows that $\lambda_{lk} \Delta t$ can be treated as a transition probability in time Δt (for small Δt).

As in the case of the discrete Markov processes, assuming that the process has a regeneration state, we can define steady-state probabilities

$$Q_k = \lim_{t \to \infty} P\{z(t) = k | z_0 = l\}, \quad k \in Z \tag{B.12}$$

which are independent of initial state l. In this case the equilibrium (balance) equations and the normalizing equation are defined by

$$\nu_k Q_k = \sum_{l \neq k} \lambda_{lk} Q_l, \quad k \in Z \tag{B.13}$$

$$\sum_{k \in Z} Q_k = 1 \tag{B.14}$$

The balance equation says that the rate of state k departure equals the rate of arrival in this state. In fact this principle applies to any subset of states, $z' : z' \neq z$ and can be stated as

$$\sum_{l \in z'} Q_l \sum_{k \notin z'} \lambda_{lk} = \sum_{k \notin z'} Q_k \sum_{l \in z'} \lambda_{kl} \tag{B.15}$$

Product form and recursive solutions

One of the basic objectives of the stationary Markov process analysis is to find the steady-state probabilities. In general this can be done by solving the set of equilibrium (balance) and normalizing equations as defined by Equations (B.5) and (B.6) for the discrete case and Equations (B.13) and (B.14) for the continuous case. Nevertheless in some cases the solution can be achieved in a less complex manner. For example, let us consider a single-link loss system, the same as the one analyzed in Chapter 8, with capacity L which is offered several connection classes, each of them characterized by Poissonian arrival rate λ_j, mean holding time μ_j and bandwidth requirement, d_j. In this case the steady-state probabilities can be calculated from the following product form equation:

$$Q(\mathbf{x}) = \frac{1}{G(X)} \prod_{j \in J} \frac{(\lambda_j/\mu_j)^{x_j}}{x_j!} \tag{B.16}$$

where $\mathbf{x} = [x_j, j \in J]$ denotes the system state, x_j is the number of class j connections in the system, and $G(X)$ is the normalization constant defined as

$$G(X) = \sum_{\mathbf{x} \in X} \prod_{j \in J} \frac{(\lambda_j/\mu_j)^{x_j}}{x_j!} \tag{B.17}$$

where X denotes the set of all feasible system states.

The solution for the above system can be further simplified if proper bandwidth discretization is applied and if only the probabilities of global number, x, of occupied bandwidth units are of interest. Namely, when all bandwidth allocations, d_j, can be expressed as integer numbers of the largest common denominator, the steady-state probabilities can be found from the following recursive formula:

$$Q(x) = \frac{Q'(x)}{\sum_{x \in X} Q'(x)} \tag{B.18}$$

where unnormalized state probabilities $Q'(x)$ are evaluated from

$$Q'(x) = \frac{1}{x} \sum_{j \in J} Q'(x - d_j) d_j \frac{\lambda_j}{\mu_j} \tag{B.19}$$

assuming, for example, $Q'(0) = 1$.

Uniformization technique

The uniformization technique transforms an arbitrary continuous-time Markov process, which in general can have different average sojourn times in different states, into an equivalent continuous-time Markov process where the average time between transitions is constant. This is done by introducing fictitious transitions between the same states. The rate of transition in the equivalent system, ν', can have any value which satisfies

$$\nu' \geq \nu_l, \quad l \in Z \tag{B.20}$$

In the equivalent process the state transition instances form a Poisson process with rate ν' and the state transitions are described by a discrete-time Markov process with the following one-step transition probabilities:

$$p'_{lk} = \begin{cases} \frac{\nu_l}{\nu} p_{lk} & \text{for } k \neq l, \\ 1 - \frac{\nu_l}{\nu} & \text{for } k = l, \end{cases} \tag{B.21}$$

It can be shown that the equivalent Markov process is probabilistically identical to the original process (Tijms, 1986).

The uniformization technique is very powerful. For example, it can provide a closed form equation for evaluation of the time-dependent transition probabilities $p_{lk}(t)$, where t is the time during which transition from state l to state k occurs, or it can serve for calculation of the distribution of the first-passage time into a state or a set of states [see e.g. (Tijms, 1986)]. Also, it can be used for transformation of a continuous-time Markov process into a discrete Markov chain without loosing the information about the state sojourn times.

B.2 Markov Decision Processes

In the preceding description of Markov processes we assumed that the transition probabilities and the transition rates are given. Nevertheless, in many cases the state transitions can be controlled by the system owner or the system user. In such situations a question arises: What are the optimal control decisions? Markov decision theory provides powerful tools for analysis of the probabilistic sequential decision processes. In this appendix we concentrate on stationary processes with infinite planning horizon. In general a Markov decision model can be formulated from the system cost or the system reward perspective. Both formulations are basically equivalent since maximization of the system reward under given potential maximum reward corresponds to minimization of the system cost. Also, the corresponding decision models are essentially the same. In this appendix we use the reward maximization formulation. The objective of Markov decision models is to find an optimal control policy which maximizes the long-time average reward per unit time or, equivalently, which minimizes the long-time average cost per unit time.

B.2.1 Discrete-time Markov decision models

Consider a dynamic system whose evolution is described by a sequence of system states $\{z_t, t = 0, 1, ...\}$ belonging to a discrete state space Z. The time t points can be interpreted as the system review instances and are assumed to be equidistant. When the system arrives at time t it is classified into state $l \in Z$. Then a decision is made to which state the system should be moved. For each state a set of possible actions is defined, A_l, $l \in Z$ assumed to be finite. Each action, a, results in a certain reward, $r_l(a)$, which is given to the system once the decision is executed. Then, in the next time epoch, the system moves to a state which is a function of the transition probabilities $p_{lk}(a)$ being a function of the action taken. The transition probabilities satisfy

$$\sum_{k \in Z} p_{lk} = 1, \quad l \in Z \tag{B.22}$$

Such a system is called a discrete-time Markov decision model if rewards $r_l(a)$ and transition probabilities $p_{lk}(a)$ are independent from the past history of the system. It should be emphasized that the reward given to the system at the decision moment can also represent the expected reward collected until the next decision moment.

A prescription for making decisions in each time instance is called *policy*. In general that policy can depend on the system history. Nevertheless, for the Markov decision processes with infinite planning horizon we need only consider stationary policies where action depends only on the current system state (Derman, 1970).

Let us assume that for each possible stationary policy π a state z^r exist which can be reached from any other state (*unichain assumption*). Under this assumption for each stationary policy we can define the associated Markov process with steady-state probabilities Q_k satisfying the set of balance and normalizing equations

$$Q_k(\pi) = \sum_{l \in Z} p_{lk}(\pi) Q_l(\pi), \quad k \in Z \tag{B.23}$$

$$\sum_{k \in Z} Q_k(\pi) = 1 \tag{B.24}$$

Then the long-time average reward per time unit under policy π is given by

$$\overline{R}(\pi) = \sum_{k \in Z} r_k(\pi) Q_k(\pi) \tag{B.25}$$

If

$$\overline{R}(\pi^*) \geq \overline{R}(\pi) \tag{B.26}$$

for each stationary policy π, the stationary policy π^* is *average reward optimal*. Unfortunately finding the *average reward optimal* policy by computing the average rewards for each stationary policy is not feasible in most cases due to the enormous number of possible stationary policies. Nevertheless, the optimal policy can be found by applying efficient algorithms which construct a sequence of improved policies. *Relative values* play a central role in some of these algorithms.

Relative values

Let us assume that bias values $w_l(\pi)$ exist such that

$$w_l(\pi) = \lim_{n \to \infty} [R_n(l, \pi) - n\overline{R}(\pi)] \tag{B.27}$$

where $R_n(l, \pi)$ is the total expected reward over n decision instances. Then we can define the difference in total expected rewards starting in state k rather than in state l, over a long period of time, as

$$w_k(\pi) - w_l(\pi) = \lim_{n \to \infty} [R_n(k, \pi) - R_n(l, \pi)] \tag{B.28}$$

B.2 Markov Decision Processes

Intuitive analysis of this relation suggests that having the bias values one can improve the current policy by choosing as the next state the one with maximum positive difference $w_k(\pi) - w_l(\pi)$, $k \in Z_l$ where Z_l denotes the set of states which can be reached from state l. Obviously the improvement will be achieved when the current policy selects a state with a smaller difference. Note that the above policy improvement analysis indicates that the optimal policy is *deterministic* in the sense that, for a given process state and new connection class, the action is always the same (rejection or acceptance on a given path).

Since we are only interested in a difference between the bias values it is sufficient to know the relative values defined as

$$v_k(\pi) = w_k(\pi) + c, \quad k \in Z \tag{B.29}$$

where c is an arbitrary constant. It can be proved, e.g. (Tijms, 1986), that under a stationary policy a particular set of relative values (defined by a particular value of constant c) can be evaluated from the following set of linear equations

$$v_l(\pi) = r_l(\pi) - \overline{R}(\pi) + \sum_{k \in Z} p_{lk}(\pi) v_k(\pi), \quad l \in Z \tag{B.30}$$

by assuming $v_s = 0$ for an arbitrary state s. The equations (B.30) are called *value determination equations* for policy π.

Policy-iteration algorithm

It can be proven (Howard, 1960) that the following relation holds:

- If for a new policy π'

$$r_l(\pi') - \overline{R}(\pi) + \sum_{k \in Z} p_{lk}(\pi') v_k(\pi) \geq v_l(\pi), \quad l \in Z \tag{B.31}$$

then π' is an improved policy with $\overline{R}(\pi') \geq \overline{R}(\pi)$

Based on this relation, the policy-iteration algorithm is formulated as follows:

1. Choose an initial stationary policy π.

2. For given policy π, solve the set of value-determination equations

$$v_l(\pi) = r_l(\pi) - \overline{R}(\pi) + \sum_{k \in Z} p_{lk}(\pi) v_k(\pi), \quad l \in Z \tag{B.32}$$

by setting the relative value $v_s(\pi)$ for an arbitrary reference state s to zero.

3. For each state $l \in Z$ find an action a_l producing the maximum in

$$\max_{a \in A_l} \left\{ r_l(a) - \overline{R}(\pi) + \sum_{k \in Z} p_{lk}(a) v_k(\pi) \right\} \tag{B.33}$$

The improved policy π' is defined by choosing $\pi'_l = a_l$ for all $l \in Z$. If the improved policy equals the previous policy π the algorithm is stopped. Otherwise go to step 2 with π replaced by π'.

It can be proven (Howard, 1960) that this procedure converges to π^* in a finite number of iterations.

It is important that the number of iterations in the policy-iteration algorithm is practically independent from the number of states. According to (Tijms, 1986) this number typically varies between 3 and 15.

Value-iteration algorithm

The value-iteration algorithm avoids solving a set of linear equations. Instead it uses the recursive solution approach from dynamic programming. In particular the value-iteration algorithm evaluates recursively the value function $V_n(l)$, where $n = 1, 2, ...$, from

$$V_n(l) = \max_{a \in A_l}\{r_l(a) + \sum_{k \in Z} p_{lk}(a) V_{n-1}(k)\}, \quad l \in Z \tag{B.34}$$

The value function $V_n(l)$ can be interpreted as the expected reward from the system within n transition periods assuming state l at the beginning of the considered time horizon and terminal reward of $V_0^s(k)$ at the end of this time; for proof see (Dernardo, 1982; Derman, 1970). This interpretation indicates that the difference $\overline{R}(\pi^n) = V_n(l) - V_{n-1}(l)$ approaches the maximum average reward from the system for large n:

$$\overline{R}^* = \lim_{n \to \infty}[V_n(l) - V_{n-1}(l)] \tag{B.35}$$

It is important that for a finite n one can evaluate bounds on the average reward under the optimal policy, $\overline{R}^*$. These bounds are defined by

$$m_n = \min_{k \in Z}\{V_n(k) - V_{n-1}(k)\} \tag{B.36}$$

$$M_n = \max_{k \in Z}\{V_n(k) - V_{n-1}(k)\} \tag{B.37}$$

and it can be proven (Tijms, 1986) that

$$m_n \leq \overline{R}(\pi^n) \leq \overline{R}^* \leq M_n, \quad n \geq 1 \tag{B.38}$$

It can be also shown that if for each stationary policy the resulting Markov process is aperiodic, the following holds:

$$\lim_{n \to \infty} m_n = \lim_{n \to \infty} M_n = \overline{R}^* \tag{B.39}$$

provided that $\overline{R}^*$ is independent from the initial state (Bather, 1973; Schweitzer and Federgruen, 1979).

Based on the presented discussion the value-iteration algorithm can be defined as follows:

1. Determine initial values $\{V_0(l) : l \in Z, 0 \leq V_0(l) \leq \min_a r_l(a)\}$ and let $n = 1$.

2. Evaluate the value functions

$$V_n(l) = \max_{a \in A_l}\left\{r_l(a) + \sum_{k \in Z} p_{lk}(a)V_{n-1}(k)\right\}, \quad l \in Z \qquad (B.40)$$

and find policy π^n which maximizes the right side of Equation (B.40) for all $l \in Z$.

3. Compute m_n and M_n. If $0 \le M_n - m_n \le \epsilon m_n$, where ϵ determines a required relative accuracy, stop the algorithm with policy π^n.

4. Set $n = n + 1$ and go to step 2.

The convergence of the value-iteration algorithm can be accelerated by using a dynamic relaxation factor. In this case step 4 is modified as follows:

- Determine states u and o for which

$$V_n(u) - V_{n-1}(u) = m_n \qquad (B.41)$$
$$V_n(o) - V_{n-1}(o) = M_n \qquad (B.42)$$

Then compute the dynamic relaxation factor

$$\omega = \frac{M_n - m_n}{M_n - m_n + \sum_{k \in Z}[p_{uk}(\pi_u) - p_{ok}(\pi_o)][V_n(k) - V_{n-1}(k)]} \qquad (B.43)$$

and modify $V_n(k)$ as

$$V_n(l) = V_{n-1}(l) + \omega[V_n(l) - V_{n-1}(l)] \qquad (B.44)$$

Set $n = n + 1$ and go to step 2.

According to (Tijms, 1986), numerical experiments indicate that the dynamic relaxation factor may significantly increase the speed of convergence.

Linear programming approach

The linear programming formulation approaches the policy optimization problem explicitly in the sense that the key element of the algorithm is a linear program maximizing the average reward. To simplify presentation we consider the case where each stationary policy induces an aperiodic Markov chain. In this case we can define $Q_l(a)$ as the steady-state probability of being in state l and choosing action a.

The first step of the linear programming algorithm is to solve the following linear program

$$\max_{\{Q_l(a):\, a \in A_l,\, l \in Z\}} \left\{\overline{R} = \sum_{l \in Z}\sum_{a \in A_l} r_l(a)Q_l(a)\right\} \qquad (B.45)$$

subject to

$$\sum_{a \in A_k} Q_k(a) - \sum_{l \in Z}\sum_{a \in A_l} p_{lk}Q_l(a) = 0, \quad k \in Z \qquad (B.46)$$

$$\sum_{l \in Z}\sum_{a \in A_l} Q_l(\pi) = 1, \quad Q_l(\pi) \ge 0 \qquad (B.47)$$

The solution, $\{Q_l^*(a) : a \in A_l,\ l \in Z\}$ can be found by applying the simplex method. Note that the constraints correspond to the balance equations and the normalizing equation, respectively.

In the second step of the linear programming algorithm the closed set of states S is identified first

$$S = \left\{ l \ \Big|\ \sum_{a \in A_l} Q_l^*(a) > 0 \right\} \tag{B.48}$$

Then for each state $l \in S$ the optimal decision is defined as $\pi_l^* = a$, $\{a : Q_l^*(a) > 0\}$.

Finally the algorithm verifies whether there are states outside S (the system can be in one of these states at time $t = 0$ only). If $S = Z$, the algorithm is stopped. Otherwise for each state $l \notin S$ the policy is defined as $\pi_l^* = a$, $\{a : p_{lk}(a) > 0,\ k \in S\}$.

Comparison of algorithms

Each of the presented algorithms for the optimal policy calculation has some advantages and the choice depends on particular application and requirements. The main advantage of the policy-iteration algorithm is a very small number of iterations.

The value-iteration algorithm is in general the best approach for solving large-scale Markov decision problems. It is easy to implement since it does not require a solution of a linear equation set. It is also a simple method to evaluate average reward in a single Markov chain.

The linear programming formulation allows to incorporate additional constraints. In particular these constraints can be probabilistic, which results in a *probabilistic* policy where the action can be chosen randomly from a predefined set according to predefined probabilities which determine the policy; for more details see, e.g. (Tijms, 1986). Although it requires solving a set of linear equations, one can use available general purpose codes.

B.2.2 Continuous-time Markov decision models

In many cases the times between the consecutive decision instances are not identical and random. The semi-Markov decision model can be used to analyze such systems. In this case most of the characteristics of the semi-Markov decision model are the same as in the discrete case except for the addition of the description of the time between two decision epochs. In particular when the system arrives at the review time t it is classified into state $l \in Z$. Then a decision is made. For each state a set of possible actions is defined, A_l, $l \in Z$ assumed to be finite. Each action, a, results in a certain reward, $r_l(a)$, which is given to the system until the next decision epoch. The reward usually consists of a lump reward given at the moment of the decision and a reward rate given continuously in time. In the next time epoch, the system moves to a state which is a function of the transition probabilities $p_{lk}(a)$ which depend on the taken action in state l. Such a system is called a semi-Markov decision model if rewards $r_l(a)$, transition probabilities $p_{lk}(a)$ and time until the next decision $\tau_l(a)$ are independent from the past history of the system.

In the following we describe three algorithms for semi-Markov decision models which calculate an optimal policy maximizing the long-time average reward per time unit. These algorithms are equivalents of the algorithms for discrete-time Markov decision processes and their derivation is very similar. Therefore we will concentrate mainly on the presentation of the algorithms and differences with their discrete equivalents.

Policy-iteration algorithm

1. Choose an initial stationary policy π.

2. For given policy π, solve the set of value-determination equations

$$v_l(\pi) = r_l(\pi) - \overline{R}(\pi)\tau_l(\pi) + \sum_{k \in Z} p_{lk}(\pi)v_k(\pi), \quad l \in Z \tag{B.49}$$

by setting the relative value $v_s(\pi)$ for an arbitrary reference state s to zero.

3. For each state $l \in Z$ find an action a_l producing the maximum in

$$\max_{a \in A_l}\left\{r_l(a) - \overline{R}(\pi)\tau_l(a) + \sum_{k \in Z} p_{lk}(a)v_k(\pi)\right\} \tag{B.50}$$

The improved policy π' is defined by choosing $\pi'_l = a_l$ for all $l \in Z$. If the improved policy equals the previous policy π the algorithm is stopped. Otherwise go to step 2 with π replaced by π'.

In a manner analogous to the discrete case this procedure converges to π^* in a finite number of iterations.

Observe that by dividing Equation (B.49) by $\tau_l(\pi)$ and some simple transformations we can arrive at

$$\overline{R} = q_l(\pi) + \sum_{k \in Z} \lambda_{lk}(\pi)[v_k(\pi) - v_l(\pi)], \quad l \in Z \tag{B.51}$$

where $q_l(\pi) = r_l(\pi)/\tau_l(\pi)$ is the rate of reward. This form of equations can also be used in the policy-iteration algorithm, resulting in the more explicit maximization of the average reward in the policy-improvement step:

$$\max_{a \in A_l}\left\{q_l(a) + \sum_{k \in Z} \lambda_{lk}(a)[v_k(\pi) - v_l(\pi)]\right\} \tag{B.52}$$

Application of this form for optimization of the CAC & routing in ATM networks is described in Chapter 5.

Value-iteration algorithm

A direct formulation of the value-iteration algorithm for the semi-Markov decision model is not possible since the recursive relation (B.34) for the expected reward takes advantage of the identical transition times. To cope with the problem one can apply the uniformization technique presented in the first part of this appendix. Having identical average transition times $\tau' = 1/\nu'$ allows application of the value iteration-algorithm derived for the discrete system. This approach results in the following algorithm:

1. Determine the initial values $\{V_0(l) : l \in Z, 0 \leq V_0(l) \leq \min_a q_l(a)\}$ and let $n = 1$.

2. Evaluate the value functions

$$V_n(l) = \max_{a \in A_l}\left\{\tau' q_l(a) + \tau' \sum_{k \in Z} \lambda_{lk}(a)[V_{n-1}(k) - V_{n-1}(l)] + V_{n-1}(l)\right\}, \quad l \in Z \tag{B.53}$$

and find policy π^n which maximizes the right side of Equation (B.53) for all $l \in Z$.

3. Compute m_n and M_n. If $0 \leq M_n - m_n \leq \epsilon\, m_n$, where ϵ determines a required relative accuracy, stop the algorithm with policy π^n.

4. Set $n = n + 1$ and go to step 2.

As in the discrete case the convergence of this algorithm can be accelerated by using the dynamic relaxation factor.

Linear programming approach

By applying the uniformization technique, the linear programming formulation for the discrete case can be used for the semi-Markov model with the following modification to the linear program:

$$\max_{\{Q_l(a):\, a \in A_l,\, l \in Z\}} \left\{\overline{R} = \sum_{l \in Z} \sum_{a \in A_l} q_l(a) Q_l(a)\right\} \tag{B.54}$$

subject to

$$\sum_{a \in A_k} Q_k(a)/\tau_l(a) - \sum_{l \in Z}\sum_{a \in A_l} \lambda_{lk} Q_l(a) = 0, \quad k \in Z \tag{B.55}$$

$$\sum_{l \in Z}\sum_{a \in A_l} Q_l(\pi) = 1, \quad Q_l(\pi) \geq 0 \tag{B.56}$$

B.3   Discussion and Bibliographic Notes

Markov decision theory is an outgrowth of Markov process theory and dynamic programming. The concept of dynamic programming was popularized by (Bellman, 1957). (Howard, 1960) combined the concepts of Markov process theory and dynamic programming which resulted in development of the policy-iteration algorithm. Since then Markov decision theory has expanded, providing a powerful and versatile tool for analyzing probabilistic sequential decision processes with an infinite planning horizon. There is a substantial literature on Markov decision processes. In this appendix we followed (Tijms, 1986). Besides providing an excellent explanation of the underlying concepts, this book provides a wide variety of realistic examples illustrating the basic models and solution algorithms.

In this appendix we limit the presentation to models with the optimality criterion based on long-time average reward (or cost) per unit time. The other possibility is to apply the criterion based on the total expected discounted reward (or cost) where the value of the incurred reward (or cost) is a function of time, see e.g. (Howard, 1960). Nevertheless, for most applications of Markov decision theory the first criterion is believed to be more appropriate. We presented three basic models: policy iteration algorithm, value-iteration algorithm, and linear programming formulation. Each of these models has some strengths and limitations and the choice depends on the particular application and modeling objectives. Application of policy-iteration and value-iteration algorithms to traffic control in telecommunication networks is presented in Chapter 5.

References

Bather, J. 1973. Optimal decision procedures for finite Markov chains. *Adv. Appl. Prob.*, 5:521–540.

Bellman, R. 1957. *Dynamic Programming.* Princeton University Press, Princeton.

Derman, C. 1970. *Finite State Markovian Decision Processes.* Academic Press, New York.

Dernardo, E.V. 1982. *Dynamic Programming.* Prentice-Hall, Englewood Cliffs, New Jersey.

Howard, R. A. 1960. *Dynamic Programming and Markov Process* The M.I.T. Press, Cambridge, Massachusetts.

Kelly F.P. 1979. *Reversibility and Stochastic Networks*, Wiley, New York.

Schweitzer, P.J., and Federgruen A. 1979. Geometric convergence of value-iteration in multi-chain Markov decision problems. *Adv. Appl. Prob.*, 11:188–217.

Tijms, H.C. 1986. *Stochastic Modelling and Analysis: A Computational Approach.* New York: Wiley.

Appendix C

Cooperative Game Theory

IN THIS appendix we focus on a particular aspect of cooperative game theory associated with arbitrated solutions and issues of fairness and efficiency. Nevertheless, to position these issues in the game theory domain we start with an introduction of basic concepts, some of them applicable to all games. Then we discuss briefly the difference between the cooperative and non-cooperative games followed by a general description of different aspects of cooperative games. Finally we describe the arbitrated solutions in the context of fairness and efficiency.

C.1 Basic Concepts

C.1.1 Players

Every game has a set of players which can be considered rational decision makers. In general the set of players, $J = \{1, 2, ..., n\}$, is finite and each player knows how many players take part in the game. A random element of the game can be introduced by adding player 0, which is called nature of chance. Player 0 makes its decision based on a probability distribution which is known to all the other players. In the following three forms of the games are described: extensive, strategic, and coalitional.

C.1.2 Extensive form

The extensive form describes a full tree of possible decision sequences. Consider, for example, a game with two players, P1 and P2, where the first player starts the game by choosing one of the two options, L or R. Then the second player tries to guess which option was chosen by the opponent. If the choice is right, he wins one unit of money from the opponent; otherwise he pays one unit to the first player. The loosing player has right to start another matching or quit (Q). Assuming that the game is limited to two matchings, all possible game sequences and outcomes are illustrated in Figure C.1. The game moves from node to node along the tree. Although the tree form suggest that all actions are made in sequence, in fact the choice and matching are done in parallel. This is illustrated by the dashed boxes which indicate that the current information set is imperfect. In other words, the player who makes decision at this stage does not know in which particular node of the dashed box the game is located.

Utility functions

In Figure C.1, the outcomes of the game are indicated in monetary payoffs. Observe that the players' satisfaction from the game may be not proportional to the monetary payoff. For example, the most important to the players may be not to lose, while the amount of money won or lost can be of lesser importance. To take into account such a behavior the outcome of the game can be transformed into utility units by means of a utility function. In our example this transformation can have the following form: $-2 \to 0.0$, $-1 \to 0.0$, $0 \to 0.3$, $1 \to 0.9$, $2 \to 1.0$.

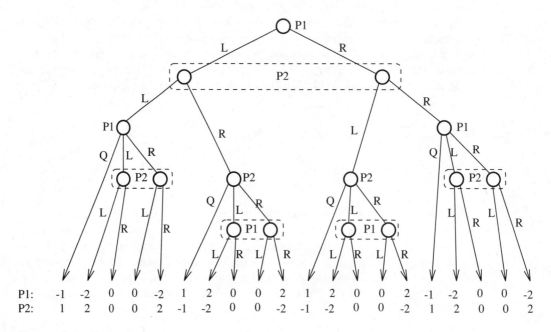

Figure C.1. *Game tree.*

Game environment

There are three basic pieces of information that a player can posses: the set of players, all actions available to the players, and all potential outcomes for all players.

In general a player can have complete or incomplete information about the game. Another aspect of the information availability to a player is the issue of forgetting. Perfect recall corresponds to a situation where the information set is consistent with a never-forgetting player. If this is not true, the game is classified as the one of imperfect recall.

The term common knowledge refers to the fact that all players have the same knowledge about the game. An example of common knowledge is a binding agreement which can be interpreted as a signed contract enforced by an outside authority. In general a binding agreement imposes restriction on the available actions on all players. A commitment is a particular case of binding agreement where one player restricts his actions and makes it common knowledge. A threat is a classical case of commitment.

C.1.3 Strategic form

The extensive form of presenting games is convenient for illustration of basic concepts from the game theory and some simple games but can be very unwieldy in many games due to a possible enormous number of nodes. In many cases this number can approach infinity. For this reason it may be convenient to describe the game in terms of each player's strategy. In this case a player's decision is a function of the information about the game available to the player. While the information set can be still very large (especially in the case of complete knowledge and perfect recall), the strategy description can be significantly simplified if only a part of information is important from the decision viewpoint.

While in general the strategic form is simpler and more natural than the extensive form, it is clear that this description suppresses the underlying game move structure. Two basic classes of strategies can be recognized: deterministic and probabilistic. In the first case the decision is a deterministic function of the available information. In the second case the choice has a random element which makes that based on the same information different decisions can be reached in different instances.

Equilibrium point

An equilibrium point is a combination of players' strategies for which each player maximizes his own utility by optimizing his own strategy, given the possible strategies of the other players. The equilibrium point was introduced by (Nash, 1950). Under certain assumptions it can be proven that all n-player non-cooperative games have at least one equilibrium point (Friedman, 1990). The solution uniqueness can be proven for some game classes (Friedman, 1990).

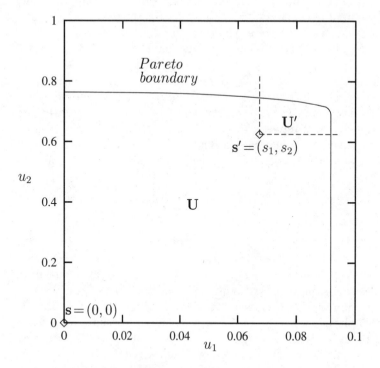

Figure C.2. *Example of bargaining domain.*

C.1.4 Coalitional form

The coalitional form assumes that any group of players can make contractual agreements. A group making a binding agreement is called coalition, C. Note that this form of game assumes certain cooperation among players. In such cooperative games the main focus of analysis are payoffs which can be achieved by each player or coalition. The strategies themselves are of lesser interest.

Characteristic function

The characteristic function describes the payoff possibilities for each coalition. This payoff, $v(C)$, represent the total utility that coalition C can obtain when their members cooperate.

Core

The core is a set of game solutions which are agreeable to all players. Here the term agreeable is strictly defined: agreeable game solution must give as much to any single player and to any potential coalition as they would get acting for themselves. The reason for this definition is that in general the game solution must be accepted by all players and potential coalitions. If any of the players or potential coalitions can achieve a better game outcome acting on their own, they will not agree to the proposed solution. In general the core can consist of many solutions, one solution, or can be empty. Nevertheless it should be indicated that the core consisting of one solution is unusual.

Bargaining domain and Pareto optimality

Since in cooperative games the final outcome is of main interest it is often convenient to analyze the domain of all possible outcomes, U, which is called bargaining domain. An outcome of the game is defined by the values of players utilities $\mathbf{u} = \{u_1, ..., u_n\} : \quad u_i \in R$. In general each utility can be expressed in different units. To simplify presentation in the following we consider the bargaining domains which are convex, closed, and bounded sets of $R^n : \quad R \geq 0$. An example of such a domain for $n = 2$ is given in Figure C.2.

A Pareto optimal solution is defined as a solution which ensures that there is no other solution in which a player can increase his utility without adversely affecting the other's. In Figure C.2 all solutions on the upper right boundary are Pareto optimal and form the Pareto boundary. The outcome of the game depends also on the starting point, $\mathbf{s} = \{s_1, ..., s_n\} : \quad \mathbf{s} \in U$, which is sometimes called the conflict or disagreement point. The starting point corresponds to a pre-game assumption that the game outcome, $\mathbf{u}^*$, cannot be worse for any player than the starting point, $\mathbf{u}^* \geq \mathbf{s}$. The starting point can limit the bargaining domain as shown in Figure C.2.

C.2 Cooperative vs. Non-cooperative Games

The principal difference between the cooperative and non-cooperative games is the introduction of binding agreements in cooperative games. For this reason the modeling and analysis of non-cooperative games focuses on optimization of players' strategies in order to achieve an equilibrium point. Due to the competitive nature of the bargaining process in non-cooperative games the outcome can be both unfair and non-efficient (below the Pareto boundary).

In the case of cooperative games, the possible agreement on the solution shifts focus on the features and properties of the solution which are often expressed in terms of axioms. The axioms try to incorporate the fairness features in the solution while efficiency is provided by required Pareto optimality. It should be underlined that there is no universal and objective fairness criteria and that there are several competing concepts, each of them having strong supporters. This indicates that the solution fairness judgment is a subjective process which may depend on a particular game formulation.

The analysis of cooperative games can also include the bargaining process. In this case the possibilities of binding agreement should be built into the game formulation. This approach was used in the original derivation of the Nash arbitrated solution for two-player cooperative games (Nash, 1950). Nevertheless, in the following we concentrate on the outcome of the cooperative games.

C.3 Different Formulations of Cooperative Games

Historically there are two formulations of cooperative games: One for two-player games and the other for n-player games with $n > 2$. The difference between the two is much deeper than the difference between the number of players. Let us consider first the n-player formulation ($n > 2$). The central issue in this formulation is the possibility of forming different coalitions with different sub-sets of players. While the final solution is achieved with cooperation of all players, the potential solutions for different possible coalitions influence the final outcome. This is evident in the already defined set of agreeable solutions (core). Concerning the choice of a particular solution from the core the potential coalitions are also important. This feature can be well illustrated by means of the Shapley value which can serve to find a unique solution.

To simplify presentation let us consider games with transferable utility, which means that the amount of utility achieved by a coalition can be divided among its members in any mutually agreeable fashion. In general the Shapley value, $\Phi(v) = \{\Phi_1(v), ..., \Phi_n(v)\}$, defines the payoff to each player. This payoff is a weighted average of the contributions that the player makes to the payoff of each coalition to which he belongs. The weight depends on the number of players, k, in each coalition. The value is given by

$$\Phi_i(v) = \sum_{C \in \mathbf{C}} [v(C) - v(C \setminus \{i\})]\frac{(k-1)!(n-k)!}{n!} \qquad (C.1)$$

where $\mathbf{C}$ is the set of all coalitions. This solution can be also characterized by four natural conditions; for details see e.g. (Friedman, 1990).

Now it can be easily understood why there is significant difference between two and more player games. Simply there is only one coalition in a two-player game, which consists of both players, and there are no potential coalitions. As a result the proposed solutions for two-player games are analogous to a conflict which is resolved by an external arbiter according to certain rules which are believed to provide fairness and efficiency. Hence, the solution algorithms are called arbitrated schemes based on axioms. As previously indicated there is no objective measure of the fairness; thus different arbitration schemes can have appeal to different users and in different applications. This feature makes the arbitration schemes similar in concept to general laws governing society which are also constructed to be fair but are perceived differently by different people.

Although historically the arbitration schemes were developed for two-player games, the notion of external arbitration without considering the potential coalitions can be extended to a general n-player games. In fact some of the arbitration schemes developed for two-player games can be easily extended to the general case. This framework fits very well some of the problems considered in this book where the network control manager can be treated as an arbiter whose objective is to provide fair and efficient access to network resources for all users. In the remainder of this appendix we concentrate on a class of arbitration schemes which can be easily extended to the general case.

C.4 A Class of Arbitrated Solutions

In this section we discuss a class of arbitrated solutions derived by (Cao, 1982) for two-player games. This class is based on maximization of the product of player preference functions. This form of the objective function is associated with the Nash arbitrated solution. Originally Nash arbitration was defined by four axioms:

- *Symmetry*: If the bargaining domain is symmetric with respect to the axis $u_1 = u_2$ and the starting point is on this axis, then the solution is also on this axis.

- *Pareto optimality*: The solution is on the Pareto boundary.

- *Invariance with respect to utility transformations*: The solution for any positive affine transformation of $(U, \mathbf{s})$ denoted by $V(U), V(\mathbf{s})$ is $V(\mathbf{u}^*)$ where $\mathbf{u}^*$ is the solution for the original system.

- *Independence of irrelevant alternatives*: If the solution for $(U_1, \mathbf{s})$ is $\mathbf{u}^*$, and $U_2 \subseteq U_1$, $\mathbf{u}^* \in U_2$, then $\mathbf{u}^*$ is also the solution for $(U_2, \mathbf{s})$. In other words this axiom states that if U_1 is reduced, in such a way that the original solution and starting point are still included, the solution of the new problem remains the same.

It is important that the Nash solution can be also expressed as maximization of the product of normalized utilities

$$\max_{u \in U}\{u'_1 \cdot u'_2\} \tag{C.2}$$

where u'_i is utility normalized by its maximum value in the utility domain U. Here the normalized utilities can be treated as the players' preference function which means that in the Nash solution each player is concerned only with his own gain. Cao extended this formulation to a family of preference functions which also takes into account the other player's gain. This family is defined as

$$v_1 = u'_1 + \beta(1 - u'_2) \tag{C.3}$$
$$v_2 = u'_2 + \beta(1 - u'_1) \tag{C.4}$$

where $\beta = [-1, 1]$ is a weighting factor. Using this definition of player's preference function one can define a class of arbitrated solutions as

$$u^* = \arg(\max_{u \in U}\{v_1 \cdot v_2\}) \tag{C.5}$$

Observe that for $\beta = 0$ the solution u^* corresponds to the Nash arbitration. For $\beta \neq 0$ the player's preference function takes into account the other player's utility. In particular for $\beta = 1$ the preference function treats with the same weight the player's gain and the other player's losses so the solution equalizes the normalized utilities of both players. For $\beta = -1$ both gains have the same weight so the solution maximizes the sum of normalized utilities. Interestingly these two solutions correspond to well known arbitration schemes. The first solution ($\beta = 1$) is equivalent to the Raiffa-Kalai-Smorodinsky (henceforth called Raiffa) solution (Raiffa, 1953; Kalai and Smorodinsky, 1975) which is based on four axioms. The first three axioms are the same as the ones for the Nash arbitration. The fourth is called *Monotonicity* and is defined as follows:

- *Monotonicity*: If $U_2 \subseteq U_1$ and $\max\{u_1 : \mathbf{u} \in U_1\} = \max\{u_1 : \mathbf{u} \in U_2\}$ and $\max\{u_2 : u \in U_2\} \leq \max\{u_2 : u \in U_1\}$, then $u_2^{2*} \leq u_2^{1*}$, where $\mathbf{u}^{j*}$ denotes the solution for $(U_j, \mathbf{s})$. In other words, for any subset of U_1 the solution for the second player cannot be improved if the maximum utility of the first player is constant.

The second solution ($\beta = -1$) is equivalent to the modified Thomson solution (Cao, 1982) which is defined by the utilitarian rule maximizing the sum of the normalized utilities.

In (Cao, 1982) it is shown that by changing β continuously from -1 to 1 one can achieve monotonically and continuously a set of solutions which relates $\beta \in [-1, 1]$ to a part of the Pareto boundary. This fact has an appealing geometrical interpretation. Namely for each β the solution is given by the tangent point between the Pareto boundary and a hyperbola from the set of hyperbolae defined by the function $v_1 \cdot v_2 = \text{const}$. As we are moving from $\beta = 0$ to $\beta = -1$ the branches of the hyperbola are widening to become a straight line normal to the line $u_1' = u_2'$ for $\beta = -1$. As we are moving from $\beta = 0$ to $\beta = 1$ the branches of the hyperbola are narrowing to become a semi-line $u_1' = u_2'$ for $\beta = 1$. This feature shows that although the Nash, Raiffa, and the modified Thomson arbitration schemes have very different characteristics they are special cases of a wider class of arbitration schemes. In this context it is also important to emphasize that in the preference space, $\{v_1, v_2\}$, all mentioned solutions become Nash solutions.

C.4.1 Extension to the general case

The simple mathematical form of the presented class of arbitrated solutions and its natural geometrical interpretation makes the extension to general n-player games relatively simple. Namely, instead of starting from axiomatic representation extensions for Nash, Raiffa and Modified Thomson solutions, one can try to generalize the mathematical representation of these solutions ensuring that the central features are preserved and that the extension also covers the case of two-player games. Generalization of Nash arbitration scheme is straightforward. In this case the solution is defined by

$$u^* = \arg\left(\max_{u \in U}\left\{\prod_{i=1}^{i=n} v_i\right\}\right) \tag{C.6}$$

where the preference function is the same as in the two-player game ($v_j = u_j'$). In (Stefanescu and Stefanescu, 1984; Mazumdar, Mason, and Douligeris, 1991) it was proved that this formulation of the Nash generalized arbitration scheme is equivalent to a formulation based on generalized axioms. This result is important for further analysis since the considered class of arbitration schemes is based on Nash solutions in preference function domains.

Assuming that the objective of the modified Thomson arbitration scheme is to maximize the sum of the normalized utilities, the generalization of this scheme is also straightforward and is defined by the following preference function:

$$v_j = \sum_i u'_i \tag{C.7}$$

The objective of the Raiffa arbitration scheme is to equalize the normalized utilities. This objective can be achieved by applying the preference function defined as

$$v_j = \sum_j u'_j + 1 - \frac{1}{n-1} \sum_{i \neq j} u'_i \tag{C.8}$$

This form indicates that in the generalized case of the Raiffa arbitration scheme the player's preference function treats with the same weight the player's gain and the average losses of the other players.

All three cases can be viewed as special instances of a set of preference functions defined by

$$v_j = u'_j + |\beta(n-1)| - \beta \sum_{i \neq j} u'_i \tag{C.9}$$

for $\beta = 0, -1, 1/(n-1)$ respectively. As in the case of two players, by changing β continuously from -1 to $1/(n-1)$ one can achieve continuously a set of solutions which relates $\beta \in [-1, 1/(n-1)]$ to a part of the Pareto boundary.

C.5 Discussion and Bibliographic Notes

Although in this appendix we introduced several basic concepts applicable to non-cooperative and cooperative game theory, the latter is the main focus of the presentation. Within the cooperative game theory one can recognize two areas of interest. The coalitional formulation of the game is important in the cases where the players have freedom to establish an agreement with any subgroup of players. Although the final solution requires consent from all players, the potential coalition agreements influence the final solution. The other formulation assumes that there is an external arbitrator who decides what the game outcome should be based on a set of axiomatic rules which can be compared to laws in a society. These rules should provide a fair and efficient game outcome and are accepted by all players. Historically this approach was developed mainly for two-player games since coalitional approach cannot be applied in this case. Nevertheless, the general n-player case is also of interest, especially in the cases where there are several players who want to maximize their outcome but because of the game nature they cannot form coalitions. In this case they have to resort to an external arbiter in order to achieve a fair and efficient payoff. This formulation fits very well into the problem of fair and efficient resource allocation to connection classes in telecommunication networks where the connection classes can be interpreted as players and the network control takes the role of an arbiter. For this reason we described a generalization of a class of arbitration schemes derived for two-player games. Application of this approach to network resource management is described in Chapter 8.

The reader interested in game theory can refer to a large literature on the subject. For example, a nice overview of the history and development of game theory is given in (Aumann, 1987). A survey of all important areas of game theory can be found in (Aumann, 1991). In (Aumann, 1985) the focus is on methodology, on the application of game theory to the real world, and on the objectives of the game theory. The presentation in this appendix follows (Friedman, 1990) except the section on the class of arbitrated solutions which is based on (Cao, 1982).

References

Aumann, R.J. 1985. What is game theory trying to accomplish? In *Frontiers of Economics,* edited by Arrow, K., and Honkapohja, S., pp. 28–100. New York: B. Blackwell.

Aumann, R.J. 1987. Game theory. In *A Dictionary of Economics: the New Palgrave,* edited by Eatwell, J., Milgate, M., and Newman, P., pp. 460–482. New York: Stocton Press.

Aumann, R.J. 1991. *Handbook of Game Theory.* New York: North-Holland.

Cao, X. 1982. Preference functions and bargaining solutions. In *Proceedings of IEEE CDC-21*, pp. 164–171. IEEE Computer Society Press.Proc.

Friedman, J.W. 1990. *Game Theory with Applications to Economics.* Oxford University Press.

Kalai, E., and Smorodinsky, M. 1975. Other solutions to Nash's bargaining problem. *Econometrica* 43:513–518.

Mazumdar, R., Mason, L.G., and Douligeris, C. 1991. Fairness in network optimal flow control: Optimality of product forms. *IEEE Transactions on Communications* 39(5).

Nash, J. 1950. The bargaining problem. *Econometrica* 18.

Raiffa, H. 1953. Arbitration schemes for generalized two-person games.. In *Contributions to the Theory of Game II*, edited by H.W. Kuhn and A.W. Tucker. Princeton.

Stefanescu, A., and Stefanescu, M.W. 1984. The arbitrated solution for multiobjective convex programming. *Rev. Roum. Math. Pure. Appl.* 29:593–598.

Von Neuman, J., and Morgenstern, O. 1953. *Theory of Games and Economic Behavior.* Princeton University Press.

Appendix D

Notation

Latin letters

a_j — class j offered traffic
a^s — traffic offered to link s
$\bar{a}^s$ — traffic carried on link s
$\bar{a}_j$ — connection class j carried traffic
$\bar{a}_j^k$ — connection class j traffic carried on path k
$\bar{a}_j^s$ — connection class j traffic carried on link s
$\mathbf{a}$ — vector of a_j
$\bar{\mathbf{a}}_v$ — vector of $\bar{a}_j$
$\bar{\mathbf{a}}_j$ — vector of $\bar{a}_j^k$
$\bar{\mathbf{a}}$ — matrix of $\bar{a}_j^k$
$\bar{\mathbf{a}}^s$ — vector of $\bar{a}_j^k$ using link s
$A(t)$ — cell rate superposition process
A_j^s — parameters and functions describing connection class j served by link s
$\mathbf{A}^s$ — vector of A_j^s
$\mathbf{A}^k$ — parameters and functions describing traffic offered to equivalent path model
b^s — link blocking probability (single-rate system)
b^k — path blocking probability (single-rate system)
b_j^s — connection class j link blocking probability
b_j — connection class j network blocking probability
B_j — connection class j loss probability
$\bar{B}$ — overall traffic loss probability
$\bar{B}^s$ — overall link traffic loss probability
$\bar{\mathbf{B}}$ — vector of $\bar{B}^s$
$\mathcal{B}$ — cell loss probability
$\tilde{B}$ — average cell burst length of on-off source
c — network cost

c^s — link cost
c_i^s — IACF link cost for class i connections
$\mathbf{c}_j$ — declaration error of class j connection
C — link residual capacity
$\mathcal{C}_i$ — link residual capacity normalized by class i connection "bandwidth" requirement d_i
d_j — equivalent bandwidth required by class j connection
d_j^e — "bandwidth" allocated to the restoration virtual paths of connection class j in state e
D — equivalent bandwidth required by a set of aggregate connections
$\mathcal{D}$ — average cell delay
$\mathbf{e}$ — model error (Kalman filter)
E — set of failure states
$E[\ ]$ — expected value
$E_c[\ ]$ — expected value as seen by connections
$E(\ ,\ )$ — Erlang B function
$\mathcal{F}_s$ — set of flows using link s
g_j — path net-gain from accepting class j connection
$g_j^s(\ )$ — link net-gain from accepting class j connection
$\overline{g}_j^t$ — time average path net-gain
G^s — "bandwidth" allocated to the sth VNL
G_p^s — "bandwidth" allocated to the sth primary VNL
G_b^s — "bandwidth" allocated to the sth back-up VNL
$G_j^k(v)$ — complementary path shadow price cumulative distribution function
$\mathbf{G}$ — vector of G^s
h^s — link efficiency factor
h^c — link efficiency constraint
h_j — "bandwidth" reservation parameter in DBR policy
h_j^k — probability of selecting path k for class j connection
$h_j^s(\mathbf{x})$ — link s "bandwidth" reserved for CAC objectives
$h_j^k(\mathbf{z})$ — probabilities of randomized policy
$\mathbf{h}$ — connection declared traffic parameters
H — network reward loss probability
H_j — "bandwidth" reserved for other connection classes in BR policy
$H_j(\mathbf{x})$ — "bandwidth" reserved for other connection classes in DBR policy
$\mathcal{H}$ — background sparseness
$\mathbf{H}$ — aggregate declared parameters of connections accepted on a link
J — set of connection classes
J — set of players (game theory)
K — buffer capacity
$\mathcal{K}$ — Kalman gain (Kalman filter)
L^s — link capacity
$\mathbf{L}$ — vector of L^s
LR — leak rate
m_j — average cell rate of class j connection
M_k — average rate of the aggregate cell process

$\bar{M}_k$ — measured average rate of the aggregate cell process
$\hat{M}_k$ — estimated average rate of the aggregate cell process
$p_j^s(\mathbf{x})$ — link shadow price
$p_j^k(\mathbf{y})$ — path shadow price
$\bar{p}_j^s$ — average link shadow price
P — peak rate of the connection cell rate process
$P_j^s(x)$ — probability of routing class j connection on a path comprising link s in state x
$\mathcal{P}$ — error covariance matrix (Kalman filter)
$Pr\{\ \}$ — probability
q — reward rate
$\mathbf{q}$ — declaration error parameters (Kalman filter)
$Q^s(\mathbf{x})$ — link state distribution probability
$\mathbf{Q}$ — vector of $Q^s(\mathbf{x})$
$\mathcal{Q}$ — covariance matrix of the model error (Kalman filter)
r_j — reward parameter of connection
r_j^s — reward parameter of link connection
$\mathbf{r}$ — vector of r_j
$\mathbf{r}^s$ — vector of r_j^s
$r_i^{s'}$ — link s connection reward parameter in COT model
$R(\ )$ — expected reward from the system (Markov decision theory)
$\overline{R}$ — time average reward from the system
$\mathcal{R}$ — link "bandwidth" reserved for the estimation error
s — instant rate sample
s_i — player utility at starting point (game theory)
$\mathbf{s}$ — starting point (game theory)
S — set of links
$\tilde{S}$ — average silence length of on-off source
$\mathcal{S}$ — background softness
t — time
t_j — threshold in DBR policy
u_i — player utility (game theory)
$u_j^{k,e}$ — "bandwidth" allocated to restoration path k serving connection class j in state e
$\mathbf{u}$ — matrix of $u_j^{k,e}$
$\mathbf{u}$ — vector of player utilities (game theory)
$\mathbf{u}^*$ — optimal solution (game theory)
$\mathbf{u}$ — measurement error (Kalman filter)
U_j — connection class throughput normalized by link capacity
$\mathcal{U}(\epsilon)$ — normalized Gaussian distribution coefficient
$\mathbf{U}$ — bargaining domain (game theory)
$v(\ )$ — relative value — continuous time (Markov decision theory)
$v_k(\ )$ — relative value — discrete time (Markov decision theory)
v_i — player preference function (game theory)
v_j — variance of the instant connection cell rate

$\tilde{V}$ — set of network nodes
$\tilde{V}'$ — set of OD nodes in the multi-casting group
$V_n(\ )$ — value function (Markov decision theory)
V_k — variance of the instant cell rate of the aggregate cell process
$\bar{V}_k$ — measured variance of the instant cell rate of the aggregate cell process
$\hat{V}_k$ — estimated variance of the instant cell rate of the aggregate cell process
$\mathbf{w}$ — Kalman filter state change
W_k — Kalman filter state
W_j — set of alternative paths for class j connections
$\mathbf{W}$ — set of W_j
x_j — number of class j connections
$\mathbf{x}$ — link state
X — set of link states
$\mathbf{y}$ — network or path state in decomposed model
$\tilde{Y}$ — measurement error parameters (Kalman filter)
$Y_s^k(x)$ — set of all possible path states for given state on link s ($s \in S^k$)
$Y_{j,nb}^k$ — set of non-blocking path states for connection class j
z_j^k — number of class j connections on path k
$\mathbf{z}$ — network state
Z — set of network states
$\tilde{Z}$ — measured aggregate cell process parameters

Upper indices

^{c} — constraint indication
^{d} — declared parameter indication
^{e} — failure state index (Chapter 10) or extrapolation index (Kalman filter)
^{k} — priority index (Chapter 3) or path index
$^s, ^o$ — link indices
^{t} — multi-hour case index

Lower indices

$_e$ — error variable index
$_h$ — high priority index
$_{j, i}$ — connection class indices
$_l$ — low priority index
$_k$ — discrete time index

Greek letters

λ — connection arrival rate
$\lambda_j^s(\mathbf{x})$ — link state-dependent connection arrival rate
$\bar{\lambda}$ — rate of accepted connections
μ — connection departure rate
π — policy

π^* — optimal policy
ρ — average cell rate
ζ — bandwidth reserved for declaration error
Δ_k — estimation error (Kalman filter)
Λ^s — vector of $\lambda_j^s(\mathbf{x})$
Ψ — vector of quality of service parameters
Ψ^c — vector of quality of service constraints
Π^s — set of link performance characteristics
Π^k — set of path performance characteristics
$\mathbf{\Pi}$ — vector of Π^s
$\mathbf{\Pi}_e$ — vector of Π^k
Υ — CAC & routing parameters
Φ_k — transition matrix (Kalman filter)
Ω — a subset of link states
$\Omega_j(a_j, \pi)$ — domain of the connection class flow distributions

Index

AAL (*see* ATM adaptation layer)
ABR:
  protocol, 21
    current cell rate (CCR), 21
    desired rate (DR), 21
  (*See also* Service categories)
ABT:
  protocol, 21
  (*See also* Service categories)
ABT/DT:
  protocol, 22
  (*See also* Service categories)
ABT/IT:
  protocol, 22
  (*See also* Service categories)
Access link, 2
ACR (*see* Connection traffic contract)
ADM (*see* Digital hierarchy)
Aggregate equivalent bandwidth, 68–91
  actual, 85
  estimated, 85
  estimation, 75–76
    error analysis, 82
Asynchronous Transfer Mode (ATM), 1
  adaptation layer (AAL), 5
  cell, 2
  layer, 10
    service categories, 10
  links, 3
  switch, 3
    input port, 3
    output port, 3, 9, 20
  transport network, 5
ATM (*see* Asynchronous transfer mode)
ATM network dimensioning, 245–255
  adaptation, 255–256
  link bandwidth modularity, 253
  link dimensioning functions, 247
  multi-hour case, 254–255

ATM network dimensioning (*Cont.*):
  optimization variables:
    flow distribution, 249–253
    link capacities, 247
    link performance, 248

"Bandwidth," 19
  aggregate, 69, 70
  quantization, 134
  requested, 69
  reserved:
    for accepted connections, 70
    for declaration error, 84
    for estimation errors, 69, 70, 75, 85, 87
    for new connection, 83, 84
"Bandwidth" allocation:
  to connections, 23, 40–63, 97
    equivalent bandwidth (*see* Equivalent bandwidth)
    minimum rate, 43
    peak rate, 43
  to virtual networks, 30, 214
    complete sharing, 222
    enforcement, 31, 225
    limited sharing, 222
    separation, 223
  to virtual paths, 26, 31, 218
B-ISDN (*see* Broadband integrated service digital network)
BR [*see* CAC (link GoS)]
Broadband integrated service digital network (B-ISDN), 1
BT (*see* Connection traffic contract)
Burst, 20, 34
  layer, 34
    performance, 72, 84

CAC (*see* Connection admission control)
CAC & RM (*see* CAC & routing manager)

INDEX

CAC & routing, 26, 95–132
  bandwidth constrained problem, 97
  classification, 98
  fairness, 27
  generic formulation, 102
  implementation:
    centralized, 128
    distributed, 130
    hierarchical, 131
  link and path metrics, 98
  manager, 33
  multi-point connections, 140, 149–155
    dynamic, 140, 142, 154
    implementation, 143
    link cost based strategies, 141
    point-to-point path based strategies, 141
  objectives:
    access fairness, 102
    network revenue maximization, 103
    reward maximization, 102
    traffic maximization, 103
  optimal operating point, 97
  optimization criteria, 97
  path recommendation, 98, 103, 128
  strategies:
    adaptive, 99
    adaptive load sharing, 99
    decomposition of MDP (MDPD), 107, 125, 150
    Dynamic Alternative Routing (DAR), 99
    fixed, 98
    inverse of residual link capacity (IRCF), 150
    least loaded path (LLP), 99, 125, 150
    load sharing, 98
    macro-state-dependent, 99
    reward maximization, 27, 102
    sequential (SEQ), 99, 123, 125
    state-dependent, 99
CAC cell, 68, 84
CBR (see Service categories)
CBS (see Policing mechanism)
CC [see CAC (link GoS)]
CCR (see ABR protocol)
CDV (see Quality of service)
CDVT (see Connection traffic contract)
Cell:
  flow control, 20
    credit-based schemes, 20
    explicit rate schemes, 20
    fast reservation protocols, 20
    rate-based schemes, 20
    window mechanism, 20

Cell (*Cont.*):
  header, 2
    format, 3
    loss priority (CLP), 6
  layer, 18, 20, 97
    measurements, 20
Circuit-switched networks, 2, 23
  multi-rate, 98, 246
CLP (*see* Cell header)
CLR (*see* Quality of service)
Congestion control, 20
Connection:
  admissible region, 44–63
    boundary, 44–63
    boundary convexity, 45
    cross-section, 59
    linear approximation, 46, 48
    modified linear approximation, 46
    non-linear approximation, 47, 48
  admission control, 24, 96
  admission control (link GoS):
    "bandwidth" reservation policies (BR), 203
    complete sharing policy (CS), 202
    coordinate convex policies (CC), 202
    decomposition of link problem (DBRA), 206
    dynamic "bandwidth" reservation policies (DBR), 203
    heuristic model (DBRH), 206
    optimal policies, 192–199
    simplified policies, 202–206
  admission control (link QoS), 70
    adaptive, 68–91
    analysis, 83–87
    conditions, 40, 56, 63
  admission control (path GoS):
    advanced reservation, 100
    classification, 100
    delayed admission, 100
    loss system, 100
    multi-point, 143
  admission control (path QoS), 44, 97
  background:
    process, 46
    softness, 46, 47
    sparseness, 46, 47
  cell rate:
    average, 72
    variance, 72
  class:
    many-to-many, 140, 143, 154
    many-to-point, 140
    multi-point, 140

INDEX

Connection, class (*Cont.*):
  narrow-band (NB), 101
  point-to-many, 140
  point-to-point, 140
  wide-band (WB), 101
 layer, 18, 23, 97
  measurements, 28, 108
 routing, 96
  path (CRP), 217
  (*See also* CAC & routing)
 setup, 5
  "bandwidth" reservation, 220
Connection parameters (cell layer), 40, 41
 declaration, 42, 68, 69, 72
  aggregate, 69, 70
  error, 42, 70, 74, 80
  error generator, 84
  modification, 90
Connection process description, 164
 link state dependent, 164
 one moment, 164
 two moment, 164
  Bernoulli-Poisson-Pascal distribution, 163, 164
  Equivalent Random Theory (ERT), 164
  Pascal distribution, 172
Connection superposition process parameters (cell layer), 67, 69
 estimation, 68, 73–75
  error, 74
  error analysis, 80–82
 instant cell rate, 72
  autocorrelation, 72
  mean, 72
  variance, 72
 measurement, 68, 70
  error, 74
  error analysis, 76–79
Connection traffic contract, 6
 allowed cell rate (ACR), 11
 burst tolerance (BT), 10
 cell delay variation tolerance (CDVT), 10
 minimum cell rate (MCR), 9
 peak cell rate (PCR), 7, 41
 sustainable cell rate (SCR), 10
Connectionless:
 servers, 13
 services, 13
COT (*see* Link performance models)
Cross-connect switch, 4, 217
CRP (*see* Connection routing path)
CS [*see* CAC (link GoS)]
CTD (*see* Quality of service)
CTP (*see* Service categories)

DAR (*see* CAC & routing strategies)
DBR [*see* Service categories; CAC (link GoS)]
DBRA [*see* CAC (link GoS)]
DBRH [*see* CAC (link GoS)]
DCS (*see* Digital hierarchy)
Demultiplexer, 2
DH (*see* Digital hierarchy)
Digital hierarchy, 31, 243
 add-drop multiplexers (ADM), 243
 digital cross-connect switches (DCS), 243
 layers, 243
  transmission rates, 244
 plesiochronous (PDH), 243
 self-healing capabilities, 257
 synchronous (SDH), 243
 Synchronous Optical NETwork (SONET), 243
 terminal multiplexers (TM), 243
Direct-link-path, 99
DJM (*see* Multi-cast tree design heuristics)
DLP (*see* Link performance models)
DMUX (*see* Demultiplexer)
DR (*see* ABR protocol)

EPM (*see* Network performance models)
Equivalent bandwidth, 42–63
 allocation, 42
  worst case, 90
 evaluation:
  approximate, 75
  required, 68
Estimation theory, 269–276
 covariance matrix, 273
  estimation error propagation, 274
 linear system model, 270
  discrete representation, 271
  transition matrix, 270
 observability and controllability, 272

Fair queuing, 9, 64
Fast packet switched networks, 1
FCM (*see* Flow control manager)
FIFO (*see* First-in-first-out)
Filters recursive discrete, 71
 linear, 71
  Kalman, 74
 non-linear, 71
First-in-first-out, 9
Flow control, 33
FPS (*see* Fast packet switched networks)

Game theory, 186, 293–301
　arbitration schemes, 188, 299–301
　　loss equalization, 193
　　modified Thomson, 188, 194, 300
　　Nash, 188, 299
　　Raiffa, 188, 193, 300
　　traffic maximization, 194
　bargaining domain, 186
　coalitional form, 296
　　characteristic function, 296
　　core, 296
　cooperative, 187, 297, 298
　cooperative vs. non-cooperative, 297
　extensive form, 293
　　utility function, 294
　fairness criteria, 188, 297
　　axioms, 188, 299
　non-cooperative, 187, 297
　Pareto boundary, 187, 297
　　evaluation, 190–192
　Pareto optimality, 187, 297
　players, 186, 293
　preference function, 189, 299
　　product, 189, 299
　starting point, 187, 297
　strategic form, 295
　　equilibrium point, 295
　utility, 186, 294
General processor sharing, 9
GoS (*see* Grade of service)
GPS (*see* General processor sharing)
Grade of service, 23, 160
　distribution control, 125
　requirements, 97

Kalman filter, 73–75, 274–276
KMB (*see* Multi-cast tree design heuristics)

LCR (*see* Policing mechanism)
Leaky bucket (*see* Policing mechanism)
Link capacity:
　residual, 40
Link performance models:
　cell layer:
　　fluid approximation, 47, 56, 63
　　Modulated Markov Poisson Process (MMPP), 48
　connection layer, 108–115
　　aggregation into uni-dimensional link state (UDL), 115
　　class oriented transformation (COT), 114

Link performance models, connection layer (*Cont.*):
　　decomposition of the link process (DLP), 114
　　steady arrival rates (SAR), 113
LLP (*see* CAC & routing strategies)

Markov decision process, 103
　optimal policy:
　　randomized, 190
Markov decision process (MDP), 103, 283–290
　continuous-time, 288
　decomposition, 106
　discrete-time, 283
　linear programming approach, 287, 290
　optimal policy, 104
　　deterministic, 104, 285
　policy iteration algorithm, 104, 107, 285, 289
　　relative value, 104, 284, 285
　　value-determination equations, 104, 285
　value-iteration algorithm, 110, 286, 290
　value function, 110
Markov process, 105, 190, 279–283
　continuous-time, 280
　decomposition, 106
　discrete-time, 280
　near complete decomposability, 114
　solution, 282
　　equilibrium equations, 160, 282
　　product form, 282
　　recursive solution, 282
　uniformization technique, 110, 282
MCR (*see* Connection traffic contract)
MDP (*see* Markov decision process)
MDPD (*see* CAC & routing strategies)
MMPP [*see* Link performance models (cell layer)]
Modulated Markov Poisson Process (MMPP), 63
MPC1 (*see* Multi-cast tree design heuristics)
MPC2 (*see* Multi-cast tree design heuristics)
MST (*see* Multi-cast tree design heuristics)
Multi-cast:
　server, 141
　switch, 141
　tree, 141
　tree design heuristics, 144–149
　　complexity order, 146, 148
　　DJM, 148
　　KMB, 147
　　MPC1, 144

INDEX

Multi-cast, tree design heuristics (*Cont.*):
    MPC2, 144
    MST, 146
    SPT, 145
    TAK, 145
    virtual channel (MVC), 141
Multi-link-path, 100
Multiplexer, 2
MUX (*see* Multiplexer)
MVC (*see* Multi-cast virtual channel)

NB (*see* Connection class)
NCTP (*see* Service categories)
Network:
    directed cost, 142
    resources (*see* "Bandwidth")
    undirected cost, 142
Network performance (connection layer):
    decomposition into link problems:
        fixed-point equations, 247
Network performance models, 27, 159–179
    decomposition into link problems, 161
        fixed-point equations, 161, 170
        hysteresis effect, 165
        link independence assumption, 161, 166–168
        link loading function, 161, 170, 172
        link performance function, 161, 171, 172
        multiple solutions, 165
        solution uniqueness, 165
    decomposition into path problems, 162
        background traffic, 162
        Equivalent Path Model (EPM), 168
        path loading function, 164
        path performance function, 163
    decomposition methods, 160–169
        accuracy, 168
        damping factor, 162
        repeated substitution, 162, 164
    reward maximization, 169–179

OAM (*see* Operation and maintenance cell)
Operation and maintenance cell, 6
OPT1 (*see* Virtual network design)
Overlay packet switched network, 13

Packet switched networks, 4
Payload type, 6
PCR (*see* Connection traffic contract)
PDH (*see* Digital hierarchy)

Physical link:
    resource manager, 34
Physical network, 215
    layer, 18, 31
    resource manager, 33
    self-healing, 31
    survivability, 31
PL-RM (*see* Physical link resource manager)
PN (*see* Physical network)
PN-RM (*see* Physical network resource manager)
Policing mechanism, 7, 87
    leaky bucket, 7
        cell burst size (CBS), 7
        leak cell rate (LCR), 7, 90
Protocol reference model, 5
    control plane, 5
    management plane, 5
    user plane, 5
PT (*see* Payload type)

QoS (*see* Quality of service)
Qualified bandwidth, 19
Quality of service, 6, 69
    cell delay variation (CDV), 6
    cell loss rate (CLR), 6
    cell transfer delay (CTD), 6
    constraints, 40–58, 83
    guarantees:
        deterministic, 64
        statistical, 40, 64

Reactive traffic control, 20
Real-time applications, 10
Resource management:
    architecture, 32
    cell, 6, 11, 20, 25, 68
    and traffic control, 17
        layered structure, 18
    VN design, 218
    VP_{ee} design, 218
    VP_{nn} design, 218
Resources (*see* "Bandwidth")
Reward:
    average from network, 102, 103
    connection parameter, 102, 103
        division, 106, 109
    link connection parameter, 106
    link process, 106
    maximization, 102
        average link shadow price, 109, 116

INDEX

Reward, maximization (*Cont.*):
   link independence assumption, 105, 106, 120
   link net-gain, 106
   link shadow price, 103, 106
   path net-gain, 103, 105, 106, 121
  network process, 106
  rate from connection, 102
  rate from network, 104
RM (*see* Resource management cell)
RM & TC (*see* Resource management and traffic control)
Routing (*see* CAC & routing)

SAR (*see* Link performance models)
SBR (*see* Service categories)
Scheduler, 9
Scheduling, 9
SCR (*see* Connection traffic contract)
SDH (*see* Digital hierarchy)
SEQ (*see* CAC & routing strategies)
Server:
  asynchronous, 51, 56
  synchronous, 51, 56
Service categories (ATM layer), 10
  ATM block transfer (ABT), 12
    with delayed transmission (ABT/DT), 12
    with immediate transmission (ABT/IT), 12
  available bit rate (ABR), 11
  constant bit rate (CBR), 10
  controllable traffic parameters (CTP), 12
  deterministic bit rate (DBR), 11
  non-controllable traffic parameters (NCTP), 12
  statistical bit rate (SBR), 12
  unspecified bit rate (UBR), 11
  variable bit rate (VBR), 10
Shaping, 7
Simulated allocation, 265
SONET (*see* Digital hierarchy)
Source:
  on-off, 41
    burst length, 41
    silence length, 41
  policing, 42, 43, 70
  worst case, 24
Source parameters [*see* Connection parameters (cell layer)]
SPT (*see* Multi-cast tree design heuristics)
Statistical multiplexing:
  cell layer, 39, 42, 50
    gain, 2, 42

Statistical multiplexing (*Cont.*):
  connection layer, 62
  of errors, 69
  of measurement errors, 70
STM (*see* Synchronous Transfer Mode)
Survivability, 256–264
  failure scenarios, 256
  restoration, 257
    connection, 258
    GoS, 258
    hitless path switching, 259
    link, 257
    objectives, 259–261
    path, 236, 258
  self-healing network design, 261–264
Synchronous Transfer Mode (STM), 2

TAK (*see* Multi-cast tree design heuristics)
Tariff, 27, 103, 216
  design, 266
    elasticity function, 266
Telecommunications Information Network Architecture (TINA), 35
Telecommunications Network Management (TNM), 35
TINA (*see* Telecommunications Information Network Architecture)
TM (*see* Digital hierarchy)
TNM (*see* Telecommunications Network Management)
Traffic:
  layered entities, 18
Transport layer, 242
  dedicated, 242
  link bandwidth modules, 242
  primary link, 242
  stand-by link, 242
  transmission facilities, 242

UBR (*see* Service categories)
UDL (*see* Link performance models)
UNI (*see* User network interface)
UNI connection setup manager, 33
UNI-CSM (*see* UNI connection setup manager)
UPC (*see* Usage parameter control)
Usage parameter control, 7
User network interface, 7

VBR (*see* Service categories)
VC (*see* Virtual channel switch)

VCC (*see* Virtual channel connection)
VCI (*see* Virtual channel identifier)
VCL (*see* Virtual channel link)
Virtual channel (VC), 3
   connection (VCC), 3
   identifier (VCI), 3
   link (VCL), 3
Virtual network (VN), 18, 28
   applications, 215–217
      interrelation, 217
   definition, 214
   design, 30, 225–233
      economic constraints, 226
      equality loss constraints, 228
      inequality loss constraints, 227
      link efficiency constraints, 228
      optimization algorithm (OPT1), 228
   identifier (VNI), 214
   layer, 18, 28
   link (VNL), 18, 28, 214
      end-to-end, 28
      identifier (VNLI), 28, 214
      node-to-node, 28
      resource manager, 33
      vs. VPC, 217–221
   management oriented, 30, 216
      backup VN, 30, 216, 236
   nested, 62, 215
   parent, 215
   resource manager, 33
   service oriented, 29, 61, 215
      QoS VN, 97
   subnetworks, 217

Virtual network (VN) (*Cont.*):
   update, 233–236
      link marginal costs, 235
      measurements, 233
   user oriented, 29, 216
      multi-point connections, 30, 216
      private VN, 29, 216
Virtual path (VP), 3
   applications, 217–221
   connection (VPC), 3
      vs. VNL, 217–221
   end-to-end (VP_{ee}), 26
   identifier (VPI), 3
   layer, 34
   link (VPL), 3
   node-to-node (VP_{nn}), 26
VN (*see* Virtual network)
VNI (*see* Virtual network identifier)
VNL (*see* Virtual network link)
VNLI (*see* Virtual network link identifier)
VNL-RM (*see* Virtual network link resource manager)
VN-RM (*see* Virtual network resource manager)
VP (*see* Virtual path switch)
VPC (*see* Virtual path connection)
VPI (*see* Virtual path identifier)
VPL (*see* Virtual path link)

WB (*see* Connection class)
Worst case, 64
   design, 69

ABOUT THE AUTHOR

Zbigniew Dziong (Quebec, Canada) is a professor at INRS-Telecommunications, a research institute at the University of Quebec. He has carried out contract research in advanced and broadband networks for Ericsson Telecom and Bell Northern Research (Nortel) and has over 44 articles published in his field.